The Urban Book Series

The Urban Book Series is a resource for urban studies and geography research worldwide. It provides a unique and innovative resource for the latest developments in the field, nurturing a comprehensive and encompassing publication venue for urban studies, urban geography, planning and regional development.

The series publishes peer-reviewed volumes related to urbanization, sustainability, urban environments, sustainable urbanism, governance, globalization, urban and sustainable development, spatial and area studies, urban management, transport systems, urban infrastructure, urban dynamics, green cities and urban landscapes. It also invites research which documents urbanization processes and urban dynamics on a national, regional and local level, welcoming case studies, as well as comparative and applied research.

The series will appeal to urbanists, geographers, planners, engineers, architects, policy makers, and to all of those interested in a wide-ranging overview of contemporary urban studies and innovations in the field. It accepts monographs, edited volumes and textbooks.

Indexed by Scopus.

Salomón González-Arellano · Bruno Gandlgruber
Editors

Cities as Anticipatory Systems

Springer

Editors
Salomón González-Arellano
Departament of Social Sciences
Universidad Autónoma Metropolitana
Unidad Cuajimalpa
Mexico City, Mexico

Bruno Gandlgruber
Departamento of Institutional Studies
Universidad Autónoma Metropolitana
Unidad Cuajimalpa
Mexico City, Mexico

ISSN 2365-757X ISSN 2365-7588 (electronic)
The Urban Book Series
ISBN 978-3-031-78164-3 ISBN 978-3-031-78162-9 (eBook)
https://doi.org/10.1007/978-3-031-78162-9

This Springer imprint is published by the registered company Springer Nature Switzerland AG
The registered company address is: Gewerbestrasse 11, 6330 Cham, Switzerland

Foreword

In just over 10,000 years—a blink of an eye in the history of *homo sapiens*—we have gone from nomadic micro-groups to megacities. Apart from megacities—a predominantly Asian and African phenomenon—and sprawls—the widespread cities typical of the United States—most people live in small and medium-sized cities—let's say between 30,000 and 1 million inhabitants. These cities are neither too big nor too small. They offer the advantages of living in an urban context without imposing the additional costs linked to abnormal urbanization developments, whether intensive as megacities or extensive as in sprawls.

Even if few noticed it, 2004 was a turning point: for the first time in the history of humanity, the number of citizens exceeded the number of rural people. The trend towards a further increase in people living in urban contexts continues, and the United Nations expects that by 2050, three-quarters of the world's population will live in urban contexts. This single fact—there are and will be more and more citizens—explains why it makes sense to write books about cities: because it concerns most of us. There is much more, though. Perhaps the most interesting question is why more and more people decide to live in cities. The answers are different, but if we look for a trait that can unite them, perhaps the most plausible explanation is that in the city, you feel free, and perhaps you are. With its apparently unavoidable pendant, we feel alone and often are.

The dominant megatrend of the constant increase in citizens is contrasted by the secondary countertrend of those fleeing the city and deciding to return to the mountains and woods. This is a fascinating phenomenon that presents original aspects. Firstly, those who return to live 'in nature' are not the same people who abandoned the more peripheral mountains and valleys. The latter had abandoned the places of their childhood and had gone to the city to have a hope of life, a job, and the possibility of a career. The former return to nature having a job; they are often high-level managers, and even with significant international experience, they are perfectly integrated into their work contexts. A good internet connection is all they need to continue their work. The move towards cities and the counter move towards nature are two different trends; each of them, in its way, is relevant. What matters is not to confuse them and not to confuse their dimensions. It is no coincidence that I defined the first (towards the city) as a 'megatrend', that is to say, as a critical process, while I called the second (return to nature) a 'secondary countertrend'. The first is dominant; the second is remarkably smaller.

Cities are and will continue to be essential for pure numerical data and other reasons. Cities are becoming internal frontiers, where important dividing fault lines are accumulating. Infrastructures that are unable to stay up to date, the impact of the climate crisis and in particular of heat waves, ageing of the population, administrations incapable of understanding the needs of citizens, progressive decline in birth rate, flight of young people who increasingly move from sleepier cities to more active ones are all factors that are already visible now and which promise to become increasingly impactful in the coming decades. Cities must be able to manage the many transformations taking place and, to do so, a new ruling class is needed, a ruling class capable of looking far ahead, understanding the transformations taking place, implementing innovative projects, and involving citizens in designing their futures.

A city is also a multiplier for its territory. A sleepy city also puts the surrounding area to sleep. A lively city generates activities and resources in its context.

There is no shortage of financial resources today, but they are necessary to generate critical, innovative projects. Then, it is also a question of knowing how to obtain the needed resources and the rules of international competition. Only some people will make it because they have no ideas, don't know where to knock, or are shortsighted.

The difference between cities that will be at least partially able to manage the maturing transformations (and gather the necessary resources) and those that will not will grow progressively.

This book focuses on anticipation, trying to read possible changes that may occur in the coming decades. This book does not offer universally valid recipes; one-size-fits-all recipes don't work. Everyone must find their way and use their imagination and their organizational ability. This book provides some essential reflections that can help administrators, architects, urban planners, and indeed anyone who cares about their city to begin a reflection on its context, on the problems that they will have to manage, on the opportunities that could emerge, in a word on the future of your city.

Roberto Poli
UNESCO Chair in Anticipatory Systems
University of Trento
Trento, Italy

Contents

Contributors

Marion Bonhomme National Institute on Applied Sciences (INSA), Laboratoire Matériaux et Durabilité des Constructions (LMDC), Toulouse III University Toulouse, Toulouse, France

Rafael Calderón-Contreras Department of Social Sciences, Universidad Autónoma Metropolitana-Cuajimalpa, Mexico City, Mexico

Bikramaditya K. Choudhary CSRD/SSS, Jawaharlal Nehru University, New Delhi, India

Marco Córdova FLACSO Ecuador, Quito, Ecuador

Marcos Tavares de Arruda Filho Instituto de Energia E Ambiente—Universidade de São Paulo, São Paulo, Brasil

Brais Estévez Departamento de Xeografía, Universidade de Santiago de Compostela, Santiago de Compostela, Spain

Manuela Á. Fernández-Borrero COIDESO-Universidad de Huelva, Huelva, Spain

Mamdouh Ibrahim Zaky Mohamed Galil College of Engineering and Architecture, University of Nizwa, Nizwa, Oman

Bruno Gandlgruber Department of Institutional Studies, Universidad Autónoma Metropolitana-Cuajimalpa, Mexico City, Mexico

Farzaneh Gharaati Department of Architecture, Tarbiat Modares University, Tehran, Iran

Salomón González-Arellano Department of Social Sciences, Universidad Autónoma Metropolitana-Cuajimalpa, Mexico City, Mexico

José Luis González Madrigal Universidad Autonoma Metropolitana, Mexico City, Mexico

Julia Hidalgo National Centre for Scientific Research (CNRS), Laboratoire Interdisciplinaire Solidarités, Societés, Territoires (LISST), Toulouse II University, Toulouse, France

Marco Jaso Department of Institutional Studies, Universidad Autónoma Metropolitana-Cuajimalpa, Mexico City, Mexico

Daniela Losiggio CONICET, Universidad Nacional Arturo Jauretche, Buenos Aires, Argentina

Mohammadjavad Mahdavinejad College of Engineering and Architecture, University of Nizwa, Nizwa, Oman;
Department of Architecture, Tarbiat Modares University, Tehran, Iran

Blanca Miedes-Ugarte COIDESO-Universidad de Huelva, Huelva, Spain

Nora A. Morales Zaragoza Department of Design Theory and Processes, Universidad Autónoma Metropolitana-Cuajimalpa, Mexico City, Mexico

Diana Paz FLACSO Ecuador, Quito, Ecuador

María Pozzio CONICET, Universidad Nacional Arturo Jauretche, Buenos Aires, Argentina

Laura E. Quiroz Rosas Department of Social Sciences, Universidad Autónoma Metropolitana-Cuajimalpa, Mexico City, Mexico

Pedro Roberto Jacobi Instituto de Energia E Ambiente—Universidade de São Paulo, São Paulo, Brasil

Mónica Velázquez Téllez Universidad Autonoma Metropolitana-Cuajimalpa, Mexico City, Mexico

Part I
Introduction

Chapter 1
Introduction

Salomón González-Arellano and Bruno Gandlgruber

Abstract The imaginary future of cities has been the subject of science fiction writers, architects, illustrators, filmmakers, videogame designers, fashion designers, restaurant chefs, etc. Futures exploration has been slowly expanding in recent years by using broader approaches to imagine alternative future scenarios and better understand existential risks and opportunities, the potential of municipal governments and citizens. One of these approaches focuses on the inherent anticipatory capacity of biological and social systems. Anticipation theory postulates that social systems create images of themselves in the future that play an essential role in decision-making processes in the present. These processes are embedded in broader social dynamics that frame the construction of futures in urban contexts. This book aims to characterize cities as evolving anticipatory systems with diverse implicit and explicit uses of images of the future. Each contribution explores specific aspects of anticipation in cities worldwide, particularly in the Global South. The chapters of this book are related to specific aspects of anticipatory systems and some of their conceptual components, such as anticipatory assumptions, anticipatory capabilities, or anticipatory governance.

Keywords Anticipatory systems · Cities · Urban futures · Future literacy

1.1 Introduction

Urban settlements have always been important for human development, particularly in the modern epoch. As the current trends of urbanization will foreseeably continue in this century, cities will play a crucial role in coping with the major challenges for

S. González-Arellano (✉)
Department of Social Sciences, Universidad Autónoma Metropolitana-Cuajimalpa, Mexico City, Mexico
e-mail: sgonzalez@cua.uam.mx

B. Gandlgruber
Institutional Studies Department, Universidad Autónoma Metropolitana, Mexico City, Mexico
e-mail: bgandl@cua.uam.mx

S. Gonzalez-Arellano and B. Gandlgruber (eds.), *Cities as Anticipatory Systems*, The Urban Book Series, https://doi.org/10.1007/978-3-031-78162-9_1

human societies in our days in strategic areas like health, energy, equality, etc. Many features of the current physical, political, social, and economic configurations have proven to be limited in resolving many of these issues. The systems of organizing housing, transport, or nutrition established in the last century show severe tensions like the excessive use of resources, the creation of climate-sensitive emissions, and other problems that have led to diminishing levels of sustainability, quality of life, and equality.

There is a vast production on the future of cities that can be traced at least back to the nineteenth century. The imaginary future of cities has been the subject of science fiction writers, architects, illustrators, filmmakers, videogame designers, fashion designers, restaurant chefs, etc. Alongside these visions of futures, urban studies have recently generated a series of future scenarios that are widely accepted by the expert community despite often narrow pathways to the future (Gonzalez-Arellano 2024). These "hegemonic" futures are meta-narratives that indicate not only how probable futures of cities are projected but also how to imagine diverse possible futures that are compared to specific selected preferable futures and specific trajectories to approach them. The established forms of institutionalization of city and urban futures are, in our opinion, dominated by a small group of guiding assumptions that can be synthesized into the following components:

- Toward a planetary urbanism: One of the main transformations of cities is updating the very conception of the city and the urban. For some authors, the category "city" is becoming obsolete in the face of the new conditions of urbanization, a process that has taken on an unprecedented dynamic of reconfiguration in recent years. It is not only a demographic and quantitative issue; "the urban" is surpassing the socio-spatial reality of the "city" to establish itself in the most diverse spaces of the planet: alpine resorts, transatlantic routes, deserts, even in and out of the terrestrial atmosphere (Brenner and Schmid 2014; Brenner 2018).
- Toward new geographies: One of the effects of this planetary urbanization is the generation of new geographies. Cities and "the urban" will take on new configurations, some in concentration dispersion, others in corridors or constellations of settlements or infrastructures. In addition to new urban forms, the future of "the urban" will be marked by new places, not only by the emergence of architectural or infrastructural innovations that will respond to new socio-cultural, political, and technological realities but through the adaptation of "old" places and structures of the built environment. Rethinking urban geography also means rethinking the coexistence between cities and nature as part of scenarios of "nature futures" (Mansur et al. 2022).
- Toward new temporalities: Future imaginaries of the city often emphasize socio-spatial transformations and ignore the other dimension of social reality: time. The city of the future will not only be polychronic (multiple temporalities), it will also experience a gradual temporal extension (24-h city), the acceleration of urban rhythms, and the synchronization of activities with distant places will be accompanied by new temporalities such as asynchronous and simultaneous ones (Madanipour 2017). If space were the dominant dimension in the industrial city of

the nineteenth and twentieth centuries, in the city of the future new temporalities (and chronotopes), would be the lever of innovations of the urban and the city (Zeiderman 2008).

- Toward new urban metabolisms: Urban crises associated with environmental problems mark the need to reconfigure urban metabolism. Current flows of resources and energy in cities are unsustainable, and the presence of specific "weak signals" already suggests a paradigm shift. Initiatives such as local production and consumption of food, clothing, water, energy, and raw materials are adopted in new urban lifestyles. The urban footprint that currently exceeds tens of thousands of kilometers from urbanites will, in the future, tend to be more selective and introduce aspects of social, economic, and ecological responsibility (Zhang et al. 2015). The transformation of sociotechnical and socio-ecological systems opens windows of opportunity for preferable futures, which is crucial for successful climate mitigation and adaptation (Egerer et al. 2021; McPhearson et al. 2016).
- Toward new forms of governance: The social production of the future of the urban and of cities requires an updating and reconfiguration of the governance of cities and territories. The way of imagining the future of cities is based on processes of vertical institutionalization of the future constituted by regulatory, normative, and cognitive instruments that often adopt path dependencies and, therefore, are not very innovative. The mismatch between the institutionalized governance of the city and a new socio-ecological reality characterized, among other things, by the aspects mentioned above will favor the emergence of innovations and new forms of participative governance (Sheikh et al. 2023).

All these dimensions of city futures can be associated with specific processes of creating images, narratives, or models of the future on different levels of individual and collective decision-making (UN-Habitat 2022). This process depends on diverse capacities to build and compare different futures in new ways. The existing forms of representation of the city of the future on a global level (like within the C40 group of cities for climate leadership, Davidson et al. 2019) or the images related to particular cities that are expressed, for example, in foresight exercises or strategic planning processes (Ravetz and Miles 2016), are often somewhat limited to linear, rigid extrapolations of existing structures, systems, and mental models. The reason for the persistence of narrow pathways to the future is the underlying paradigms and methods that are used to formulate narratives that reflect practices associated with the more traditional future studies and that rely heavily on quantitative techniques with anticipatory solid assumptions about the continuity of past and current configurations in the tradition of forecast models. Although the scope of the exploration of alternative futures has been slowly widened in recent years by using broader strategic foresight techniques to imagine alternative futures scenarios (Inayatullah 2011) and get a better understanding of the existential risks and opportunities, the potential of city governments and citizens to imagine and apply alternative futures is often rather limited.

Anticipation theory, one of the more recent approaches of contemporary futures studies, has developed a framework to characterize the anticipatory capacities of all kinds of social systems (Miller et al. 2018; Poli 2019). They part from the idea that social systems create images of themselves in the future that play an essential role in the diverse decision-making processes of all actors in the present. Even if the prevailing images of the future contain no changes, the system uses these images by orienting its resources and policies toward a reproduction of itself, avoiding variations. The creation and diffusion processes of future images express specific forms of "using the future" to guide processes of continuity and change. These processes are embedded in broader social dynamics that frame the construction of futures in urban contexts.

This book aims to characterize cities as evolving anticipatory systems with diverse implicit and explicit uses of images of the future. Each contribution explores specific aspects of anticipation in cities worldwide, particularly in the Global South, with recent contributions reflecting on the decolonization and construction of futures (Miraftab 2017). All the chapters of this book are related to specific aspects of anticipatory systems and some of their conceptual building blocks, like anticipatory assumptions, anticipatory capacities, or anticipatory governance; there is no homogeneous perspective or methodology behind the different efforts. The participating authors explore specific characteristics of cities as anticipatory systems in specific urban contexts, mostly in different parts of the Global South.

The guiding thread of this work is based on the postulate that the city, as a complex socio-spatial system, contains an inherent attribute, the capacity for anticipation. Cities and, even more so, metropolises are a set of coordinated systems at different scales. Urban studies have explored the analysis of cities from the perspective of systems theory from various approaches, for example, as socio-ecological systems, urban service systems (health, energy, transportation, etc.), physical systems (morphological of the built environment), socio-economic systems, political systems, etc. Each of these urban systems constructs its vision of the future that allows for sharing projects with other actors and actors of the various systems of the cities. This set of shared visions allows a set of values, regulations, and urban knowledge to be oriented toward the future, which is what we call the institutionalization of the future of cities (González-Arellano 2022). The institutionalization of the future of cities has a pragmatic component: anticipation. While the future scenarios of cities are collectively constructed and updated, it is through anticipation that these imaginaries are translated into action. As we will see throughout the chapters that make up this book, the anticipation of cities is embedded in each city system. In different ways, urban systems update their present state in the face of the different future scenarios they can imagine. This inherent attribute of anticipation of systems is diverse and uneven among urban systems, inhabitants, and places of cities.

This book is divided into four sections. Section I, "New Conceptual Frameworks for Cities as Anticipatory Systems," comprises two chapters that propose theoretical and conceptual frameworks for anticipatory systems in cities. Salomón González explores in this Chapter under the title "Cities as anticipatory systems" how anticipatory systems apply to cities, decomposing anticipation into future disposition, predictive capacity, and transformative updating. Due to their complexity and decentralized nature, cities harbor diverse anticipatory regimes, forming urban anticipatory communities shared among actors. Urban futures' production shapes policies rooted in diverse visions. Institutionalized futures highlight cities' anticipatory nature. In Chap. 2, titled "Hopeful Determination as an Urban Anticipatory System. An Enactive Approach," Blanca Miedes Ugarte and Manuela Á. Fernández-Borrero delve into urban development dynamics through a lens of hopeful determination. It intricately analyzes foresight, adaptability, citizen engagement, and multi-stakeholder cooperation, highlighting their role in fostering urban resilience and proactive change. Enactive cognition theory underpins the argument, emphasizing urban actors' active role in shaping their environment, thus underscoring the significance of "hopeful determination" in urban planning and development.

Section II of "Building Anticipatory Systems in Urban Sectors (contexts) in the Global South" includes four chapters that explore social configurations of anticipation processes in four different cities in the Global South on topics like mobility, care, health, and construction. In Chap. 3, titled "Sustainable Urban Mobility as an Anticipatory System: The Case of the Quito Metro," Marco Córdova and Diana Paz conceptualize sustainable urban mobility as an anticipatory mechanism within urban development. Future projections of socially inclusive, environmentally sustainable, and economically efficient urban configurations inform present policy and planning. This chapter analyzes through a qualitative case study the role of the metro system in Quito as a sociotechnical system that anticipates growth, climate change mitigation, and socio-cultural dynamics. Daniela Lossigio and Maria Pozzio wrote Chap. 4, titled "Sustainability of Life and Care in the Metropolitan Area of Buenos Aires, Argentine: Feminism and Anticipation." The chapter advances a discourse blending future studies and feminism, elucidating the slogan "*Se va a caer*" (Patriarchy will fall) to explore future narratives amid threats and the ascent of the ultra-right. The authors trace theoretical genealogies to conceptualize care and its anticipatory implications for practical proposals. In Chap. 5 ("Inclusive Urban Health Systems and Universal Well-Being: Mirage or a Real Future?"), Bikramaditya K. Choudhary relates the survival of humanity to the dependence on health systems for preventive and curative measures. Urbanization will require an understanding of urban complexities and livability. Future cities are deemed synonymous with "healthy" systems and economies, in plain contradiction to reigning neoliberal regimes. The concept of an "anticipatory city" is crucial. Despite visions for an unknown future, the paper addresses inequality and unchecked consumption in future cities. Mohammadjavad Mahdavinejad and Farzaneh Gharaati look at anticipatory processes related to construction paradigms in the Global South in Chap. 6, titled "Anticipating the future of Iranian cities: High-tech vs. nature-based solutions." In this chapter, the authors employed the discipline of anticipation and a co-design methodology involving experts in urban planning to envision future scenarios, drawing from future literacy approaches. Anticipated urban futures encompass high-technology, capital-intensive developments contrasted with nature-based, socio-economically empowering, and eco-friendly solutions. The study advocates for a future focused on happiness metrics, circular economies, and novel value creation, culminating in a proposed scenario for the Global South's urban future.

Section III comprises four chapters that analyze different levels and subsystems of cities as anticipatory systems in Mexico. In Chap. 7, Nora Morales and Laura Quiroz examine under the title "Anticipating asbestos-free futures—participatory mapping as elements to envision transitions based on hope" the co-creation of future visions within an asbestos-affected community on the outskirts of Mexico City, underscoring the significance of employing Transition Design principles in addressing complex social-ecological challenges. Through participatory mapping, community members generated alternative scenarios challenging normative urban futures. Expressions of hope related to preferable futures emerge as a motivational factor fueled by community narratives, driving action toward envisioned outcomes. Mónica Velázquez revises in Chap. 8 ("Anticipation in three restaurants of Mexico City: Imagining edible fungi in urban food futures") how three mushroom-based restaurants in Mexico City anticipate urban food transitions amid the current eco-social crisis. It discusses international food scenarios, urban food dynamics, mushrooms' emergence, their role in the Mexican food system, and the anticipatory capacities of these restaurants. Another contribution by José Luis González in Chap. 9 ("Anticipatory Systems and Security in Mexican Cities. A case-based review") reflects on urban security anticipation, examining attempts by various urban actors and their influence on the city's anticipatory capacity. Considering conflicting discourses and actions, it questions whether cities can constitute effective security anticipation systems. In Chap. 10, Bruno Gandlgruber and Marco Jaso explore zero-plastic futures under "Anticipating zero plastics futures in urban contexts in Mexico." Based on a brief overview of the consequences of plastics contamination and some general characteristics of the plastics economy, they analyze different images, narratives, and models of the future that are related to the plastics futures and their applications to cities in the Global South. In the second part of the chapter, they give an overview of regulation processes related to plastics in Mexico to describe some elements of an anticipatory system in construction in this area.

Section IV includes anticipatory elements for urban climate policies. Pedro Roberto Jacobi, Marcos Tavares de Arruda Filho, and Rafael Calderon compare in Chap. 11 ("Recife and the anticipation of extreme weather events—paths to advance climate justice") debates on climate change governance, especially in urban areas as crucial elements of public management. Anticipatory governance requires long-term vision changes to prevent crises. Recife, highly vulnerable to climate change, illustrates the importance of anticipation in local agendas. Finally, Brais Estévez, Julia Hidalgo, and Marion Bonhomme, in Chap. 12 with the title "Anticipatory action and future urban living in a context of increasing temperatures: an analysis from the Barcelona Climate Shelter Network" show how in recent years, global uncertainties, stemming from ecological and social crises linked to human-induced climate change, pose existential threats to cities and the planet. Local governments respond with anticipatory initiatives like Barcelona's Network of Climate Shelters, engaging citizens in collaborative resilience-building efforts to address urgent climate challenges.

References

Brenner N (2018) New urban spaces: urban theory and the scale question. Oxford University Press, New York

Brenner N, Schmid C (2014) Planetary Urbanization. In: Brenner N (ed) Implosions/explosions: towards a study of planetary urbanization. Jovis, Berlin, pp 160–163

Davidson K, Coenen L, Gleeson B (2019) A decade of C40: research insights and agendas for city networks. Global Pol 10(4):697–708

Egerer M, Haase D, McPhearson T, Frantzeskaki N, Andersson E, Nagendra H, Ossola A (2021) Urban change as an untapped opportunity for climate adaptation. Npj Urban Sustainability 1(1):22

González-Arellano S (2022) The future scenarios of cities: an analysis of their institutional construction. Int J Urban Reg Res 46(5):865–873. https://doi.org/10.1111/1468-2427.13103

González-Arellano S (2024) Tipología de escenarios futuros de ciudades. Revista EURE - Revista De Estudios Urbano Regionales, 50(151). https://www.eure.cl/index.php/eure/article/view/4911

Inayatullah S (2011) City futures in transformation: emerging issues and case studies. Futures 43(7):654–661

Madanipour A (2017) Cities in time: temporary urbanism and the future of the city. Bloomsbury Publishing

Mansur AV, McDonald RI, Güneralp B, Kim H, de Oliveira JAP, Callaghan CT, Pereira HM (2022) Nature futures for the urban century: integrating multiple values into urban management. Environ Sci Policy 131:46–56

McPhearson T, Iwaniec DM, Bai X (2016) Positive visions for guiding urban transformations toward sustainable futures. Curr Opinion Environ Sustain 22:33–40

Miller R, Poli R, Rossel P (2018) The discipline of anticipation: foundations for futures literacy 1. Routledge, Transforming the future, pp 51–65

Miraftab F (2017) Insurgent practices and decolonization of future (s). The Routledge Handbook of Planning, Routledge, pp 276–288

Poli R (2019) Handbook of anticipation: theoretical and applied aspects of the use of future in decision making. Springer

Ravetz J, Miles ID (2016) Foresight in cities: on the possibility of a 'strategic urban intelligence.' Foresight 18(5):469–490

Sheikh H, Mitchell P, Foth M (2023) Reparative futures of smart urban governance: a speculative design approach for multispecies justice. Futures 154:103266. https://doi.uam.elogim.com, https://doi.org/10.1016/j.futures.2023.103266

United Nations Human Settlements Programme (2022) World cities report 2022: envisaging the future of cities. United Nations Research Institute for Social Development

Zeiderman A (2008) Cities of the future? Megacities and the space/time of urban modernity. Critical Plann 23–39

Zhang Y, Yang Z, Yu X (2015) Urban metabolism: a review of current knowledge and directions for future study. Environ Sci Technol 49(19):11247–11263. https://doi.org/10.1021/acs.est.5b03060

Part II
New Conceptual Frameworks for Cities as Anticipatory Systems

Chapter 2
Cities as Anticipatory Systems

Salomón González-Arellano

Abstract The production of urban futures significantly influences public policies and decision-making. Futures emerge from a diversity of visions that are continually updated. However, amid this diversity, many urban actors share an institutionalization of futures. The institutionalization of futures contributes in a certain way to the anticipation of cities understood as an inherent property of complex systems. This chapter aims to demonstrate how anticipatory systems can be applied to cities. This anticipatory property of cities is neither homogeneous nor static in the various subsystems that make up cities. In this chapter, the notion of anticipation is decomposed into two dimensions: (a) the future disposition and predictive capacity of the systems and (b) the transformative or updating capacity of the system as a response to possible evolution. Because of their complex nature and decentralized system, cities present a diversity of complementary future and anticipatory regimes, and the diversity of future and anticipatory regimes of cities are shared among urban actors, forming urban anticipatory communities.

Keywords Anticipatory systems · Urban systems · Future regime · Anticipatory communities · Urban modeling

2.1 Introduction

There is an international consensus that the world will be increasingly urban and that cities have gained relative influence as they face the central governments of the nation-state (Herschel and Newman 2017; Johnson 2018). This idea attributes a hegemony to cities due to their demographic and economic dimension and identifies them as strategic spaces for triggering innovation processes. Hence, cities are conceived as decision-making spaces and promoters of change while having a central place in producing imaginaries of futures (European Commission 2019; Ferraris et al. 2018).

S. González-Arellano (✉)
Department of Social Sciences, Universidad Autónoma Metropolitana-Cuajimalpa, Mexico City, Mexico
e-mail: sgonzalez@cua.uam.mx

S. Gonzalez-Arellano and B. Gandlgruber (eds.), *Cities as Anticipatory Systems*, The Urban Book Series, https://doi.org/10.1007/978-3-031-78162-9_2

The production of urban futures significantly influences public policy design and decision-making (van Dorsser et al. 2018). Recent literature points to the importance of anticipation in public policy conception and implementation. These are practices that some authors refer to as "long-term policies," as they are conceived for at least one human generation and handle deep uncertainty both for the generation of the problem they address, as well as for the conception of responses to it (Koskimaa et al. 2021; Meissner 2012; Organization for Economic Cooperation and Development [OECD] 2019; Sprinz 2012).

From a strategic rational perspective, the construction of future scenarios can be considered a tool that influences the development of cities. This capacity to decide why and how to use imagination to introduce the non-existent future into the present is known as Futures Literacy and consists of a key competence in the development of collective intelligence of cities and territories (González Arellano 2014; Miller 2018). Thus, scenario production plays a communicative role by focusing on strategic and sensitive issues and synthesizing diverse visions and knowledge. In addition to the practical utility based on strategic rationality, imagining the future touches dimensions of culture, science fiction, religion, and even superstition (MacKay and McKiernan 2018).

There are many visions of the future of cities. Futures emerge from a diversity of visions that are continually updated with the arrival and consolidation of events in the present, new information, or changes in values. However, amid this diversity, concrete futures are institutionalized, identifiable, and shared by many urban actors. The institutionalization of city futures is influenced by images of hegemonic futures that hold particular agendas, such as those related to climate change, the notion of smart cities, the UN Sustainable Development Goals (SDGs), or global events such as a pandemic, terrorism, or mega-events (Olympic Games, concerts, summits of transnational organizations).

The process of institutionalization of the future identifies three "pillars" that interact in the construction of future images of cities: (a) the regulatory pillar, which guides (forces) governments and other city actors to generate future scenarios, often through the creation of laws, international commitments, urban planning agenda; (b) the cultural-cognitive pillar, which contributes to the construction of images of the future through shared knowledge and information of the various dimensions of the city; and (c) the normative pillar, which, with the set of values and norms common among urban actors, guides the construction of what the futures of cities should be (González-Arellano 2022). The institutionalization of futures contributes in a certain way to the anticipation processes of cities by incorporating predictive and prospective models of itself or the external environment into the city's formal and informal functioning and coordination.

This chapter aims to demonstrate how the notion of anticipatory systems, first developed by Rosen (1985) and later by other authors (Poli 2010; Miller 2018), can be applied to cities. This proposal is based on the following postulates: (a) cities can be understood as systems with properties such as self-organization, bifurcation, and adaptation, along with others that characterize complex systems; (b) anticipatory theories consider that systems (biological and social) can modify today (update) based on a predictive model being this an inherent property of anticipation (Rosen et al.), (c) based on the above ideas, cities can be understood as anticipatory systems, (d) because of their complex nature, and because they are a decentralized system, cities present a diversity of complementary future and anticipatory regimes, and (e) the diversity of future and anticipatory regimes of cities are shared among urban actors, forming urban anticipatory communities.

In this sense, it is relevant to better understand cities as anticipatory systems, especially in periods of unprecedented socio-ecological transformations. The current socio-ecological transition offers signals (weak and strong) of future changes that should trigger anticipatory processes of a very diverse nature in the cities' lives (González-Arellano 2024). This chapter proposes that understanding cities as anticipatory systems can open new perspectives on the new nature of the urban world and the actors and institutions that produce the city.

2.2 The City as an Anticipatory System (AS)

This section discusses the foundational concepts of anticipation that allow us to propose the cities as anticipatory systems. First, we discuss how the city has been conceived from the theory of systems and, in particular, how it can be modeled as a complex system. In a second moment, Rosen's proposal presents the concept of anticipation, followed by other authors. This section explains the two dimensions of anticipation (future imagination and transformative agency) and the anticipatory system (AS) notion. Finally, it is shown how the general notion of an anticipatory system can be an intrinsic attribute of the city.

2.2.1 The City as a Complex System

The city has been analyzed from many perspectives: historical, geographical, sociological, economic, political, and ecological. Its multidimensional nature is the origin of the emergence of the interdisciplinary field of urban studies. Among the lengthy repertoire of perspectives for the study of cities, approaches based on systems theory stand out. For more than half a century, cities and regions have been studied from systems theory. Synthetically, this approach involves identifying elements or components of the city or a region, their interrelationships, and the functions-structure that makes it a system (Pumain 2018). Indeed, this approach presents at least two scales that have dominated the specialized literature on analyzing urban systems. The first deals with urban systems resulting from interrelationships between cities at varying scales, such as regional, national, or international. These pioneering studies developed a theoretical framework of urban systems, which involves viewing cities not as isolated entities but as interconnected elements within broader regional or global systems. Urban systems theory considers the flow of people, goods, and information between cities to see how these interactions contribute to the overall dynamics of urban areas (Berry 1964; Blumberg 1971).

The second approach comprises urban systems within cities, such as road systems, transportation, energy grid, city services, urban food systems, and health systems. There are numerous versions and types of models of these urban systems (For a review of the evolution and diversity of urban systems modeling, see Batty 2005, 2013). For example, Blumberg (1971) presents a vision of the city from a series of subsystems: subsystems of spaces and structures of land uses, subsystems of functional elements of the city in social and economic terms, a third subsystem is transportation, and finally a system of services such as water, electricity, hospitals.

With the developing understanding and modeling of urban systems, concepts from complex systems theories were gradually incorporated into the study of cities. The contributions of complex systems theories to city studies improved the understanding of city analysis more broadly. Hence, the city as a complex systems approach adopts a definition as a system "in which large networks of components with no central control and simple rules of operation give rise to complex collective behavior, sophisticated information processing, and adaptation via learning" (Mitchell 2009 cited in Fabrizio Li Vigni 2020).

From French geography and with the theoretical contributions of complex systems and the theory of self-organization, geographers proposed remarkable advances in modeling cities and regions. Pumain and Sander have promoted a school of spatial modeling since the late 1980s and, in particular, of the self-organizing capacity of cities. These works show an excellent example of applying complex systems concepts to explain the behavior and evolution of cities. With the development of the new quantitative geography and the development of computer science, a path from the spatial modeling of cities allowed significant advances in the theoretical and applied reflection on the city as a system. An outstanding exponent is Batty (2005, 2013) and his team from the Center for Advanced Spatial Analysis (CASA) in London, who have contributed to the reconceptualization of the city, emphasizing its dynamic, open, adaptive nature, networks, flows, interactions, scales, places, and demonstrating the relevance of understanding the city as a complex system.

Although there are different approaches to analyzing cities from the complex systems theories, it is possible to identify common attributes that allow us to understand cities as complex systems. First, cities must be conceived as systems of systems, in other words, as a set of components that create subsystems constituting networks through interconnections. Second, interconnections facilitate flows of energy, matter, information, people, goods, and money, creating feedback loops that are more or less intense depending on the dynamism of places at different scales and distances. Third, this process is a property of dynamic and adaptive systems, giving cities emergent properties that translate into innovations of many different kinds: cultural, technological, social, etc.

City Components and City Subsystems. A city's physical infrastructure comprises various tangible structures, encompassing buildings, roads, bridges, utilities, parks, and other elements that constitute the city's physical environment. The human population residing and working within the urban area represents a pivotal component, exerting significant influence over the city's cultural milieu, economic landscape, and social dynamics (Pumain 2018). The economic systems inherent to cities are characterized by intricate networks encompassing businesses, industries, and employment opportunities, collectively contributing to the overall economic vitality of the urban entity. Concurrently, social systems within cities are shaped by the complex interplay of relationships, community structures, and cultural dynamics, collectively defining the city's social fabric and exerting substantial influence over individual behaviors and interactions among residents.

Urban Interconnections and Networks. The transportation networks within urban areas encompass a myriad of infrastructural components. This infrastructure plays a pivotal role in facilitating the seamless movement of individuals and enhancing accessibility throughout the urban expanse (Urry 2007). In tandem with physical transportation networks, cities are intrinsically linked by communication networks, constituted by digital infrastructures such as the Internet and telecommunications. These technological frameworks play a critical role in fostering the exchange within the city itself and extending beyond its geographical confines (Batty 2013).

Furthermore, the economic vitality of cities is intricately tied to supply chains, wherein the flow of goods and services assumes paramount significance. Cities rely extensively on the inflow of commodities from external sources, necessitating the establishment and maintenance of complex supply chain networks to meet the diverse needs of their populace. This dynamic interdependence underscores the multifaceted nature of urban systems and the integral role of urban networks in sustaining the functional coherence of contemporary cities.

City Feedback Loops. Economic feedback mechanisms within urban contexts denote the reciprocal relationship between alterations in economic conditions and ensuing effects on employment rates, thereby exerting consequential influence over consumer spending patterns and business activities. Fluctuations in economic variables contribute to a dynamic interplay that shapes the intricate network of economic interactions within the urban milieu. Similarly, environmental feedback processes delineate the reciprocal interactions between urban development activities and their impact on the local environment. The alterations brought about by urbanization can instigate environmental changes, thereby influencing the overall well-being of city residents. The intricate interdependence between urban development and environmental dynamics underscores the need for a comprehensive understanding of these feedback loops that help inform sustainable urban planning practices and policies prioritizing the holistic welfare of the urban populace and the surrounding environment.

Cities Emergent Properties. A distinctive amalgamation of individuals, traditions, and historical narratives constitutes cultural identity within an urban context. This unique cultural identity of a city materializes through the dynamic interactions among diverse individuals and communities. The resulting synthesis of these components contributes to forming a city's distinct cultural character, encapsulating the shared values, practices, and expressions that collectively define its identity. Furthermore, cities commonly function as focal points for innovation, fostering an environment where diverse ideas and talents converge. This convergence catalyzes the generation of novel technologies, artistic endeavors, and cultural trends. The intricate interplay of varied perspectives and skill sets within urban settings engenders a fertile ground for creativity to flourish. Therefore, cities emerge as dynamic hubs wherein synthesizing diverse intellectual and creative inputs engenders a continuous innovation cycle, influencing technological advancements and artistic and cultural expressions.

City Adaptability. Urban planning plays a pivotal role in the adaptive evolution of cities, as it involves formulating and implementing strategic development approaches to contend with prevailing challenges while proactively anticipating future needs. This process encompasses a comprehensive examination of the urban landscape, considering population growth, economic dynamics, and environmental sustainability to inform judicious planning and development initiatives. Simultaneously, the resilience of cities assumes paramount importance in the face of multifaceted challenges. Urban environments are susceptible to shocks, encompassing economic downturns, natural disasters, and public health crises. Navigating such adversities necessitates establishing adaptive systems and formulating contingency plans to mitigate the impact of disruptive events. The imperative for resilience underscores the significance of strategic foresight, robust infrastructure, and proactive governance in fortifying cities against unforeseen challenges and ensuring their sustained functionality and well-being.

Understanding cities as systems provides a holistic perspective that can inform urban planning, policy-making, and sustainable development efforts. It acknowledges the interconnectedness of various elements and the need for coordinated efforts to create and maintain a thriving urban environment.

2.3 Anticipatory Systems (AS)

As discussed in the introductory chapter, future studies have devoted relevant attention to anticipation. The relevance of recent contributions from the anticipation studies community points to the gradual consolidation of anticipation as a discipline (DoA) in its own right (Miller 2018). An excellent example of this dynamic field is the two volumes of the collective work Handbook of Anticipation, Theoretical and Applied Aspects of the Use of Future in Decision Making (Poli 2019) to realize the broad spectrum that anticipation theories offer as a legitimate research topic.

In a highly synthetic way, anticipation is understood as a process when "the future is used in action." In other words, anticipatory processes allow the translation of future scenarios or images into present behaviors. This process radically differs from reaction, which implements action after an event. A reaction should be seen as a complementary process to anticipation. Anticipatory processes can respond to predictive models outside or inside the system (Poli 2019). For example, a city can trigger an anticipatory process in the face of dangerous meteorological scenarios, such as launching preventive actions in the face of a hurricane whose extrapolation of its trajectory represents a threat. Alternatively, anticipatory actions can be developed in the face of demographic scenarios of its population, such as the creation of care policies or pension funds in the face of an aging population. From this approach, anticipation is composed of two dimensions: (a) a willingness and ability to imagine future scenarios, and (b) integrating this vision of the future to take some action.

The distinction between these two dimensions is important because, from this perspective, predictive or prospective exercises are not, strictly speaking, a form of anticipation since they are limited to generating visions of the future and do not lead to action in the present. As mentioned by Poli, these two dimensions of anticipation are consistent with Rosen's definition: "An anticipatory system is a system containing a predictive model of itself and its environment, which allows the system to change state at an instant in accord with the model's predictions about a later instant" (Rosen 2012:8, 313, originally published in 1985).

Considering the above, we can synthetically construct an anticipatory system for the city from other anticipatory system modeling exercises. Suppose we have the system of a particular city that we can consider as system S. City S can be understood as a dynamic system for which we can make predictions of states of S based on the knowledge of an initial state and input of the system. For example, a city's food system is dynamic with some degree of behavioral knowledge: production, distribution, sales, consumption, and waste. These are known with some degree of confidence.

However, let us explicitly incorporate a predictive model of city S into another system we will call M. As a dynamic system, the city system S evolves in real-time. At the same time, the possible trajectories for the predictive model M follow a time variable faster than the real-time of S. Therefore, any change observable in the model of M serves as a predictor of the behavior of some observable attributes of the city system S at a later instant. Regarding the example of the urban food system, the predictive model of system M will be able to generate trajectories of changes in the food supply due to a prolonged drought or war thousands of kilometers from the city system S (see Fig. 2.1).

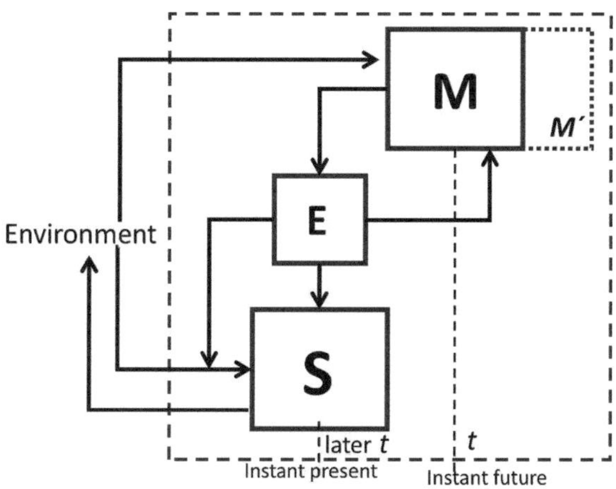

Fig. 2.1 Anticipatory system (adapted from Rosen 1985)

Now, let us establish that the predictive system M and the city system S are linked, allowing them to maintain specific interactions. From this operation, we can observe that the output of the observable in M becomes an input to the city system S. Since M is a model of the future state of the city system S, M then influences the present state transition in the city system S. Precisely, this operation is what we have characterized earlier as the anticipatory behavior of a system. An example of the interaction between M and S can be the predictive model of the population demographics of S. Specifically, M's future scenarios of population aging are articulated with the care and health systems of the city S.

Thus, the interactions between the system M and the city system S will be mediated by a set of operators, denoted as E, which give it the capacity to exert influence directly on S or on the environmental inputs that affect S. Thus, M exerts through E an influence to modify the dynamic properties inherent to the city system S. For M to qualify as a coherent model, the actions executed on S must be executed correspondingly on S. The schematic representation of this system is shown in Fig. 2.1 Similarly, we can build a system M with outputs incorporating predictions regarding inputs to the city system S. In this case, the present change of state of the city S will not depend on the predictive model of M, but instead on information related to future inputs to the city system S. An example would be to observe the evolution in the arrival of immigrants to the system S; this information feeds on the model M to generate a predictive model of the requirement of shelters and immigrants' dinners (canteens) of the system of the city S (see Fig. 2.1).

The schematic representation of this system is shown in Fig. 2.1. The result is an adaptive system in which anticipation of future behaviors (M) serves as the stimulus for alterations to the city system's (S) state. By encapsulating this arrangement in a unified framework, the resulting construct manifests as an adaptive system in which anticipation of future behaviors serves as the stimulus for alterations to the system's state. Accordingly, we will designate this system as an anticipatory system of the city S.

A system is more than any model thereof, as a model is, by definition, incomplete and is a reduction of reality. Consequently, the behavior predicted by a model will exhibit discrepancies from the system's actual behavior. The discrepancy between the predictive model M and the behavior of the city system S can be considered an error of the anticipatory system. This error should be considered when modeling the anticipatory system of city S (see Fig. 2.1).

Anticipatory systems are an inherent attribute in socio-ecological systems, which are complex systems that involve the interaction between social and ecological components. In this context, anticipatory systems refer to the ability of these systems to predict and respond to changes or disruptions in their environment, often to maintain sustainability and resilience. In socio-ecological systems, anticipation is essential for promoting sustainability, reducing vulnerabilities, and enhancing the resilience of communities and ecosystems. By predicting and responding to changes and disruptions, these systems contribute to the long-term well-being of human societies and the natural environment.

Anticipatory systems for urban planning involve using data, technology, and predictive models to proactively address various challenges and optimize the management of urban environments. These systems aim to enhance the quality of life for city residents, improve resource efficiency, and promote sustainable development. Some critical aspects of anticipatory systems in urban planning are Traffic and Public Transportation Management, Energy Efficiency, Urban Development (land use planning), Climate Resilience, Waste Management, and Urban Resilience (natural disasters and pandemics). Overall, anticipatory systems in urban planning are essential for creating more efficient, sustainable, and resilient cities. They leverage data-driven insights and advanced technologies to anticipate future challenges and opportunities, enabling cities to make proactive decisions and improve the quality of life for their residents.

2.4 From Future Regimes to Anticipatory Communities

As mentioned above, anticipation contains two dimensions: (1) the disposition of actors to imagine the future and (2) incorporating the future into the transformative capacity of the system. Both dimensions behave differently according to the capabilities of each system: on the one hand, its capabilities or disposition toward some future regime, and on the other hand, its capacity and disposition to transform the system based on the future vision of itself and the external environment. In a way, these two dimensions correspond for the model in Fig. 2.1 to the predictive model system M (dimension for modeling the future), and the component E, a set of operators that, based on M, influence the system S (transformational dimension of the system S).

To better understand the first dimension of anticipation, let us introduce the concept of "regimes of the future" described by French sociologists Chateauraynaud and Debaz (2017). The authors propose the notion of the future regime as the disposition to imagine and practice the future by actors, individuals, and organizations: How do actors conceive of time? Is it linear or non-linear, open or closed, slow or accelerated? Table 2.1 synthesizes its main characteristics under three axes and identifies future regimes: urgency, expectation, anticipation, prediction, foresight, promise, prophecy, and science fiction. A future regime is defined as a set of shared beliefs and practices around a specific scenario and representation of the future, a particular type of project of understanding evolution, and a specific model of science, policy, or even fictional interface.

For the second dimension of anticipation, the transformative capacity, i.e., the actions triggered by the predictive model of M on S, depends on several variables. First, as already mentioned, the predictive model generated by M should contain aspects that are relevant for S. Second, the interface E that allows the deployment of operators that make M coherent over S may present biases of at least four types: (a) the normative bias, (b) the shared beliefs about time and causality, (c) the agency

Table 2.1 Future regimen adapted from Chateauraynaud and Debaz (2017)

	Time modeling	Logic of action
High-frequency	Continuous flow	Visual navigation; uncertain anticipation in a very short time
Urgency	Time is short	Race against time to save a situation
Expectation	Suspended time	Continual displacement of expectation time
Optimization	Adaptation to an open future	Automatization of the learning process; creative randomness
Anticipation	Accelerated time	Action over an ongoing process; preparedness
Prediction	Calculated and linearized time	Calculation space planning
Prospective	Time is non-linear and open	Scenario production
Promise	Delay logic	Engaging in a promise, proving one's credibility
Prophecy	Eschatological time	Announcing an inevitable future

available to transform both the system and the environment, (d) the diversity and quantity of material and immaterial resources available.

From the intersection of these two dimensions, it is possible to identify diverse ways of practicing anticipation at the individual and collective levels. That allows us to use the notion of anticipatory communities developed by Li Ving. For this chapter, we understand as anticipatory communities the group of actors who share a specific anticipatory practice, i.e., they share both a future regime and anticipatory policies as their transformative capacity (i.e., a prescriptive vision of the environmental governance we need to implement). Therefore, communities of anticipation share a vision of the future, a prescriptive ideology, and a practical investment in a city and governance project with which to face the future.

We can expect diverse actors, lifestyles, interests, functions, and interactions in a complex social system like cities. Amid this diversity, there are affinities of all kinds: political, professional, ideological, religious, residential, food, sports, clothing, music, and others. These affinities can take the form of formally or informally recognized communities. In addition to this diversity, the actors have unequal transformative capacity to transform their system and influence their environment.

Figure 2.2 shows an adaptation of the proposed future regimes grouped by their similarity in interweaving the two dimensions of anticipation: (a) temporal dimension, frequency of change patterns, and (b) transformative capacity or updating of the system. As shown in the figure, it is possible to identify five groups corresponding to the relationship between the two dimensions of anticipation: future regimes and transformative capacity to update the present.

The first group of anticipatory practices (C1) responds to high-frequency, random, with little time of anticipation, and highly changeable (or with the flow of time stopped in the case of expectation) situations. Here, we find the predictive modes

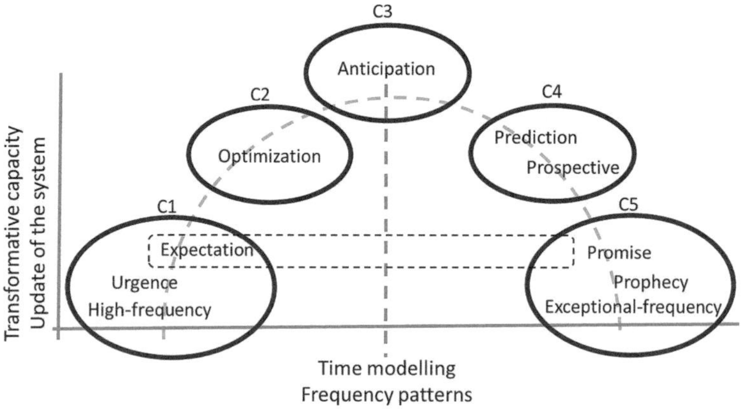

Fig. 2.2 Future regimes and transformative capacity of the system. *Source* Prepared by the author

of high-frequency events, urgency, and expectation, which urban actors treat with a low transformative capacity of the system. This group of anticipatory practices is very close to reactive systems because they are very close in time to events (almost "too late"). In this group, as in C4, we have included the expectation of maintaining a poorly defined short and long temporal range.

The practices we have named optimization (C2) start from a future regime with high-frequency and regularity event patterns, which allows a high degree of predictability. Actors adopting optimization as anticipatory practices modify the system, relying on this high degree of predictability to maximize benefits and reduce costs. It is a regular process of learning and adaptation with a low degree of transformative change aimed at improving the current functioning of the system.

Anticipation (C3) is the ideal way of updating the system in response to a predictive model of the system or its environment. It is more potent than optimization, prediction, and foresight because it adopts a greater transformative capacity and updates the state and functioning of the system. It is similar to prediction and foresight in generating scenarios in that it is less interested in contingency and more interested in medium- and long-term scenarios and is open to alternative scenarios. This practice is the one that, in the strictest sense, corresponds to an anticipatory system.

The prediction and prospective practices of the group (C4) often adopt a medium and long-term future regime. The objective is to elaborate scenarios in the realm of the possible under a linear or tendential vision for prediction and a non-linear and more creative, alternative, and exploratory one for prospective. In both cases, their primary purpose is to stimulate the imagination and is less about acting on the system's present state. Although they are inputs for anticipation, their primary purpose is not to update the present.[1]

Finally, promise and prophecy (C5) are, in the extreme, not very anticipatory practices because they mobilize few actors due to their distant time horizon. Some of these visions of futures go beyond the present generations and move into the realm of utopia, collapse, or, at best, hope. In this group, as in C1, we can include the expectation of maintaining a poorly defined temporal range.

2.5 Anticipatory Communities

We can understand an anticipatory community as a collective of actors that share (a) a future and anticipatory regime, (b) standard causal models of how the world works, and (c) shared future policies. From the above, it is possible to identify four types of anticipatory communities of cities: (a) the Repairers, (b) the Managers, (c) Anticipators-Strategists, the Visionaries-Counselors, and the Utopians-Prophets.

2.5.1 Repairers

This community corresponds well to the future and anticipatory regime C1 in Fig. 2.1 Urban actors with care and alert functions (anticipation) in the face of high-frequency events, emergencies, and the expectation of events with very low predictability. This type of anticipatory community group has actors who have identified recurring patterns of high, random frequency and have little time to anticipate. Because of the quasi-random or sudden nature, the causality of events is often known, although the actors in the system have little control. Security, civil protection, and surveillance systems adopt this anticipatory mode. Part of the anticipatory actions of urban traffic management work on "sudden events with little reaction time."

The anticipatory policies of this community are characterized by the design and practice of protocols that allow, through protocols, the automatization of a series of appropriate behaviors for events (accidents) that follow familiar patterns. These communities are in many urban systems, in charge of governmental and non-governmental actors. Earthquake protocols have been an established practice in Mexico City for several decades. Unfortunately, in recent years, in some cities, the practice of protocols against attacks by organized crime or terrorist actors has been established.

---

[1] Although constructing visions of the future can transform present consciousnesses, these approaches do not, in their pure version, aim to transform the functioning and present state of the system.

2.5.2 Managers

This group of urban actors forms a community that is a suitable adopter of the anticipatory regime based on optimization practices. Like the repairers, they do not seek a transformative change in the state or functioning of the system but rather an incremental improvement based on learning predictable behavioral patterns. Their relationship with the future is usually short and medium-term, with a reliable internal predictive model and low uncertainty. Indeed, unlike the group of repairers, managers focus on anticipation in the face of predictive models internal to the system, and therein lies their most excellent agency in establishing modest changes to improve the current performance of system functions.

The anticipatory policies of this community can come from strategic management practices, such as continuous improvement. Some examples of this can be urban systems such as transportation, energy, or food waste management, which, based on regular patterns and low uncertainty, try to maximize benefits and minimize costs, such as the optimal design of service routes and energy consumption schedules. This anticipatory community can be found in the governmental sector, companies, and social and political organizations whose logic of maximizing benefits differs from the economic one and translates into social or political impact with the least effort.

2.5.3 Anticipators-Strategists

The community of anticipators is composed of urban actors with a high level of agency, capable of influencing the state and functioning of the system based on internal or external predictive models. They are actors with a good understanding of the complexity of urban systems, able to identify relevant interactions between components, redundancies, and bifurcations. Actors of this anticipatory community share more complex causal models than the previous groups, are familiar with medium and long-term future regimes, have a creative and open perspective on predictive models of the future of cities, and are alert to interactions with the environment as external to the system.

The anticipatory policies of this community are characterized by adopting both a pragmatic and creative stance toward the construction of future scenarios often related to relevant problems such as climate change, food, population aging, and others. Predictive models of these problems require putting transformative actions beyond improving (optimizing) the current state and functioning of the systems. For this reason, these communities will generally be outside the government or organizations' management routine (managers). Some of the profiles of the members of these communities are artists, creatives, activists, consultants, politicians, entrepreneurs, or academics. We can distinguish three types of anticipators: (a) internal anticipation, oriented to prepare for the predictive model of the future; (b) deep anticipation, the transformers of the future scenario; and (c) open anticipation, which is the anticipation of the future.

2.5.4 Visionary-Advisors

This group of actors share future visions and anticipatory regimes based on prediction and prospection. In urban systems, these communities often present themselves as experts capable of constructing medium- and long-term images of a parcel in urban reality. They often share techno-scientific rhetoric for constructing medium and long-term scenarios based on rational, objective, and more or less transparent methods.

Their anticipatory practices are limited to the construction of predictive models that offer clarity of the future. The members of this anticipatory community may be close to or distant from decision-makers and may present a diversity of methods and very diverse topics. From forecasting a linear and trend-based vision of the future, we find actors who model the territorial growth of cities, demographic scenarios, and even some climate scenarios. On the other hand, prospection is also oriented toward creating scenarios, but in a less deterministic and trendy way than prediction. Emphasis is placed on the diversity of alternative futures.

This community is essential as it represents many actors involved in the institutional construction of city futures. Although, in principle, they do not seek to modify the current state and functioning of the system since their function is to generate future scenarios. Their influence lies in the production of futures. Institutional processes, such as producing urban development plans, demographic or climate scenarios for programs or laws, and more, often endorse it. In addition, this group can contribute to the Anticipator-strategist community by providing inputs to create their predictive models.

2.5.5 Utopian-Prophets

At the extreme side of the dimension for time modeling are actors who share concerns about the long and very long-term future. This anticipatory community comprises visions of the future that fall into utopia or collapse. Some scenarios surpass the present generation's horizon, the city's existence, or even humanity. In this group, we also include anticipatory practices of the "promise" or "hope" type, in the sense that they share a mobilizing force and expectation that presents a low transforming agency of the current state and functioning of the system.

Their anticipatory policies are characterized by low or no agency over the system's present state. Their purpose is to warn, orient, and make the value systems of the current social and ecological system visible. In these communities, hopeful, utopian, dystopian, and catastrophic positions can be observed simultaneously. In this community, we can find actors of diverse profiles, from artists, creative people, science fiction writers, religious people, philosophers, and activists.

2.6 Anticipatory Operators

These typologies of anticipatory communities are ideal models; in the real life of cities, the actors share more than one community, and some of the anticipatory practices are superposed and adapted according to the situation and the personal and institutional resources available. None of these communities is better or worse than the others, as they are all products of culture and the shared futures literacy in the incessant production of cities.

So far, the examples of anticipatory systems in cities suggest a process focused on actors, i.e., on anticipatory actions performed by individuals or organizations. It is possible to broaden the notion of an anticipatory system by adopting a broader system view, going beyond human actors. Actor-network theory (ANT) allows, at least in a suggestive way, to imagine an anticipatory system of (human and non-human) actors. ANT approach emphasizes the importance of non-human entities, or "actants," in shaping social interactions and networks. This approach can help model anticipatory systems, especially because, as demonstrated by Rosen (1985) and numerous examples (Poli 2019), anticipation is an inherent attribute of both human and non-human systems. Lussault (2007) adopts the notion of actant and the main postulates of ANT to reflect on the role of spatial entities (human and non-human) in organizing societies. Lussault replaces the term actant with that of the spatial operator and, similarly to ANT, stable human and non-human spatial operators as actants with a spatial agency.

Therefore, following Lussault and ANT, anticipatory operators can be human and non-human actors that maintain a network of interactions in an anticipatory system. From this point of view, an anticipatory system of cities can incorporate the notion of anticipatory operators formed by persons and organizations, and other types of operators such as non-human living organisms (animals and plants), geophysical operators such as the weather, or active volcano, immaterial operators such as laws, information, values, or collective beliefs about the future. Human and non-human anticipatory operators establish networks by interactions, and in the case of cities, these networks are mainly decentralized and multi-scale (neighborhood, city, region, country, world).

The anticipatory function of cities is a distributed property, not centralized in a subsystem specialized in that function. However, it can be observed that in some cities, both in the government bureaucracy and in the organization of companies, departments, or teams formally dedicated to foresight, anticipation is an inherent attribute in each part of the system. That allows anticipatory practices to be developed at any scale of the system, to a greater or lesser extent, and from different approaches.

As already mentioned in another text, the institutionalization of the futures of cities is built from regulatory, normative, and cognitive arrangements (González Arellano 2022). We consider that the above is valid for forming anticipatory regimes and communities. This process establishes a hegemonic future that imposes conservative imaginaries and a narrow margin for imagining scenarios of cities "outside the box." That takes on even greater importance in a socio-ecological transition phase characterized by disruptive regime shifts (energy, climate, food, geopolitical) that occur with high uncertainty, complexity, and innovation. Faced with this unprecedented scenario, we fear that traditional planning and future construction methods are becoming less and less adequate.

2.7 Conclusion

As discussed above, the central idea of this chapter is to understand anticipation as an inherent attribute of socio-ecological systems, that is, as a property intrinsic to the functioning of systems. In this sense, and as developed in the first section of this article, if cities are understood as complex systems, then anticipation is a functional property of cities.

Cities as complex systems have the inherent property of anticipation. This anticipatory property of cities is neither homogeneous nor static in the various subsystems that make up cities. This chapter has reviewed the notion of anticipation, starting from two dimensions: (a) the future disposition and predictive capacity of the systems and (b) the transformative or updating capacity of the system as a response to possible evolution.

Following Rosen's founding proposal, we have adapted the general model of an anticipatory system for the city system in this chapter. This proposal was based on three aspects: (a) cities are complex systems, (b) anticipation is an inherent attribute of complex systems, and (c) therefore, cities present diverse anticipatory policies corresponding to the varied future regimes and the unequal transformative capacity of urban actors. In this chapter, we identified five modalities explained by the future regime adopted and the transformative agency of the system's actors: repairers, managers, anticipators-strategists, visionary advisors, and utopian prophets. These anticipatory types are not mutually exclusive nor exclusive to human actors; non-human actors also participate in the anticipatory systems of cities.

The anticipatory policies of cities are the result of the institutionalization of the future that cities develop. As mentioned, besides being diverse, anticipatory practices follow differentiated institutionalization processes. On the one hand, we can find formal or informal regulations that reward or punish certain actors for developing anticipatory processes based on very preset models or scenarios. An excellent example of this is laws and other types of incentives at all scales that address global climate change scenarios. In addition to regulatory instruments that incentivize cities' anticipatory practices, certain actors also share the normative dimension of what the city should be.

An example is the Charter for the Right to the City signed by Mexico City, which constitutes the collective construction of a city model to which an important group of actors adhere. This model of the future of the city takes the form of anticipatory practice in concrete and performative actors, such as the signing of the charter by the government of Mexico City's government in a public act and accompanied by stakeholders of the city, which had even influenced the drafting of the first Political Constitution of Mexico City in 2017. In addition to the regulatory and normative dimensions, a third dimension contributes to institutionalizing the future and anticipatory practices in cities. The cognitive and cultural dimension is a source of incentives for actors to build anticipatory models and anticipatory actions in cities. Access to information on the internal urban systems or the external environment of cities can trigger anticipatory initiatives. A recent and familiar example for us is the global pandemic of COVID-19. The information about the pandemic that quickly circulated the world allowed many governmental and non-governmental actors to take various anticipatory actions. A great example is governments initiating negotiations with pharmaceutical companies to purchase vaccines for the public and private sectors.

These three dimensions, regulatory, normative, and cognitive, interact and together contribute to the construction of coordination, conventions, and agreements, formal and informal, between actors on future and anticipatory regimes, more or less consensual between local or external urban actors. Here, such anticipatory regimes and communities are operating at various scales. International humanitarian aid can be an example of this.

Finally, the understanding and development of anticipatory systems often involves a combination of technological innovation, data-driven decision-making, collaboration, and a commitment to addressing specific challenges or needs. Public institutions, private organizations, and communities may employ different strategies and approaches to create and implement these systems depending on the context and goals. Comparing urban systems as anticipatory systems involves assessing how different cities or urban areas use future scenarios and transformative capacities supported by data, technology, institutional, and planning strategies to anticipate and address various challenges and opportunities.

References

Batty M (2005) Cities and complexity: understanding cities with cellular automata, agent-based models, and fractals. The MIT Press, Cambridge, MA, p 565

Batty M (2013) The new science of cities. The MIT Press, Cambridge, MA

Berry BJL (1964) Cities as systems within systems of cities. Pap Reg Sci Assoc 13:147–163

Blumberg DF (1971) The city as a system. Simulation 17(4):155–167. https://doi.org/10.1177/003754977101700404

Chateauraynaud F, Debaz C (2017) Aux bords de l'irréversible. Sociologie pragmatique des transformations. Éditions PETRA

European Commission (2019) Cities as innovation hubs. European Commission, the future of cities: opportunities, challenges and the way forward. Publications Office, pp 105–112. https://doi.org/10.2760/375209

Ferraris A, Santoro G, Papa A (2018) The cities of the future: hybrid alliances for open innovation projects. Futures 103:51–60. https://doi.org/10.1016/j.futures.2018.03.012

González-Arellano S (2022) The future scenarios of cities: an analysis of their institutional construction. Int J Urban Reg Res 46(5):865–873. https://doi.org/10.1111/1468-2427.13103

González-Arellano S (2014) Inteligencia territorial y la observación colectiva territorial. Espacialidades 04(02):89–109. http://espacialidades.cua.uam.mx/vol/04/2014/02/04_Gonzalez.php

González-Arellano S (2024) Tipología de escenarios futuros de ciudades. Revista EURE—Revista De Estudios Urbano Regionales 50(151). Recuperado a partir de https://www.eure.cl/index.php/eure/article/view/4911

Herrschel T, Newman P (2017) Cities as international actors. Urban and regional governance beyond the Nation State. Palgrave Macmillan UK. https://doi.org/10.1057/978-1-137-39617-4

Johnson C (2018) The power of cities in global climate politics. Saviors, supplicants, or agents of change? Palgrave Macmillan London/City Leadership Laboratory, University College London (UCL). https://doi.org/10.1057/978-1-137-59469-3

Koskimaa V, Rapeli L, Hiedanpää J (2021) Governing through strategies: How does Finland sustain a future-oriented environmental policy for the long term? Futures 125:102667. https://doi.org/10.1016/j.futures.2020.102667

Li Vigni F (2020) Five anticipation communities in complex systems sciences: complexity science and its visions of the future. Futures120:102551. https://doi.org/10.1016/J.FUTURES.2020.102551

Lussault M (2007) L´Homme spatial : la construction sociale de l'espace humain. Seuil

MacKay RB, McKiernan P (2018) Scenario thinking: a historical evolution of strategic foresight. Cambridge University Press

Meissner D (2012) Results and impact of national Foresight-studies. Futures 44(10):905–913. https://doi.org/10.1016/j.futures.2012.07.010

Miller R (2018) The discipline of anticipation. Exploring key issues. In: Miller R (ed) Transforming the future: anticipation in the 21st century. UNESCO, Routledge

Mitchell M (2009) Complexity. A guided tour. Oxford and New York, Oxford, University Press

Organization for Economic Cooperation and Development (OECD) (2019) Strategic foresight for better policies. https://www.oecd.org/strategic-foresight/ourwork/Strategic/Foresight/for/Better/Policies.pdf

Poli R (2010) The many aspects of anticipation. Foresight 12(3):7–17

Poli R (2019) Handbook of anticipation. Theoretical and applied aspects of the use in decision making. Springer

Pumain D (2018) An evolutionary theory of urban systems. In: Rozenblat C, Pumain D, Velasquez E (eds) International and transnational perspectives on urban systems, Springer

Rosen R (1985) Anticipatory systems. Philosophical, mathematical, and methodological foundations. Oxford, Pergamon Press (2nd edn Springer 2012)

Rosen R (2012) Anticipatory systems philosophical mathematical and methodological foundations. Springer, New York

Sprinz DF (2012) Long-term environmental policy: challenges for research. J Environ Dev 21(1):67–70. https://doi.org/10.1177/1070496511435667

Urry J (2007) Mobilities. Polity

Van Dorsser C, Walker WE, Taneja P, Marchau VAWJ (2018) Improving the link between the futures field and policymaking. Futures 104:75–84. https://doi.org/10.1016/j.futures.2018.05.004

Chapter 3
"Hopeful Determination" as an Urban Anticipatory System—An Enactive Approach

Blanca Miedes-Ugarte and Manuela Á. Fernández-Borrero

Abstract This chapter offers a compelling exploration into the dynamics of urban development through the lens of hopeful determination. This concept is illuminated through the case study of Plan Integral del Distrito 5 de Huelva (PID5H), showcasing how urban communities, equipped with a vision for the future, navigate and shape their developmental trajectories. The chapter intricately examines the interplay of foresight, adaptability, citizen engagement, and multi-stakeholder cooperation in fostering urban resilience and proactive change. It underscores the transformative power of shared narratives in rallying collective efforts toward inclusive and sustainable urban futures. By anchoring its argument in enactive cognition theory, the chapter argues that urban actors do not merely react to their environment but actively participate in its creation and evolution, embodying hopeful determination in their approach to urban challenges. This innovative perspective offers a nuanced understanding of how urban areas can anticipate and adapt to change, making hopeful determination a vital component of urban planning and development.

Keywords Hopeful determination · Enactive urbanism · Resilient cities · Participatory development · Anticipatory urban systems

3.1 Introduction

In this day and age, cities are facing challenges like never before. These challenges include socio-economic issues, such as the pursuit of prosperity by ensuring decent living conditions for the entire population and integrating different ways of life. They also have environmental challenges around managing urban metabolism, the effects of climate change, pollution, and biodiversity loss. Cities also face rapid

B. Miedes-Ugarte (✉) · M. Á. Fernández-Borrero
COIDESO-Universidad de Huelva, Huelva, Spain
e-mail: miedes@uhu.es

M. Á. Fernández-Borrero
e-mail: manuela.fernandez@dstso.uhu.es

S. Gonzalez-Arellano and B. Gandlgruber (eds.), *Cities as Anticipatory Systems*, The Urban Book Series, https://doi.org/10.1007/978-3-031-78162-9_3

cultural and technological changes that affect how people think, behave, and live together. These tensions are most acutely felt in the most deprived urban areas. Against this backdrop, there is a need for approaches to urban design that address existing problems and anticipate and prepare cities for an uncertain future. Here is where the concept of "hopeful determination" as an urban anticipatory system comes into play. The authors conducted an ethnographic study of the *Plan Integral del Distrito 5 de Huelva* (PID5H), a multi-actor, multi-level project initiated in 2000 to regenerate a deprived urban area of approximately 15,000 inhabitants in southwestern Spain. Through the lens of the enactive cognition approach, this chapter explores how the combination of long-term vision, adaptability, active citizen participation, multi-actor collaboration, and a multi-dimensional approach can build cities that not only confront but proactively anticipate and transform themselves in the face of emerging challenges, provided that actors are willing and able to at least partially articulate their visions of the future.

The chapter is divided into four sections. It introduces the enactive approach (Varela et al. 2017) as the basic conceptual framework, followed by discussing the concept of "hopeful determination" in the urban context using the three-horizon model (Sharpe 2013; Sharpe et al. 2016). The authors, coordinators of the Plan's scientific and technical team for over a decade, examine the evolution over time of the future horizons of the actors involved, as well as the conflicts that arose and the commitments made during the different phases of the Plan. This section also discusses the competencies developed by those involved in regenerating the urban space and co-constructing a more inclusive and resilient future.

More than a mere methodology or strategy, "hopeful determination" is a praxis, an integrated way of being, doing, and knowing in which actors and environment mutually model each other in an ongoing sense-making process in which mobilizing narratives play a defining role (Liveley 2017). The concept calls for engagement and pro-activity in urban design and reinvention. By recognizing the link between urban challenges and the inherent capacity of communities to innovate and adapt, this chapter explores the opportunities and constraints on the ground for building more resilient, inclusive, and sustainable cities.

3.2 An Enactive Approach to "Hopeful Determination"

The world can be considered a "place for things" or a "field for action." In the former, modern societies took a scientific approach to interpreting the world to define these things' verifiable and agreed-upon properties. We also explore how to use them efficiently for various purposes where appropriate. However, the world as a forum for action—an older view that includes the world as a place for things—is a place of "value, a place where everything makes sense," a sense that tells us who we are, that is shaped in social interaction, and that serves as an interpretive framework for action (Peterson 2002). Knowing what something is involves recognizing its particular attributes and motivational significance, i.e., what this means for its behavior.

That "something" must evoke an emotional response to garner enough interest for examination and analysis. Scientific methods try to remove affect from perception so that experience can be described universally. They are, therefore, woefully unsuited to the task of guiding action involving those all-pervasive emotional aspects of the human process.

The future can also be seen from the point of view of things to be controlled or instrumentalized. What we forecast or foresee will happen (predict) in the probable, plausible (explored) future. It can also be seen in terms of action, what we want to happen, or what we think should happen—desirable, normative, and prospective futures (Miller et al. 2013). All our ideas about the future, which we are more or less aware of, are a yardstick by which we judge the present. The envisioned, desired, and imagined future is a guiding framework for evaluating the present and informing action in the here and now, whether reacting to whatever is unfolding or taking transformative action to create new life possibilities. Our way of being in the world is anticipatory.

Understanding how we come to know is fundamental to understanding our relationship to the future. Recent advances in neuroscience suggest that the human way of making sense of the world is inherently predictive—indeed, it may be the brain's primary mode of operation. In a recent exploration of the science of consciousness, Seth (2021) posits that consciousness is a hallucination of the brain, akin to a controlled dream in which the brain makes its best prediction about what is going on in the outside world in response to incomplete and ambiguous sensory cues. That implies that our conscious experience of the world is a personalized internal simulation and that our experience of "reality" is an interpretation crafted by the brain. The science of emotion further explores this pivotal role of prediction in human existence (Barrett 2017). Emotions are presented as predictions that prime the body to act in line with learned expectations and environmental demands and are heavily influenced by cultural context and social norms. Concepts and language help us to construct this social reality.

In this vein, O'Hara and Leicester (2012) include such *epistemic literacy* among the three critical competencies required for navigating the VUCA (vulnerable, uncertain, complex, and ambiguous) landscape of the twenty-first century, along with an understanding of one's agency (*psychological literacy*), and the ability to read and facilitate cultural processes (*cultural literacy*).

The enactive approach to cognition is an approach that emphasizes the vital role of prediction in cognition, offers a holistic perspective that connects being, doing, and knowing, and provides a good account of how knowledge, agency, and cultural processes are intertwined. This approach has been gaining ground in recent decades, is naturalistic but not reductionist, and draws on sources from phenomenology, biology, complexity sciences, and cognitive sciences (Varela et al. 2017; Thomson 2007; Stewart et al. 2010; Di Paolo 2021; Di Paolo and De Jaegher 2012, 2022; Di Paolo et al. 2023; Sepúlveda-Pedro 2023). It distinguishes itself from older approaches that view cognition as informational processing based on internal representations. In contrast, the enactive approach sees cognition as an emergent process in the interaction between the organism and the environment, giving greater prominence to the whole body, in which the mind is embodied in perception and thought. Anticipation, viewed from this vantage point, is deeply embedded in action, in lived experience. Living beings react to the world around them and anticipate it in ways that allow them to adapt to and navigate their environment proactively. This anticipation operates both at the cognitive and physical levels through sensorimotor coupling. Organisms anticipate as they actively interact with the environment through patterns of behavior that have proven effective in the past. Anticipatory action generates consequences, which in turn inform and shape future actions.

The enactive approach recognizes the precarious nature of these processes, which are invariably prone to error, a certain amount of uncertainty, and emergent phenomena. Anticipation does not necessarily produce a predetermined action or expected outcome; instead, it guides the organism in its ongoing interaction with the world. From this perspective, anticipation is rooted in action dynamics (*praxis*). It is part of a continuous process of adjustment and adaptation involving the whole organism with its environment, as opposed to isolated brain processes. Here, mind and body are intertwined, and the whole organism is intertwined with the world in a co-creation process. Enactive cognition suggests that living organisms actively forge and maintain their identities through interacting with the world around them. Our way of knowing the world is based on something other than anticipating it, as the theories of consciousness and emotion mentioned above suggest. Instead, we are co-creators of the world, "bringing forth" only what is relevant to us.

Thus, doing and knowing are recognized as ways for organisms to "be" alive. Organisms "bring into being" or "make real" their world through their interactions with it. Perception is not merely a passive representation of reality; we perceive the world through our potential actions. We perceive things not only according to their attributes but also by what we can do with them.

A crucial element of this approach is the link between anticipation and "sense-making," whereby the organism interprets its environment in the context of its "affordances." These affordances arise from the interplay between the environment and the organism's capabilities, influenced by biological requirements and the accumulated experience of interactions. From an affective or valuational point of view, this perspective holds that the world has no intrinsic meaning independent of organisms. Organisms are embodied agents that create meaning (*enact*) through these interactions with the world. What is vital to one being may be of no consequence to another. Sense gives meaning and relevance, guides behavior, and creates a set of norms. Therefore, action is intentional and directed toward perpetuating one's life and realizing one's goals by what happens in the surrounding environment.

Enaction emphasizes continuity and innovation among natural processes, life, the mind, language, and human communities. For human beings, agency manifests itself in deeds, intentions, emotions, thoughts, relationships with others, speech, community involvement, and work. *"Embodied agents* enact *a world of significance, an ongoing achievement done in conjunction with other agents. Acts, moreover, are processual; they respond to existing potentialities, trigger or block processes of individuation, and have rippling repercussions in a meshwork of other acts past and future"* (Di Paulo and De Jaegher 2022).

Human beings are constantly formed, whose identity and understanding of the world are forged through an interactive process of attributing meaning, or "sense-making." It is not a purely individual process but one deeply rooted in social and cultural dynamics, where emotions are interwoven and shared and can, in turn, be in harmony or conflict, with unequal effects on our life chances. Through language, our actions are transformed into forms of knowledge and power, encompassing practical skills, ethical reasoning, and making and breaking habits and norms. These human actions are an integral part of the values that are the basis for the success or failure of our social interactions. They are also connected to a more extensive network that includes the subtleties of behavior, such as attitude, style, or mood. In this way, they form the ethical reality that must be continually updated.

This framework can accommodate phenomena such as hope or commitment and complex and profound states beyond the conscious anticipation of future events. In this light, hope is not a definite expectation as it is a guiding sense toward something inherently good that cannot be fully understood or anticipated. We are inspired by those who exhibit hope or courage, awakening previously hidden potentialities and influencing our actions from a pre-rational stage. Hope and commitment can transform what might be called "pre-individual potentialities," charging them with an energy that has not yet manifested itself but is already directing our moral compass and, thereby, our actions. Similarly, acts of participation bring about a shift in collective consciousness, which involves not only a cognitive shift but also a change in ethical possibilities, giving life to new practices and thus to new realities (Di Paulo and De Jaegher 2022).

Here is where the concept of "hopeful determination" makes sense as a way of being doing knowing that implies an active commitment to shaping a desired future. It transcends the individual and weaves its way into webs of interconnectedness and affect. As Motti (2017) notes, this resonates deeply with the rhythms of Deleuze (1994) and involves the co-creation of meaning with others and the world in a context of uncertainty. "Hopeful determination" perseveres in the face of adversity through adaptability and resilience in an ongoing effort to transform potentialities into concrete realities (Coleman 2017; Cook and Cuervo 2019; Doe 2020; Benton 2019). It emerges as a vital component in all human endeavors, cities being among the most complex, playing a decisive role in their current and future configuration.

3.3 Hopeful Determination in the Urban Context

Cities, seen through the lens of "things," emerge as hyper-complex systems, a mesh of deterministic inertias and emergent phenomena. To the outside observer, they represent a confluence of chance and necessity (Prigogine and Stengers 1979), whose trajectory could be more or less reliably anticipated using various methods (Miller et al. 2013). That would depend on the magnitude of this inertia (demographic patterns, economic and technological infrastructure, communications, institutions, social and cultural habits) and the stability of their operating context (the surrounding political, social, and environmental trends).

Looking at the city from an action perspective, however, adds a crucial element to the interplay between chance and necessity: the intricate dynamics of the conscious will of human actors and their efforts to steer the future (Godet and Durance 1997) toward a desired destination, guided by their values and expectations, or what we refer to here as their "hopeful determination." That will is crucial in urban planning, as it stems from prevailing visions of the future. These visions, in turn, help to mitigate uncertainty by enabling the actions of urban actors to be coordinated, fostering stability and predictability, and allowing for a certain level of foresight in the short and medium term. Thus, the co-creation of the future in urban environments emerges from the interplay between the capabilities (knowledge and power) of the different actors and groups involved in urban governance, all striving to realize these visions of the future.

A particularly heuristic model in this regard is the three-horizon model (Sharpe 2013; Sharpe et al. 2016), which aims to identify the underlying patterns in actors' behavior, i.e., their value systems and attitudes toward continuity and change, or, in the author's words, their "patterns of hope."

The three-horizon model shines a bright light on how three essential visions unfold at every unique intersection of actors in space and time. Horizon 1 reflects the perspectives of those actors who, clinging to control and management of the current system, resist change. Nevertheless, they are propelled into a future that may be unsustainable, undesirable, or both, and face the dichotomy of radically transforming themselves or fading into oblivion. Horizon 3, on the other hand, represents the visions of the more farsighted agents of change, those who are inclined to create favorable scenarios based on principles of fairness and environmental sustainability but who are hampered by a lack of resources to make a substantial impact in the short, medium, or long term. Finally, Horizon 2 embodies the visions of those pragmatic enough to combine Horizon 1 resources and Horizon 3 aspirations to effect significant change toward a preferred course. However, they risk being overwhelmed by the inertia of those in Horizon 1, who may allow change only if they maintain their dominant position.

In periods of relative stability, when urban socio-technical systems and their environment are fully embedded but unsustainable in the long term, challengers to the system face more significant obstacles. However, in internal or external turbulence threatening to destabilize existing relationships, Horizon 3 and 2 players can see a window for change. Such moments form the basis of a multi-level perspective on change and innovation that clarifies how transition occurs across different levels: the global landscape or context, socio-technical systems, and innovation niches (Geels 2002, 2007, 2010, 2011; Geels and Schot 2007).

In transformative processes of this kind, over and above the availability of material resources, narratives (Liveley 2017) play a critical role in inspiring and galvanizing people, giving meaning to individual and collective actions, and injecting "hopeful determination." As Donella Meadows (2008) observes, the most effective lever for changing a system is changing the underlying paradigm, the meta-narrative that frames the stories of the various actors. Examples of urban narratives include the Creative City, the Business City, the 15-Minute City, the Mediterranean City, and the City of Care. Such narratives spring from a profoundly uneven field, where perhaps the most telling indicator of the power of urban actors is their ability to make their vision and narratives prevail, whether through persuasion or imposition (González-Arellano 2022). Horizon 1 elites, with operations and resources that extend beyond city boundaries on many levels, often wield tremendous influence over information and communication systems. These systems play a pivotal role in generating and perpetuating meaning and making sense of life and the individual and collective values that, in turn, influence urban behaviors and decisions. They are, therefore, in a position to influence who and what “will have a future” in the city, what will remain unchanged, and what needs to evolve. These patterns uniquely shape each urban context, reflecting the complexity of particular circumstances and multi-actor, multi-level processes.

Therefore, the “narratives of hope” map an uneven future vision across the urban fabric. In underprivileged neighborhoods, not only are current conditions less than ideal but there is also the sense of the future being at a dead end. These environments, aggravated by heightened tensions and acute instability, offer remarkable opportunities for bringing about change and consolidating processes of transformative “hopeful determination.” The uniqueness of each context adds value to case studies such as the one presented below. With them, we can unravel how horizons manifest and converge within distinct landscapes while also uncovering insights that help to identify environments and practices capable of expressing the will of different actors in different contexts. The result is a climate of inspiration and unwavering hope that catalyzes the activation of their skills.

3.4 Patterns of Hope on the Ground: The Integrated Plan for Huelva’s District 5

The Integrated Plan for District 5 of the City of Huelva (PID5H) is a paradigmatic case of urban foresight planning worthy of study, both for its complexity in terms of the issues addressed and its participatory process, as well as the fact that it has lasted for more than two decades. This case provides an opportunity to analyze, in an evolutionary context, the interplay of “hopeful determinations” of different actors (their active commitment to co-create a desired future). It illustrates how these are translated into collaborative courses of action of varying intensity, depending on contextual factors, individual circumstances, visions, and collective sense-making processes organized around an overarching goal. It also makes it possible to identify the main competencies acquired by the actors in their actions, developing their individual and collective agency in an active process in which being, doing, and knowing are inextricably linked.

3.4.1 Origin and Characteristics of the PID5H

Huelva (population 141,824) is located in southwest Spain. The PID5H[1] was established in 2000 as a comprehensive urban renewal plan in the fight against poverty and social exclusion. It aims to eradicate discrimination and safeguard the dignity, quality of life, and autonomy of all the people living in seven neighborhoods (approx. 15,000 inhabitants). The area was characterized by extreme insecurity and vulnerability, having been neglected by the public authorities, but there was a certain, albeit fragmented and uncoordinated, network of social initiative organizations.

The initiative was conceived by social organizations working in the area in close collaboration with a research center at the University of Huelva. Together, they crafted a narrative about the urgent need to strengthen and coordinate efforts to bring about social change in this neighborhood. In doing so, they garnered the support of local neighborhood associations and persuaded the public authorities to get involved and allocate resources. The primary vehicle used was a multi-actor platform involving public bodies from education, health, housing, employment, social services, culture and social initiatives, businesses, neighborhood associations, state-subsidized schools, and sports clubs. This multi-level platform—national, regional, and local—involved more than 50 entities (now 73). Its participatory governance system is still an internationally recognized benchmark for partnership building, holistic visioning, multi-sectoral coordination, and citizen participation in the regeneration of neighborhoods in need of social transformation (Miedes-Ugarte and Fernández-Borrero 2010; De Paz et al. 2017).

As the very name implies, the methodological pillars of PID5H are its multi-dimensionality (holistic approach), multi-actor and multi-level partnerships, citizen participation, and long-term visioning. Its goals, lines of action, and *modus operandi* have been set out in various strategic plans (2000, 2008, 2016, 2023), each time informed by preparatory anticipatory work carried out collaboratively. This work combines a thorough diagnosis of the current situation with a forward-looking analysis of expected trends and desired scenarios in the chosen time horizon.

The governance model is similar to a *holocratic* organization's (Robertson 2007), where decision-making power is pooled and shared. Activities are organized into six thematic participation forums (citizenship and planet; employment and prosperity; health and healthy habits; education and training; children and youth; family, equality and integration). These autonomous, self-organizing work units create ad hoc commissions for specific projects as needs or opportunities arise and for as long as they last. Each forum, which meets quarterly, defines the role of the coordinator, who in turn liaises with other forum coordinators and representatives of regional and local public authorities on the Steering Committee at monthly follow-up meetings.

Decisions in the forums are made by consensus. Discussions continue until resolutions are ready to be proposed to the Steering Committee, where a new consensus must be reached. Decision-making power is decentralized throughout the organization and is not in the hands of a single person or group. All the organizations

1 All the information regarding the PID5H in this chapter can be found at www.distrito5huelva.org.

and individuals involved in the forums come together at the annual meeting of the General Assembly, the supreme representative and participatory body, which ratifies the membership of the various partners. Information is shared openly and readily available to all organization members. That is a very dynamic model of governance that has been slightly adapted over the years as circumstances dictated.

The Plan's entire operation is orchestrated by a technical office with two facilitators responsible for coordinating, communicating, planning, and sustaining the participatory process of collective responsibility. In addition, a technical team led by two members of the University (the authors, since 2010) provides scientific and technical advice throughout the process. The Plan has no legal status, and aside from a few small joint projects, the technical office's sole function is to coordinate the visions and actions of the various entities. Each entity develops its projects and manages its funds: public grants and private contributions from competitive tendering fund the technical office.

3.4.2 Phases of PID5H: Patterns of Hope

Over the past two decades, the PID5H has weathered many ups and downs, and it is possible to distinguish between different phases in its evolution based on several criteria. These include the changing needs of a rapidly aging population and new challenges such as climate change; leadership changes and vacuums within and between entities; changes in the people involved and in the very lives of the long-serving participants; changes in funding bodies; shifts in public policy approaches; and the evolving landscape of policy direction at the national, regional, and local government levels, with the attendant agreements and conflicts. Of particular value to this analysis is the aforementioned three-horizon model for its ability to heuristically identify the meta-narratives underpinning the ways how different actors engage with the PID5H, or in the author's own words, "their patterns of hope" (Sharpe 2013). How the meaning and significance of the PID5H has evolved for the various actors goes a long way toward explaining the progress and setbacks observed throughout the collaborative process.

Initial Phase (2000–2009): Horizon 3 with Support from Horizon 2

A typical pattern emerges among all the actors involved in this initial phase, clearly recognizable as Horizon 2. This pattern serves as a major impetus for innovation for Horizon 3 actors, who have been striving for years to create a better future in living conditions for the population but with little support and scarce resources. In this phase, significant changes are negotiated and approved, including establishing the participatory process and large-scale projects such as constructing a state-of-the-art health center or rehabilitating and bringing up to code part of the housing stock. The visions of the actors (government, neighborhood associations, and other entities) are strongly aligned during this phase. Naturally, conflicts of varying intensity and nature exist, especially as the entities get to know each other and learn to work together. The prevailing mood, however, can be described as one of shared "hopeful determination." The ultimate proof is that despite the challenges, the organizations remain in the game (Kahane 2017), with all renewing their commitment to participate in the Plan every year and more and more players coming on board.

Second Phase (2010–2020): Horizon 1 Reels in Horizon 2

The financial crisis in Spain worsened from 2010 onwards, and this has had a significant impact on all PID5H actors. Social organizations are left reeling by the crisis, turning all their attention inward to weather the storm. Some of them disappear. The public sector, mainly social services, has fallen victim to drastic funding cutbacks. In a context of great uncertainty, the gravity of the financial crisis has prompted a return to more welfare-oriented policies to combat poverty (Horizon 1) at the expense of the PID5H's more proactive and anticipatory policies (Horizon 2). Neighborhood associations focus on strengthening their fragile networks of community solidarity to respond to the worst neighbors' situations. They see the PID5H as an indispensable resource for further empowerment, reinforcing these networks and stepping up efforts when needed. Under these circumstances, public authorities, whose leaders are overwhelmed by the circumstances and in constant flux either need help to understand the operating logic of the PID5H or perceive the participatory process to be a threat due to increasing neighborhood demands. The neighborhood associations, with their increasingly aging representatives, point out the shortcomings and the lack of response to issues about which, after ten years of constant attention and collaboration, they have created a collective memory, have expert knowledge, and have grown weary of calling for concrete solutions. Partnership working, when it occurs, continues to bear fruit, but tensions are rising.

That is compounded by financial difficulties in maintaining the technical office, which has gone months without funding for the first time in a decade. The sustainability of the mechanism itself becomes a significant concern for some Plan members. Others, faced with existential threats of their own, do not see it as a priority. All of this weakens the participatory process, causing the Horizon 2 narrative to recede and the process itself to retreat to a more management-oriented Horizon 1 in an attempt to keep the process alive under adverse conditions.

Third Phase (2020–2023): The Return of Horizon 3 (or the Jedi)

Despite the progress and setbacks experienced by the actors in the area, there has always been a latent critical mass that can be identified with Horizon 3. Their determination sees greater potential in the environment for transitioning to a better future. In a paradoxical twist, the COVID-19 pandemic brings these actors closer together, prompting a joint reflection on the need to focus on the local population's reality and to renew the Plan's long-term, forward-looking vision. New challenges, including but not limited to energy transition, climate change, population aging, and technological adaptation, join the list of existing problems to be solved. This renewal breathes new life into the transformation narrative and is deeply meaningful to the various actors, rallying visions, resolve, and resources around the common goal of reigniting the transformative innovation of Horizon 2. Building on this momentum, a new participatory foresight and adaptive planning process called Roadmap 2030 seeks to recapture the most valuable lessons of the past two decades. The process has involved the creation of five citizen laboratories inspired by the aforementioned three-horizon approach. Combined with a mission-oriented innovation methodology (Mazzucato 2021), this has given concrete form to the future visions of the actors by identifying 37 projects organized into six major missions (one for each forum). The aim is to mobilize the attention and energy of the alliance around meaningful and hopeful common goals.

This new narrative, co-constructed by more than 200 people and endorsed by the 2022 Assembly, will reinvigorate the participatory process. The first step will be to find new funding sources to sustain it while introducing fresh approaches to how actors share enabling roles and forging new alliances between organizations, neighbors, and public bodies following a period marred by misunderstanding. To this end, the projects have also been aligned with the 17 sustainable development goals of the UN 2030 Agenda and the ten goals of the Spanish Urban Agenda, defined by the local government for the city as a whole (Miedes-Ugarte and Fernández-Borrero 2022).

The complexity of the PID5H's evolution through these phases shows that the journey to the desired future is non-linear, with the tapestry of visions reconfiguring along the way. This journey spans the period of initial alignment and innovation, where Horizon 2 drove development and participation, through the struggle for survival and the resurgence of welfare policies during the financial crisis (Horizon 1), to the period of reflection and the search for a new, forward-looking direction in the post-pandemic era (Horizon 3). All of this profoundly influenced the development of the actors' competencies, which have co-evolved throughout the process in terms of their anticipatory capacity and ability to work in partnership.

3.4.3 The Competencies Developed by the Actors Through an Enactive Lens

In each of these three phases, opportunities have arisen for actors to learn how to enhance their foresight skills and, by extension, their ability to meet future challenges. In the initial phase (2000–2009), Horizon 2 provided a transformative impulse that strengthened the actors' anticipation by constructing major infrastructure and high-community impact projects. This period was critical in laying a foundation of cooperation and trust that, by removing some of the uncertainty, proved essential to improving individual and collective anticipation. Despite the tensions, aligning actors' visions helped develop a proactive approach to identifying and addressing future problems.

In the second phase (2010–2020), the financial crisis forced actors to prioritize short-term survival over innovation. That diminished their anticipatory capacity for an extended period, as resources and attention were focused on immediate needs rather than long-term planning. Nevertheless, dealing with the crisis also provided valuable lessons in risk management and the need for flexibility in existing plans and structures.

The third phase (2020–2023), marked by the pandemic and a renewed focus on Horizon 3, was an opportunity to revisit and enhance anticipatory capacity. The now more seasoned actors have successfully incorporated emerging issues such as climate change and energy transition, reflecting their evolving ability to anticipate and build a shared vision of the future.

Developing these anticipatory skills is part of a broader set of competencies needed for the ongoing practice of "hopeful determination," learning to work together to meet future challenges with flexibility and adaptability, depending on goals and circumstances (Castañer and Oliveira 2020). Four main groups of competencies have been identified:

1. Proactive collaboration: alliance building, conflict management, effective communication.
2. Adaptation and change management: adaptability and flexibility, managing uncertainty, organizational and community resilience.
3. Innovation and transformation: innovation: systems thinking, creative thinking, adaptive planning, resource management, and fundraising.
4. Social engagement and community empowerment: civic engagement, socio-ecological awareness and responsibility.

Developing these competencies is by no means technically or formally linear. It is a complex, open-ended, never-ending process that occurs under precarious conditions and is unevenly distributed. It is a process that strongly resonates with the enactive approach to cognition. This approach views cognition not simply as a representation of the world but as a collection of embodied activities that emerge and are continually transformed through interaction with the environment. The path taken by PID5H is a real-life example of the enactive principle, where cognition and action are in constant dialogue, and every action is both a product and an enabler of understanding. The actors' acquired competencies can be seen as manifestations of their unfinished agency in continuous co-construction with the environment. In an ongoing process of sense-making, actors shape the environment and are shaped by it. Far from merely receiving or processing external information, they have developed the ability to read subtle signals (Morales and González 2021), adapt to emerging circumstances, and proactively seek to influence events in a positive direction. Anticipation here means actively participating in creating desirable futures, not merely predicting what is to come.

3.5 Conclusions

The future is in seed form right now; the will of urban actors is guided by their focus, drawn to the possibilities offered by an environment aligned with individual and collective sense-making in specific spatial and temporal contexts. We have argued that "hopeful determination" catalyzes the actors' impulse to channel their energies toward a more promising future within this sense-making framework. The enactive approach emphasizes that the future is flexible. Instead, it is an emergent entity forged by our present actions and not entirely predestined. Actors and their environment co-create worlds of meaning. These worlds steer the actors' activity toward realizing their aspirations. They transform the possibilities within the environment into tangible realities.

The analysis of PID5H illustrates a form of active and participatory anticipation that, according to Gidley's (2010, 2016) categorization, taken up by Motti (2017), falls into the "empowerment-participatory and hope-enhancing" category. It is a living process, full of uncertainties, in which all possible combinations of challenges involved in transforming neighborhoods in crisis can be found in varying degrees of intensity: the scarcity of resources, stigmatization of specific areas, limited acknowledgment and visibility of the process by the actors, uneven distribution of capacities, fragile citizen participation in continual development, resistance of some actors to change, and varying institutional support depending on whether participatory solutions conform to pre-established policies, not to mention bureaucratic barriers. It takes great determination and hope to continue working together in such a context. The PID5H exemplifies inclusive and caring governance that seeks to sustain this "hopeful determination." At the same time, it considers the emotional and affective dimensions of the participants and the community, who often face specific and ongoing challenges that affect their minds and ability to act. In a world grappling with complex socio-ecological challenges like climate change, political polarization, and post-truth, and where advances like genetic engineering and artificial intelligence are unleashing massive transformative potential at great risk of deepening inequalities, the enactive approach reminds us of the importance of reconnecting with our embodiment and with the unique textures of our interactions. That makes the sensory experience of participatory spaces highly relevant. It triggers potential connections between aspects essential for collaborative and anticipatory behavior, such as memory, foresight, voice, optimism, and yearning (Bussey 2017).

This chapter also highlights how actors construct meaning through the mobilization of narratives. As Mercier and Sperber (2017) suggest, the essence of sense-making comes from our interactions and conversations, which are the underlying catalysts for transformation. In this sense, as Kaplan and Davidoff (2014) note in their exploration of the concept of "delicate activism," the main transformative lever in many situations is to change the nature of the conversation. That is consistent with O'Hara and Leicester's (2019) proposition that the three literacies—*cognitive* (understanding how we know), *psychological* (understanding our agency), and *cultural* (understanding and facilitating cultural change)—should be strengthened as essential competencies for navigating a VUCA world with some guarantee of success. As we have shown, the enactive approach can significantly contribute to this reinforcement.

In short, anticipation in the urban context discussed here does not negate the value of analyzing expected trends or scenarios in more or less stable contexts, especially for risk prevention purposes. However, its effectiveness is inextricably linked to the actors' willingness to act accordingly and adapt their behavior. They must also activate their individual and collective agency, an ongoing process, to positively influence the future and break with the inertia and trends of the past when they are unsustainable or undesirable. It is about seeing the future as a shared legacy constantly co-constructed in an ongoing process of renewing "hopeful determination" or, as Solnit (2010) would say, as a never-ending effort to maintain and sharpen the axe with which to open emergency doors.

References

Barrett LF (2017) How emotions are made: the secret life of the brain. Pan Macmillan

Benton M (2019) Epistemological aspects of hope. In: Blöser C, Stahl T (eds) The moral psychology of hope. Rowman & Littlefield Publishers. Rowman and Littlefield International, pp 135–151

Bussey M (2017) Anticipatory aesthetics: the senses and the body in anticipatory theory and practice. In: Poli R (ed) Handbook of anticipation: theoretical and applied aspects of the use of future in decision making. Springer, pp 861–874

Castañer X, Oliveira N (2020) Collaboration, coordination, and cooperation among organizations: establishing the distinctive meanings of these terms through a systematic literature review. J Manag 46(6):965–1001. https://doi.org/10.1177/0149206320901565

Coleman R (2017) A sensory sociology of the future: affect, hope, and inventive methodologies. Sociol Rev 65(3):525–543

Cook J, Cuervo H (2019) Agency, futurity, and representation: conceptualising hope in recent sociological work. Sociol Rev 67(5):1102–1117

Deleuze G (1994) Difference and repetition. Columbia University Press

De Paz MA, Asensio MJ, Franco MI (2017) Plan integral Distrito V de Huelva: 2000/2003, vol. 87Servicio de Publicaciones de la Universidad de Huelva

Di Paolo EA (2021) Enactive becoming. Phenomenol Cogn Sci 20(5):783–809

Di Paolo E, De Jaegher H (2012) The interactive brain hypothesis. Front Hum Neurosci 6:163

Di Paolo EA, De Jaegher H (2022) Enactive ethics: difference becoming participation. Topoi 41(2):241–256

Di Paolo EA, Lawler D, Vaccari AP (2023) Toward an enactive conception of productive practices: beyond material agency. Philos Technol 36(2):31

Doe MJ (2020) Conceptual foreknowings: an integrative review of hope. Nurs Sci Q 33(1):55–64

Geels FW (2002) Technological transitions as evolutionary reconfiguration processes: a multi-level perspective and a case-study. Res Policy 31(8–9):1257–1274

Geels FW (2007) Feelings of discontent and the promise of middle range theory for STS: examples from technology dynamics. Sci Technol Human Values 2(6):627–651

Geels FW (2010) Ontologies, socio-technical transitions (to sustainability), and the multi-level perspective. Res Policy 39(4):495–510

Geels FW (2011) The multi-level perspective on sustainability transitions: responses to seven criticisms. Environ Innov Soc Trans 1(1):24–40

Geels FW, Schot J (2007) Typology of sociotechnical transition pathways. Res Policy 36(3):399–417

Gidley JM (2010) Globally scanning for 'Megatrends of the Mind': potential futures of futures thinking. Futures 42(10):1040–1048
Gidley JM (2016) Postformal education: a philosophy for complex futures. Springer International Publishing, New York, NY
Godet M, Durance P (1997) La prospective stratégique. Futuribles 45:57
González-Arellano S (2022) The future scenarios of cities: an analysis of their institutional construction. Int J Urban Reg Res. https://doi.org/10.1111/1468-2427.13103
Kahane A (2017) Collaborating with the Enemy. Berrett-Koehler Publishers
Kaplan A, Davidoff S (2014) A delicate activism: a radical approach to change. The proteus iniciative. http://socialself.org/Articles/adelicate.aspx
Liveley G (2017) Anticipation and narratology. In: Poli R (ed) Handbook of anticipation: Theoretical and applied aspects of the use of future in decision making. Springer, pp 1–120
Mazzucato M (2021) Mission economy: A moonshot guide to changing capitalism. Penguin UK
Meadows DH (2008) Thinking in systems: a primer. Chelsea Green Publishing
Mercier H, Sperber D (2017) The enigma of reason. Harvard University Press
Miedes-Ugarte B, Fernández-Borrero MA (2010) Inteligencia Territorial para la Lucha Contra la Pobreza. Aprendizajes de 20 Años Sobre el Terreno. TRABAJO. Revista Iberoamericana de Relaciones Laborales, p 23
Miedes-Ugarte B, Fernández-Borrero MA (2022) IV Plan Integral Distrito 5. Hola de Ruta 2030. from https://www.distrito5huelva.org/iv-plan-integral-del-distrito-5/
Miller R, Poli R, Rossel P (2013) The discipline of anticipation: exploring key issues. Bellagio Document 4 Working paper 1 version 0.9
Morales N, González S (2021) Cities as anticipatory systems: analyzing 'weak signals' to explore beyond the predictability of their future. Ethnographic Praxis Ind Conf Proc 2021(1):240–257
Motti VV (2017) Sources of futures studies from foresight to anticipation. Poli R (ed) Handbook of anticipation: theoretical and applied aspects of the use of future in decision-making. Springer, pp 1719–1732
O'Hara M, Leicester G (2012) Dancing at the edge: competence, culture, and organization in the 21st century. Triarchy Press
O'Hara M, Leicester G (2019) Dancing at the edge: competence, culture, and organization in the 21st century. Triarchy Press
Peterson JB (2002) Maps of meaning: the architecture of belief. Routledge
Poli R (ed) (2017) Handbook of anticipation: theoretical and applied aspects of the use of future in decision making. Springer
Prigogine I, Stengers I (1979) La nouvelle alliance: métamorphose de la science. Gallimard
Robertson BJ (2007) Organization at the leading edge: introducing Holacracy™. Integr Leadersh Rev 7(3):1–13
Sepúlveda-Pedro MA (2023) Sense-Making in the wild: the historical and ecological depth of enactive processes of life and cognition. Adaptive Behavior, 10597123231190153
Seth A (2021) Being you: a new science of consciousness. Penguin
Sharpe B (2013) Three horizons. Triarchy Press
Sharpe B, Hodgson A, Leicester G, Lyon A, Fazey I (2016) Three horizons: a pathways practice for transformation. Ecol Soc 21(2)
Solnit R (2010) Hope in the dark: the untold history of people power. Canongate Books
Stewart J, Gapenne O, Di Paolo EA (eds) (2010) Enaction: toward a new paradigm for cognitive science. MIT Press
Thompson E (2007) Mind in life: biology, phenomenology, and the sciences of mind. Harvard University Press
Varela FJ, Thompson E, Rosch E (2017) The embodied mind, revised. MIT Press, Cognitive science and human experience

Part III
New Building Anticipatory Systems in Urban Sectors in the Global South

Chapter 4
Sustainable Urban Mobility as an Anticipatory System: The Case of the Quito Metro

Marco Córdova and Diana Paz

Abstract Sustainable urban mobility is presented as an anticipatory system in urban development processes. The future projection of socially inclusive, environmentally sustainable, and economically efficient territorial configurations allows for the improvement of public policies and city planning in the present. The chapter proposes an analysis of how the Quito Metro constitutes an anticipatory system in the growth of the metropolitan district, in the mitigation and adaptation to climate change, and in the socio-cultural dynamics of the population. Methodologically, a descriptive inference exercise is proposed through a qualitative case study. The collection and processing of information are based on academic secondary sources and official documents. The case of Quito shows that the Metro is built with an anticipatory approach, arising from a more comprehensive reading of mobility as a long-term policy. However, it constitutes an implicit or prospective anticipation based on the city's planning model.

Keywords Sustainable development · Urban mobility · Metro systems · Anticipatory systems · Metropolitan district of Quito

4.1 Introduction

In recent decades, socio-ecological transitions have driven territorial strategies based on sustainable development approaches. These strategies have redefined issues such as climate change, risk due to natural hazards, natural resource management, and urban mobility. Faced with the imperative to build sustainable and resilient territories, multi-stakeholder (state, market, society) and multi-level (international, regional, local) governance dynamics have been consolidated. In this context, various state

M. Córdova (✉) · D. Paz
FLACSO Ecuador, Quito, Ecuador
e-mail: mcordova@flacso.edu.ec

D. Paz
e-mail: dmpazfl@flacso.edu.ec

S. Gonzalez-Arellano and B. Gandlgruber (eds.), *Cities as Anticipatory Systems*, The Urban Book Series, https://doi.org/10.1007/978-3-031-78162-9_4

and non-state actors have promoted a comprehensive set of public policies and planning instruments. These new forms of public action can be understood as anticipatory systems—sets of decisions made in the present based on future resources and scenarios (Poli 2017).

The contemporary notion of sustainable development has a constructivist and holistic sense. It is projected into the future regarding social equity, environmental protection, and economic efficiency, with a marked impact on territorial configuration and transformation (Gudynas 2010; Keivani 2010). Viewing development ontologically from an anticipatory discipline perspective allows understanding the planning exercise derived from sustainable development as a form of knowledge aimed at colonizing the future and giving meaning to the emerging novelty of the present by inspiring a systematically imagined future (Miller et al. 2013). This anticipatory reading of development in a sociological key implies understanding that knowledge about the future addresses the diagnosis of the trends of the times and simultaneously projects, in a speculative sense, possible scenarios and outcomes of change (Hammershøj 2017).

Sustainable development has been built on the positive and negative interpretation of the territorial transformations of the last century, explaining the trends of the transitions derived from the process, especially in contexts where structural imbalances are more evident (Acosta 2015; Bolay 2012; Winchester 2006). However, at the same time, the idea of development based on sustainability principles has led to the institutional construction of a normative, regulatory, and cultural-cognitive nature, which, through the agenda of the Sustainable Development Goals (SDGs), condenses a set of anticipatory policies that local governments have promoted to model future scenarios (González-Arellano 2022).

Analyzing the processes of public action inherent to sustainable development as a problem of anticipation allows us to understand how the utopian construction of the territories of the future is being employed in the present to enhance the design of public policies in the face of urban problems in the current context of the Anthropocene. In the field of urban mobility, from the positioning of sustainable development in the 1990s, a change is observed from previous transportation policies oriented in an engineering approach toward a new mobility paradigm that includes the relationship of the individual with space (Miralles-Guasch and Cebollada 2009). This integration addresses problems related to the unequal development of cities, later conceived from economic, social, and environmental sustainability (Quintero-González 2017; Vasconcellos 2015).

Around sustainable urban mobility in Latin America and the Caribbean (LAC), a series of policies have been promoted, focusing on accessibility, equity, integrality, and emissions reduction (Errazuriz et al. 2017; Figueroa 2001; Rivas et al. 2019). More recently, attention has turned to the link between urban development and mobility in the so-called Transport-Oriented Development (Rodriguez and Vergel-Tovar 2018). In these processes, implementing Metro systems has promoted socio-technological innovations in public transportation and urban mobility in general, serving as a central element in new ways of conceiving the city. In this sense, mobility policies are being consolidated as anticipatory systems in LAC cities based on their intention to manage uncertainty and foster the collective intelligence of cities (González-Arellano 2022). The notion of forward-looking sustainability allows for structuring comprehensive urban development models in the present, characterized by complex contexts with scarce resources, low institutionalization, and weak capacities (Paz-Gómez and Santelices 2020).

Against this background, the chapter analyzes how sustainable mobility, specifically Metro systems, configures anticipatory systems in urban development. The Metropolitan District of Quito (DMQ) is selected as a case study since it has been promoting a sustainable mobility policy that has transformed urban planning since the 1990s. Implementing the first Metro line in 2023 marks a shift in the trajectory of mobility policies, primarily redefining current strategies based on the future projection of the city's development. Methodologically, the proposed combines a contextual analysis (LAC) and a case study (DMQ) to achieve a descriptive inference level and take a qualitative approach. The collection and systematization of information are based on secondary sources related to academic publications, official documents, and reports on development and land use plans, mobility policies, and impact analysis, among others.

The chapter is structured in two parts: (i) the characterization of urban mobility and Metro systems in LAC, based on a review of the specialized literature of the region, and (ii) the analysis of the case of the DMQ in terms of the background of the incorporation of the Quito Metro as a public transportation hub, and its impact as an anticipatory system in urban development, in the mitigation and adaptation to climate change, and the socio-cultural dynamics of the city.

4.2 Sustainable Urban Mobility in LAC: From Reaction to Anticipation

Urban mobility has been associated with cities' growth and displacement dynamics. Data suggest that 80% of the population lives in cities, with LAC concentrating on highly dense cities such as Mexico City, Buenos Aires, Rio de Janeiro, Sao Paulo, Bogota, Lima, and Santiago de Chile (CEPAL 2018). The region maintained economic growth between 1995 and 2017, which incentivized the increase of the private transport fleet (CEPAL 2019). The discussion on mobility has been conceived from two main orientations. The first is related to the reaction to the urban system's road, transport, or energy dynamics, and the second involves the reconfiguration of mobility toward anticipation, derived from adopting the sustainability paradigm in urban design and planning.

Urban mobility, defined as a reactive system, was oriented by policies focused on responding to the growth of cities. This model generally supports an engineering vision that responds to the dynamics of "the greater the growth, the greater the road development." It is characterized by short-term governmental responses designed to solve the problems derived from rural–urban migration, new patterns of land occupation, and travel times in origin–destination trips (Casado Izquierdo 2008). This reactive mobility system prioritized analyses on intra-urban commuting modes (Cosacov and Di Virgilio 2018). Although traffic engineering incorporated studies on supply and demand, origin–destination surveys, vehicular flows, and transport modes to forecast future demands, mobility was not considered a response to a possible evolution. Instead, it addressed the need to move within the city with infrastructure and transport (Manheim 1979; Quintero 2011; Rodríguez Vignoli 2008).

This generation of engineering resulted in the construction of new roads and ring roads, leading to increased private transportation. Although there was an occasional increase in public transport services, these were characterized by private operations with little or no state regulation (Avellaneda 2017; CAF 2011). Within the framework of reactive responses, public transport was constituted as a tool to supply the demand for transportation in line with population and economic growth (Lupano and Sánchez 2009). Thus, in the 1990s, the boom of rapid transit buses and omnibuses did not incorporate an environmental sustainability approach. Once again, the short-term vision of implementing new transportation systems was associated with supply and demand and, to a lesser extent, with prospective planning considering the impacts of population growth and urban expansion.

This reactive urban mobility underwent its first change with the anthropocentric incorporation of displacements. In this sense, mobility focused on relating social and cultural dynamics to how individuals were transported (Levy 1999; Sheller and Urry 2006). That did not imply leaving aside the analysis tools already designed by transportation engineering but incorporated analyses on supply and service conditions (Gutiérrez 2012; Miralles-Guash 2002; Montezuma 2003), territorial, sociodemographic, and residential conditions, and travel frequencies and their distances (Cebollada and Avellaneda 2008; Lazo 2008). Furthermore, daily mobility integrates variables associated with providing goods and services, considering transportation as the planning axis. Subsequently, these analyses incorporated variables such as the environment aimed at understanding the impact of transportation use on pollution levels in cities, accessibility and proximity to goods and services, and their impact on travel costs, residential preferences, and territorial planning (Estevan and Sanz 1996; Hernández and Falavigna 2016).

The second conception of mobility is associated with incorporating the sustainability paradigm, which consolidates an approach to anticipatory systems at the urban scale (see Table 4.1). By 1992, sustainability became part of political agendas by adopting the Rio Declaration, leading governments to formulate sustainable urban mobility under an efficient transportation approach (Miralles-Guasch and Cebollada 2009; Montezuma 2003; Pozueta 2000). In 1994, the Aalborg Charter on sustainable European cities aimed to strengthen the implementation of strategies to reduce energy consumption, promote compact urban design, and foster economic development with social justice. Subsequently, the 2015 and 2030 Agenda and the SDGs consolidated three principal axes of urban mobility: social, economic, and environmental.

As an anticipation system, the economic dimension considers the efficiency of private transportation and maximizing public transportation benefits. Meanwhile, the social dimension includes issues of equity and access focused on the population of the

Table 4.1 The evolution of responses in urban mobility generation

Orientation	Reactive system	Anticipatory system
Approach	Transportation engineering	Sustainability
Vision	From Infrastructure	From mobility
Physical dimension	On supply and demand (flows and traffic)	Shape of travel (individual-space relationship)
Objectives	Connection with city spaces	Economic, social, and environmental sustainability
Hierarchy	On the vehicle	Inverted pyramid: On transport systems (Metro and buses), pedestrians and cyclists
Development oriented	Economic growth	Urban integration and transportation
Time and distance	Reduced	Accessibility and proximity

Source By the authors

lowest quintiles. On the other hand, the environmental dimension defines elements associated with reducing greenhouse gases and energy consumption and promoting clean public transportation systems as the central axis of mobility (Monzón 2005). These dimensions constitute an anticipation of urban planning that includes data, technology, and predictive models (Alcántara Vasconcellos 2020; Hernández 2017; Jirón 2017).

From these central axes of sustainable urban mobility, anticipation has been conceived in at least two primary dimensions: (1) urban development as a predictive model for land use planning and public transport, and (2) the prioritization of Metro systems in reconfiguring mobility.

4.2.1 Anticipation of Urban Development: Transit-Oriented Development

The first dimension of anticipation in sustainable urban mobility is associated with postulates on accessibility and equity, recently consolidated debates in the so-called Transit-Oriented Development (Rodriguez and Vergel-Tovar 2018). Accessibility involves considering satisfaction parameters, quality and coverage of transport systems (Metro and buses), road connectivity, physical and fare integration, and user satisfaction. This approach allows us to consider that accessibility, within an anticipatory system, considers elements related to how cities are developing and their growth patterns. In this sense, designing sustainable urban mobility requires thinking about future forms of the urban system and how to adapt cities to these scenarios to reduce or manage uncertainty. This necessity has compelled sustainable urban mobility to be conceived with a focus on the city's future locations of activities and infrastructure, the construction or expansion of transportation modes, and their connections with individuals (Hernández 2017; Montezuma 2003). The leading prospective research identifies the travel limitations of individuals based on predictive models of urban growth, considering the impacts on different population quintiles related to fares mixed with operating costs and the need for direct subsidies by governments to maintain accessibility (CAF 2017; Gutiérrez 2010).

Transit-Oriented Development (TOD) has been belatedly implemented in LAC cities. Curitiba in Brazil is considered one of the main cities in the region that has successfully integrated urban planning with mobility, particularly around BRT-type bus systems. For this city, planning must revolve around public transport as the central axis of mobility. Although it does not limit urban growth, it associates predictive capacity with responding to possible options for the evolution of urban forms through long-term policies that connect the sustainable road system with urban development. Quito, Bogota, Santiago de Chile, and Mexico City began constructing BRT systems in the 1990s and early 2000s to meet future mobility demands, while other cities sought to complement their Metro services.

However, these systems have been transformed along with the institutionalization of sustainable urban mobility policies. By the 1990s, the objective of BRT-type systems was associated with future demand, marking the first moment of anticipation. However, the exponential growth of LAC cities exceeded the capacity of these systems in some cities, forcing a transformation or update in response to the evolution of the urban system. By 2013, 63.6% of passengers were concentrated on BRT-type buses (Rodriguez and Vergel-Tovar 2013), intensifying intensified planning on the main road axes. For example, in Curitiba, the implementation of BRT buses was accompanied by the promotion of high-density land development and mixed uses, leading to the formulation of instruments related to urban management and public transport. This anticipation addressed uncertainty about urban development through long-term decision-making. Bogota and Quito have implicitly developed TOD models around BRT systems in the main transfer stations and terminals, promoting urban development, infrastructure, and accessibility (Vergel-Tovar 2019).

4.2.2 Metro Systems in the Reconfiguration of Mobility

Metro's systems have become a key element in urban mobility in terms of reaction and anticipation. In Latin America, the first Metro was built in 1913 in Buenos Aires, which by 2020 had six lines. While Brazilian cities have made remarkable progress implementing these systems, the latest Metro line extension was carried out in Brasilia, with 42 km of construction. Before the 1990s, Metro construction in Latin America was associated with the regional economic boom, coinciding with the economic growth of the main cities. However, this vision was not associated with a comprehensive understanding of mobility. In general, the provision of public transport services was considered poor quality, with low regulation by the State, resulting in quality, coverage, and road safety problems (Figueroa and Henry 1988). Through

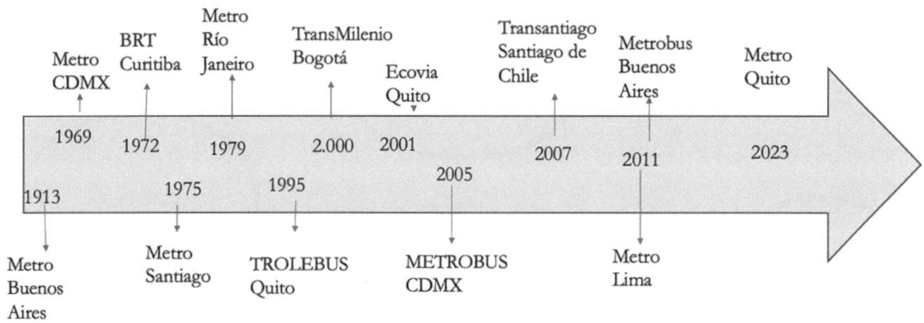

Fig. 4.1 Metro systems in LAC. *Source* By the authors

the concepts of comprehensiveness and sustainability, the Metro is considered an anticipatory system linked to environmental, socio-cultural, and economic axes (see Fig. 4.1).

The Metro de Santiago has been promoting MetroArte to link culture within the Metro stations through artistic works, dioramas, cultural diffusion, and value enhancements (De Grange 2010; Metro 2015, 2022). Artistic works have been installed inside some Metro stations, with the first one being installed in 1993 titled "Urban Interior," an oil on canvas by the artist Hernán Miranda. To date, 75 works have been exhibited, covering 11 thousand meters of permanent art. Additionally, a change in user behavior is being encouraged through the application, along with educational campaigns and the transparent Metro. That prioritizes open access to economic, technical, and financial information and regulatory frameworks.

The reconfiguration of mobility based on the subway as an anticipatory system is linked to behavioral changes, the positioning of infrastructure as a central point in travel, and the reduction of greenhouse gas emissions. The Santiago and Medellín subways, for example, have been promoting strategies such as MetroArte and Cultura Metro (Metro de Medellín 2023) through educational campaigns that consolidate the appropriation of the system while implementing changes in the use of energy, recycling, and waste from the train fleets. Santiago designed the "Huella Chile Program," which promotes the quantification of greenhouse gases within the subway infrastructure. Mexico City has promoted transparency strategies to ensure long-term policies strengthening the Metro system's predictive and transformative capabilities (Hidalgo 2022; Metro 2022; Ortuzar 1983).

In this sense, urban mobility as an anticipatory system is primarily framed in the institutionalization of long-term policies. That is associated with two main dimensions that promote the collective intelligence of cities (González-Arellano 2022) and encourage the application of predictive models in understanding present behaviors. Meanwhile, the predictive capacity dimension, environment, urban, socio-cultural, and economic development are associated with the system's capacity for transformation and its response to evolution in these areas. Considering this, the formulation of the TOD, the inclusion of climate change, social behavior change, and subsidies

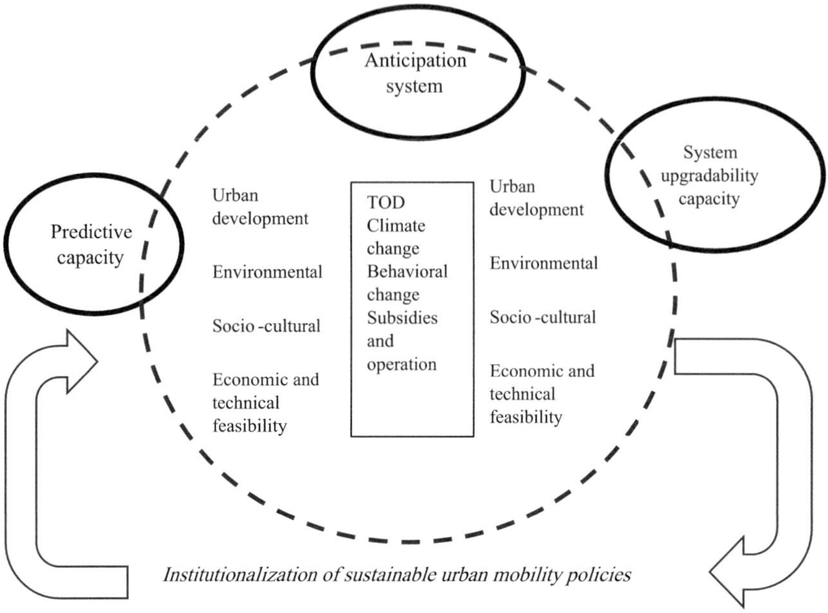

Fig. 4.2 Dimensions of urban mobility as an anticipatory system. *Source* Prepared by the authors

for operation constitute elements that enable anticipation in urban planning (See Fig. 4.2).

4.3 The Quito Metro as an Anticipatory System

With the declaration of the Law of Regime for the Metropolitan District of Quito (1993) and the decentralization reforms that Ecuador had been promoting since the 1990s to strengthen local governments, the municipality of Quito assumed a series of planning, regulation, and coordination competencies for various public services, including urban transportation. That led to a substantial improvement in public transport with the implementation of the Trolleybus in 1995 (the first BRT corridor), the modernization of the private bus fleet, and the construction of bus stops. This process was marked by a high level of conflict between the government, operating companies, and users associated with the problems of democratic governability of the emerging institutional framework of the metropolitan government of Quito (Chauvin 2006).

This background marks a change of vision from the reactive approach to transportation to the broad notion of sustainable mobility in the DMQ's policy agenda in the 2000s. Two factors marked this transformation: first, the urban development and environmental innovations incorporated by the Trolleybus project, which proposed a solution to mobility problems and was conceived as the axis of an integrated transportation system to structure a more equitable and environmentally sustainable city model. Second, the implementation of a series of instruments (laws and organizations), which underpinned a new institutional framework for urban mobility in terms of sustainable development, integrated into the Transportation Master Plan (2002) and specifically in the Mobility Master Plan 2009–2025 (Eguiguren 2015; Gordón 2011).

The formulation of urban mobility policy in the DMQ over the last three decades has been a process of innovation and political, social, and technological learning. Guided by the principles of sustainable development—environmental, social, and economic—various strategies have been implemented to assert mobility in the city as a fundamental right. Despite the limitations and shortcomings of the policies, progress is evident in areas such as improving the quality of public transportation and promoting non-motorized forms of transportation. These advancements highlight social actors' essential role, particularly activist groups and academia, whose participation and influence were crucial in reshaping the vision from a grassroots perspective, focusing on the needs of the most vulnerable. This process has redefined the culture and identity of mobility in Quito (Gordón 2011; Oleas and Albornoz 2015; Proaño 2012).

Historically, the modal distribution of trips in the city has been concentrated around 70% in public transportation (MDMQ 2015). Therefore, one of the central pillars of mobility policy has been developing an integrated mass transportation system. When the Trolleybus system faced expiration in the mid-2000s, an alternative proposal for a light rail system to enhance passenger capacity and travel speed was considered. However, this project did not materialize. Although the idea of constructing a Metro system had been previously discussed, it was dismissed at that time due to the high costs involved in the budget and the limitations of municipal management.

The proposal for a subway system resurfaced during the 2009 mayoral candidates' debate. Faced with the escalating demand to address the city's congestion issues, the construction of a mass transit system, specifically a Metro to replace the Trolleybus, gained significant traction in the political campaigns of the candidates. The mayor-elect embraced the Metro project as a key component of the new mobility policy,[1] and as the articulating element of a compact city model that optimizes land use and infrastructure in terms of rights. In the construction of the political agenda (Zahariadis 2007), the decision to promote a Metro system in Quito can be attributed to the convergence of three factors: a pronounced mobility problem that had garnered substantial attention in public opinion, and citizen demands; a redefinition of public

[1] The decision of the then mayor of Quito to discard BRT systems and promote the construction of a Metro can be explained by the logic of new public management, which seeks efficiency in the provision of the service. Additionally, it aligns with a discourse of equity and sustainability in urban mobility (Salar Khan 2016).

mobility policies based on sustainable development, advocating for the subway solution; and a political momentum reflecting consensus among various stakeholders to support the project.

Institutional restructuring was undertaken to implement Quito's new mobility policy, establishing the Secretariat of Mobility as the governing body for the policy and various operational bodies, including the company Metro de Quito. With technical support from the Metro Madrid Consortium, the conceptualization, studies, designs, construction, and equipping of the system for the First Line of the Quito Metro began in 2010. The project involved building 23 km of Metro tracks, 15 stations, and two multimodal stations in the city's north (El Labrador) and south (Quitumbe). The project was divided into two phases: Phase 1, which commenced in 2012, involved the construction of the two multimodal stations using municipal resources. Phase 2 encompassed the tunnel's construction, the remaining 13 stations, and all supporting infrastructure. The initial project budget was estimated at around US$1.5 billion, with funding expected from the municipality of Quito, the national government, and multilateral agencies (Oleas 2019).

The construction of the Quito Metro encountered various political, economic, and administrative challenges that led to delays in its completion compared to the initially established deadlines. The political instability within the DMQ's system, marked by frequent changes in mayors from different parties and movements, created a lack of continuity in city planning and policies. Each new mayor questioned the management of their predecessors, contributing to the obstruction and delay of the Metro project. Governance issues between the Metropolitan Council and mayors and tensions between the national government and the municipality further complicated matters. Financial complications arose when the national government withdrew the initially planned support, compelling the municipality to seek external resources for Phase 2 construction, thereby postponing the schedule. Challenges were encountered in financing, bidding, contracting, and operation processes. Despite these obstacles, 13 years after the project's initiation and under five municipal administrations, the Quito Metro was inaugurated in December 2023, costing US$2,017 million.

It has been noted that an anticipatory system is configured when the future is used to guide present decisions (Poli 2017). From this perspective, the Quito Metro is a relevant case to explore the anticipatory dimensions implicit in its design, both in terms of the future narrative on which the project has been delineated as the articulating axis of the sustainable mobility policy, and in the operational aspects of its implementation in different areas. That is because the Quito Metro project has just begun its operation. The design, construction, and execution processes developed in the last decade have a prospective and anticipatory character. We propose to analyze the anticipatory sense of the Quito Metro in three areas: urban development, environmental and climate change, and socio-cultural.

4.3.1 Urban Development as an Anticipatory System in Quito Metro

In its diagnosis of Quito's problems, the Metropolitan Development and Land Use Plan (PMDOT) highlights several critical issues regarding urban mobility. On the one hand, there is a constant growth in the number of vehicles, around 10% per year, which generates various adverse effects such as traffic congestion, carbon dioxide pollution, environmental deterioration, and health and quality of life problems for the inhabitants.[2] On the other hand, the supply of public transportation has failed to structure an integrated multimodal system. Ninety percent of this transportation is managed by 67 private companies, which operate disjointedly, and barely 10% of the transportation corresponds to the Integrated Public Transportation System (Trolleybus, Ecovía, North Central Corridor). Since 70% of the population travels by public transport, this generates coverage and accessibility problems (MDMQ 2021: 23, 31).

Considering this situation, the PMDOT (MDMQ 2021: 69) contemplates explicitly Strategic Objective 4: For a Connected Quito, aimed at providing reliable, quality, efficient, and safe mobility options. Under the principles of sustainability, vulnerability, and gender approaches, the policies of this objective propose planning and structuring a multimodal integrated mobility system following an urban development model based on multi-centralities. To this end, strategies and lines of action that discourage the use of private automobiles, increase non-motorized mobility, and consolidate sustainable metropolitan public transportation are proposed. In this sense, implementing the PLMQ and its extension to 2033 represents two central goals of this strategic objective.

Articulated from an integral perspective, a set of territorial systems configures the DMQ's Desired Territorial Model.[3] The city's growth has been conceived under a logic of polycentralities at different scales and influences, which takes advantage of the vocation of the territories and promotes a better distribution and territorial equity of infrastructure and services. The mobility and connectivity system structured around Quito is thus projected as the articulating axis of urban centralities and rural micro-regions, i.e., a multimodal system of Transport-Oriented Development, as a basis for promoting an inclusive, safe, resilient, and sustainable city (MDMQ 2021: 85,135). There is a sense of anticipation in the development model based on public transport, which serves the current public policy decision-making to reduce urban sprawl by consolidating centralities and more effective land use management.

Many mobility problems in the DMQ, such as vehicular congestion saturation, are directly related to imbalances in the spatial distribution of services, such as education (Astudillo-Ortega and Vásquez 2021). Hence, integrated transportation systems can

---

[2] In the case of the DMQ, market failures associated with mobility have led to, among other issues, health problems due to pollution and damage to heritage structures (Silva 2010).

[3] Environmental and Natural Risk System; Micro-regions System; Polycentric System; Mobility and Road Connectivity System; Heritage Areas System.

configure anticipatory development systems by appropriately targeting the location of metropolitan infrastructure and services, avoiding forced displacements.

The anticipatory character that the mobility and connectivity system has on the development of the DMQ is not only configured on a metropolitan dimension or scale, nor does it have a defined timeframe. The layout of the first line of the Quito Metro has generated expectations due to the impact of the different stops of the system in the sectors of the city where they are located. The location of the 15 stops was a matter of debate and dispute during its design, both in terms of the benefits and the negative externalities that the Metro circuit will have on the territory in the medium and long term.

Regarding urban implementation, two stops presented a high level of conflict between the residents of these sectors and the municipality. On the one hand, in the Solanda neighborhood in the south of the city, during the construction of the Metro stop, land subsidence problems affected the structure of several houses in the sector. This event opened a debate on the appropriateness of building a subway system in a territory with high physical vulnerability to natural hazards like the DMQ. On the other hand, some archaeological remains were found during the construction of the Plaza San Francisco stop in the city's historic center, which halted construction. At the same time, the heritage value of these vestiges was determined. Although the corresponding expertise determined that the ancient constructions found did not require special treatment, a broad academic discussion and protests from some citizen groups were generated regarding the danger of building the Metro stop in a historic sector declared a World Heritage Site. Not only was the intervention in an area of high physical fragility questioned but above all, the cultural transgression associated with its heritage status (Armijos 2021; Cabrera 2018; Plan 2016).

Based on the physical, economic, and social urban implications of the Metro's route in the territory, the system's design contemplated specific characteristics for each stop to counteract the impacts and transformations in the immediate surroundings. On the one hand, the infrastructure around the stations, through special urban-architectural programs and plans, is expected to promote urban renewal processes. However, there is also foresight regarding possible gentrification phenomena, i.e., displacement of the sector's original population and loss of identity, in high-density sectors that present physical deterioration, underutilization of land, and a deficit of infrastructure and public space (EPMMQ 2012).

The Quito Metro, a system that anticipates a series of urban transformations, mainly located in the immediate urban environment of the stops along the route, will directly impact the strategic decisions of several economic sectors. In the area of tourism, it is expected that the Metro will boost the flow of visitors to the city by increasing accessibility to sectors with tourist attractions such as the historic center (San Francisco station), but also access to sectors with a high supply of tourist accommodation and food services (El Ejido and La Alameda stations). Certainly, improving public transport in the city through a quality system can configure competitive advantages in strategic economic sectors such as tourism (Ávila 2017).

On the other hand, the anticipatory vision of the Quito Metro configures a series of urban development scenarios by modifying the current land use and occupation of those consolidated sectors through which the line passes. Depending on each typology's socio-spatial characteristics and territorial vocation, different growth hypotheses can be constructed regarding building and population density (see Table 4.2). The areas where type D stations are located have the most remarkable propensity for morphological transformation and demographic increase, owing to their commercial and public space dynamics. In a sector such as Jipijapa, located in the hyper center of the city, a high potential for urban development is projected in the medium and long term based on factors such as its level of accessibility, an average occupation of space for construction, and a tendency to change land use toward mixed activities, especially commerce. In the short term, the intensification of pedestrian flow at the Metro stop may induce a renewal of the urban infrastructure associated with pedestrian access to the area (Maldonado 2018).

The anticipatory vision of the Quito Metro has also been built in a utopian sense, that is, as an idealized projection of the functioning of the urban mobility system concerning the city's needs. While this vision is part of the narrative of an integrated multimodal system that provides accessibility and connectivity to most of the metropolitan district, this coverage has not materialized. The first line of the Quito Metro covers only part of the central urban area, and the multimodal system has problems consolidating due to conflicts in the public–private articulation of the transport management. That has meant that many district sectors, especially in the periphery, do not have access to quality public transport. An example is Calderón, a rural parish located in the north of the city, which has one of the highest demographic and urban growth rates in recent decades but was not considered within the Metro route. From a prescriptive and speculative sense, the need to extend the Metro line to new sectors has been discussed around this and other similar cases under an argument not only for mobility demand and urban development (Moreno 2020).

4.3.2 *Environmental: Anticipatory Response to Climate Change*

Many of the problems derived from the negative externalities of urban mobility in contemporary cities are related to environmental issues. The excessive circulation of motorized vehicles triggers the emission of polluting gases and high noise levels. The DMQ's vehicle fleet shows an annual growth rate of 5–10%, with a projection of 1,140,000 vehicles in 2030, and a motorization rate of around 192 vehicles per 1000 inhabitants, which could reach 450 in 2030. These values are among the highest in the region and explain the urban mobility problems experienced by the district due to motorized transport. On the one hand, there is an increasing difficulty in circulating and high levels of congestion, especially in the Hypercenter area where most activities are concentrated; travel times have increased by about 10% in the last decade, and the

Table 4.2 Types of Metro Quito stations by urban environment characteristics

Typology	Location	Characteristics	Development hypothesis
A Intermodal stations	Quitumbe Moran Valverde El Recreo La Magdalena El Labrador	Terminals connecting to the integral Mass Transportation System (SITM) with a high flow of user and commerce	Its urban environment is expected to increase its building density, mainly for purposes complementary to the transportation program
B Special stations	San Francisco Universidad Central Iñaquito	Due to their tourist, educational, commercial, and governmental characteristics, a particular study is recommended for each case	As they are in highly consolidated sectors, their functionality will directly relate to each sector's prevailing land use and occupation
C Administrative-Commercial stations	La Alameda El Ejido La Pradera La Carolina	Located within the hyper center of Quito, with intense commercial and administrative use, highly consolidated sectors where tall buildings predominate (six floors or more)	The intense flow of users to and from the stations is expected. Due to building consolidation, it is unlikely that new real estate developments will emerge to modify the environment
D Residential-Commercial stations	Solanda El Calzado Jipijapa	Residential sectors with a high percentage of sectoral commerce, where the predominant buildings are of medium and low height (less than six floors)	A possible re-densification of buildings, along with demographic densification and compaction around the stations

Source Based on Maldonado (2018), p. 20

average speed does not exceed 14 km/h. In environmental terms, because of heavy traffic and congestion, it has been determined that motorized transport, in general, is responsible for 56% of the carbon footprint in Quito (MDMQ 2015).

When faced with this problem, the DMQ's urban mobility policy has set, among its guiding principles, the preservation of the environment and the population's health. That has implied a change of vision toward sustainable mobility aligned with the Development Goals (SDG) as a response to the social, economic, and environmental impacts derived from the intensive use of motorized vehicles. Socially, the aim is for people and goods to move around efficiently and equitably. It is also expected to optimize travel times and operating costs to improve the urban economy. Environmentally, it is expected to promote transportation with renewable energy in the medium and long term. To this end, several strategies have been implemented, including vehicle controls to prevent polluting emissions[4]; promotion of non-motorized and pedestrian modes of transportation; incentives for alternative uses of private vehicles to reduce travel demand; and, mainly, the optimization and promotion of public transportation through an integrated multimodal system, where the Quito Metro[5] is the articulating axis of the system (MDMQ 2015; PMM 2009).

Environmental issues, in their broadest sense, have been on the political agenda since the 1970s, initially through specific plans (water, sewage, solid waste, and others) and later as a central axis of urban planning. As the environmental impacts associated with the development model have become more complex, new issues have been added to the agenda, including climate change. Faced with the imminent impact of climate phenomena in cities, since the early 2000s, the MDMQ assumed the responsibility and leadership to build a policy of mitigation and adaptation to climate change, which, after an important political and technical process, resulted in the Quito Climate Change Action Plan (Barrionuevo 2020). Although climate change is perceived as a problem managed from an environmental perspective, its causes and consequences are broader. Therefore, its articulation with other areas, such as sustainable mobility, is essential to understand the anticipatory nature of a system like the Metro.

The vision of the Action Plan is to make Quito a carbon–neutral and resilient territory in the face of climate change by 2050 through 19 actions related to governance, urban growth, land use, environmental services, and sustainable buildings, among others. Analyses of the DMQ situation have determined that the carbon footprint is caused by solid waste management energy consumption in the industrial, productive, and household sectors, agriculture in rural areas, land use change, and mainly urban mobility. In response to the mobility problem, the plan proposes three specific actions: (i) zero-emission public transportation, aimed at promoting the progressive replacement of conventional and public buses with clean technologies; (ii) integrated and efficient public transportation, to increase public transportation trips and reorganize routes and fares around the Metro as the central axis; (iii) active mobility, focused on increasing the network of urban cycle paths and supporting infrastructure for bicycles (MDMQ 2020).

Thus, the Quito Metro, conceived as a transversal axis of sustainable mobility, configures a system that anticipates the problems generated by climate change in at least two dimensions. Due to the Metro's accessibility characteristics, there is a real possibility of reducing the number of vehicles in transit, but also of optimizing

---

[4] In 2003, vehicle technical service centers were implemented for a mandatory annual inspection of private vehicles and twice a year for public and commercial transportation. However, the increase in pollution corresponds to the growth of the vehicle fleet, which mostly continues to use gasoline and diesel as fuels.

[5] The EPMMQ issued the Environmental Impact Study of the First Line of the Quito Metro (EPMMQ 2012). Beyond the technical details of the report, critical positions questioned the reductionist and fragmented vision of the city regarding environmental problems, suggesting specifying the scope of the impact in categories such as quality of life and the impact on biological communities (Aguilar et al. 2013).

travel times and encouraging the use of other non-motorized transportation alternatives such as bicycles, directly impacting citizens' health and quality of life. On the other hand, the energy used by this form of transportation is electrical, created by hydraulic sources. Therefore, the intensive use of this renewable energy, which does not require fossil fuels, will effectively mitigate the emission of polluting gases in the medium term. These would be local solutions to the global environmental impact produced by the negative externalities of the vehicle fleet. In this process, citizens have a fundamental role as agents of transformation toward sustainability. Thus, constructing environmental awareness, generating motivation, and empowering the population constitute social resources of an anticipatory character for Quito's Metro (MDMQ 2020).

4.3.3 Socio-Cultural Implications

The 2008 Constitution of Ecuador established a new vision of public action based on rights. That has implied that the design and implementation of public policies at different levels of government are structured in a binding manner on principles of equity, inclusion, and participation. Expressly, the notion of the right to the city has set a new course in territorial development, promoting urban planning based on sustainable and inclusive development (Pinto et al. 2023). Within this logic, DMQ's planning promotes, through its strategic objectives, a city of well-being and rights with equal opportunities for all citizens, especially for historically excluded and vulnerable groups. For this purpose, it is proposed to promote the exercise of rights by increasing access and quality of services and ensuring material conditions to reduce structural inequality gaps (MDMQ 2021).

Beyond its operational dimension, the DMQ's urban mobility policy is conceived as a critical sector for the district's progress. It promotes an inclusive, safe, and sustainable transportation and connectivity system that contributes to reducing inequity and promoting communities with a better quality of life. As part of the principles that guide the decision-making processes, there is a set of qualities that, from an anticipatory sense, focus on the social dimension of mobility in terms of rights. In this sense, accessibility guarantees adequate scope and coverage, as well as accessible costs and timely information. Equity and inclusion are focused on providing a service without discrimination of gender, age, capacity, or condition. Citizen participation seeks to guarantee a responsible inclusion of citizens within their competencies in planning mobility programs and projects (MDMQ 2015: 38).

The Quito Metro, as the axis of urban mobility, is an anticipatory element of the model of the right to the city model that guides the territorial planning of the metropolitan district. The transformative capacity of the Metro system, not only in terms of coverage and quality but also in terms of technological innovation, opens a window of opportunity in the medium and long term for the notion of sustainable mobility to consolidate and promote substantive changes in the city's structural inequities. Problems of socio-spatial segregation, urban dispersion, fragmentation, poor distribution of infrastructure and services, and excessive specialization of land use, among others, can be redefined in planning based on the advantages of accessibility and inclusion presented by the Metro.

From an anticipatory sense, the analysis of the impact of the first line of the Metro shows, on the one hand, that, in general, it will increase the accessibility of urban opportunities (Olarte 2021), depending on travel time and current levels of accessibility. On the other hand, a modest reduction of the accessibility gap is expected, with more visible benefits in long travel times (Quezada Larriva et al. 2023). However, it should be remembered that greater accessibility by connecting peripheries and centralities is not a measure that ensures the reduction of social exclusion in the city but needs to be integrated into a broader development model (Ortiz 2022).

As an anticipatory system, the Quito Metro has substantial implications for constructing a civic culture of citizen coexistence. Around the figure of the Metro, recognized as the most important urban infrastructure project in the city and the country, not only a normative dimension of the proper use of the service is being promoted, but, in a broader sense, a narrative of appropriation and respect toward common goods.[6] Certainly, the narratives configure anticipatory models, which simulate and stimulate key cognitive processes in the projection of the future (Lively 2017: 6). These strategies have been previously implemented in other municipal public transport systems such as the Trolleybus and Ecovia. For example, campaigns against the damage of stops and buses and the prevention of gender violence have been conducted, with relative results. That becomes important in a city like Quito, where urban mobility has historically been perceived as an aggressive, stressful, and constantly competitive space, inducing individual behavior of drivers and pedestrians to reproduce patterns of violence to adapt to the system and achieve the objective of mobility.[7] Although citizens know the rules governing mobility, compliance with them is partial, normalizing anarchy and impunity (Cáceres 2019).

In response to this situation, the MetroCultura strategy[8] has been conceived as a social management model based on principles of coexistence and participation,

[6] The cultural dimension of the Quito Metro had an important debate in September 2008, when a group of people broke into the Metro workshops' facilities and garages of the Metro to graffiti one of the cars. This act, categorized as vandalism and aggression to public property, generated a comprehensive discussion for and against the implications of criminalizing expressions of street art, but above all, about the new transportation system's cultural and public space dimensions (Sánchez 2021).

[7] The negative perception of urban mobility in Quito is expressed through low satisfaction (less than 50%) with the public transport service Linea Corridor Central Norte. That results in a high probability (57%) that users of this transport line will migrate to the new Metro system, indicating little identification with the current service (González 2021).

[8] The predecessor of this strategy is the Community Relations Plan included in the Environmental Management Plan (2012). Its objectives were: (i) to implement an inclusive relationship process with the different stakeholders; (ii) to promote a participatory process; (iii) to foster relationships

which is committed to the proper use and care of the transportation system. Above all, it aims to build a vision of Quito's future through the interaction of service users. To this end, the strategy proposes a system of values (respect, empathy, solidarity, tolerance, dialogue, responsibility) to govern the behavior of users in Metro facilities and spaces and to generate a sense of community among citizens. This vision has been summarized in the premise of "better users, better citizens" under three main approaches: Transportation and Public Health, Social Innovation, and Gender Equity and inclusion of priority groups. In operational terms, MetroCultura is implemented through two types of strategic plans: an information provision component on aspects related to the operation and use of the service and user service and an intervention component focused on training aspects, cultural consumption, and environmental awareness aimed at promoting new behaviors in system users (EPMMQ 2020). The strategy has a normative and prescriptive sense of regulation and control, and MetroCultura has significant communicational and pedagogical potential. It can be understood as an anticipatory system for the ethical transformation of individual and collective behavior in Quito's urban mobility.

4.4 Conclusions

Understanding sustainable mobility policies as an anticipatory system of development can be helpful in the decision-making process in the present. However, it is important not to project a reductionist narrative of a single preferred future but to consider the possibility of alternative images. That implies a systematic understanding of several possible futures (González-Arellano 2022). In this sense, subway systems are considered as anticipatory systems from different dimensions. Therefore, the case of the Quito Metro contributes to the debate of anticipatory systems around how the sustainable urban mobility policy was built and how it has been institutionalized implicitly in Quito's planning.

On the other hand, unlike the older subways created before thinking about sustainable mobility that has achieved greater consolidation of different strategies, the Quito Metro poses a substantive change in the mobility policy in the city, although still with challenges in institutional management and administration. Finally, the Metro, as an anticipatory system, provides different future scenarios in the present. Among these is a more integral reading of mobility, development, and culture-oriented to internalize the infrastructure within users and society.

These anticipatory policies of the Quito Metro stem from an articulation of anticipatory communities. In the construction and implementation process, a series of state and non-state actors with shared visions of a future and an anticipatory regime have converged. The case of Quito Metro reveals two types of communities. The first is related to a strategic anticipatory vision of political leadership, corresponding

based on trust, respect, and understanding; and (iv) to establish a dialogue relationship that prevents conflict situations during the construction and operation of the Metro (EPMMQ 2020).

to the vision of the local government, whose axis is sustainable development. This government promoted the project from a systemic perspective of the medium and long term, with interventions in the social and environmental spheres. It adopted a pragmatic and creative vision of the future.

The second community is configured from an anticipatory community of a managerial nature, centered around a new political situation designed to address the subway as a public transportation system. This community is anchored to a more operative vision of the local government, involving multilateral actors in the strategic management of the subway and in the construction process. This configuration establishes the subway as a transportation system operated by the Metro Company and other actors. However, a particular observation in the case of Quito is that once the project is concluded, it starts to take shape around the social actors. These actors contribute to building a Metro culture and new political leadership from the local government with a substantive development vision. Learning from the Quito Metro leads us to consider that communities are not exclusive to one policy area but function in response to the trajectory of decision-making.

References

Acosta A (2015) El Buen Vivir como alternativa al desarrollo. Algunas reflexiones económicas y no tan económicas. Política y Sociedad, 52(2). https://doi.org/10.5209/rev_poso.2015.v52.n2.45203

Aguilar C, Guerrero L, López E, Rodríguez M, Scarpacci M, Bustamante T, Paredes M, Perelman P (2013) Calidad de vida y comunidades biológicas: análisis del estudio de impacto ambiental del Metro en la Ciudad de Quito. Letras Verdes 13:41–52. www.flacsoandes.org/letrasverdes

Alcántara Vasconcellos E (2020) Equidad y sostenibilidad en la movilidad urbana de América Latina. In: Figueroa O, Valenzuela-Montes LM, Brasileiro A (eds) Desafíos del desarrollo urbano sostenible en el transporte y la movilidad (1st ed., vol 1). El Colegio Mexiquense A.C., pp 37–54

Armijos AM (2021) Imaginarios y discursos del patrimonio quiteño la Plaza de San Francisco y el Proyecto Metro de Quito [Tesis de maestría]. Universidad Andina Simón Bolivar

Astudillo-Ortega S, Vásquez A (2021) Detectando los desequilibrios dotacionales como generadores de problemas en la movilidad del Distrito Metropolitano de Quito. In: En Prceedings 3rd congress in geomatics engineering. Editorial Universitat Politécnica de Valéncia, pp 69–76. https://doi.org/10.4995/CiGeo2021.2021.12503

Avellaneda P (2017) Los proyectos de infraestructura y su relación con el ordenamiento territorial. In: Beuf A, Avellaneda P (eds) Ordenar los Territorios: Perspectivas y Criticas desde América Latina (1st ed., vol 1). Universidad de los Andes, pp 183–196

Ávila MS (2017) Análisis del impacto turístico del Metro de Quito en el centro histórico, una visión desde la geografía turística [Tesis de pregrado]. Pontificia Universidad Católica del Ecuador

Barrionuevo, M. de los Á (2020) La adopción de la agenda de política Pública ambiental urbana: el caso de la agenda de cambio climático en el Municipio del Distrito Metropolitano de Quito [Tesis de doctorado]. Facultad Latinoamericana de Ciencias Sociales

Bolay JC (2012) What sustainable development for the cities of the South? Urban issues for a third millennium. Int J Urban Sustain Dev 4(1):76–93. https://doi.org/10.1080/19463138.2011.626170

Cabrera S (2018) El patrimonio en disputa: la plaza vs el Metro de Quito (Paradiso Editores, vol. 1). Universidad Andina Simón Bolivar

Cáceres E (2019) Movilidad urbana en Quito: Análisis del funcionamiento del transporte público desde una perspectiva de la interacción cotidiana y la sociología de las emociones [Tesis de pregrado]. Pontificia Universidad Católica del Ecuador

CAF (2011) Desarrollo urbano y movilidad en América Latina. Banco de desarrollo de America Latina CAF

CAF (2017) Crecimiento urbano y acceso a oportunidades: un desafío para América Latina

Casado Izquierdo JM (2008) Estudios sobre Movilidad Cotidiana en México. Scripta Nova, XI I(273):741–798

Cebollada Á, Avellaneda P (2008) Equidad social en movilidad: reflexiones en torno a los casos de Barcelona y Lima. Scripta Nova, 12. https://raco.cat/index.php/ScriptaNova/article/view/115445

CEPAL (2018) Estudio Económico de América Latina y el Caribe 2018. Evolución de la inversión en América Latina y el Caribe: hechos estilizados, determinantes y desafíos de política (1st ed., vol 1). Naciones Unidas

CEPAL (2019) Estudio Económico de América Latina y el Caribe 2018. Evolución de la inversión en América Latina y el Caribe: hechos estilizados, determinantes y desafíos de política (1st ed., vol 1). Naciones Unidas

Chauvin JP (2006) Conflicto y gobierno local : el caso del transporte urbano en Quito [Tesis de Maestría, Facultad Latinoamericana de Ciencias Sociales]. http://repositorio.flacsoandes.edu.ec/handle/10469/424

Cosacov N, Di Virgilio MM (2018) Presentación del Dossier #10: Movilidades espaciales de la población y dinámicas metropolitanas en ciudades latinoamericanas. Quid 16(10):1–16

De Grange L (2010) El gran impacto del Metro. EURE 36(107):125–131

Eguiguren FJ (2015) ¿Caminando a la movilidad urbana sostenible?: desafíos de la implementación de políticas en la ciudad de Quito. [Tesis de maestría, Facultad Latinoamericana de Ciencias Sociales]. http://repositorio.flacsoandes.edu.ec/handle/10469/8420

EPMMQ (2012) Estudio de Impacto Ambiental de la primera línea del Metro de Quito

EPMMQ (2020) Metrocultura

Errazuriz M, Taddia A, Ríos R, Pérez J, Brennan P, Ortiz P (2017) Evolución de los Sistemas de Transporte Urbano en América Latina. Bid, p 67. https://doi.org/10.18235/0000828#sthash.yADhMxVk.dpuf

Estevan A, Sanz A (1996) Hacia la reconversión ecológica del transporte en España (1st ed., vol 1). Bakeaz

Figueroa O (2001) Políticas de desarrollo y políticas de transporte urbano. Coherencias y contradicciones. In: Carrión F (ed) La ciudad Construida. urbanismo en América latina (1st ed., vol 1). FLACSO, pp 377–391

Figueroa O, Henry E (1988) Diagnóstico de los Metros en América Latina. Revista Eure, XIV(42):7–17

González S (2021) Relación entre la percepción de la calidad del servicio del Corredor Central Norte de Quito y la probabilidad de migración al Metro. Neutrosphic Comput Mach Learn 16:82–95

González-Arellano S (2022) The future scenarios of cities: an analysis of their institutional construction. Int J Urban Reg Res 46(5):865–873. https://doi.org/10.1111/1468-2427.13103

Gordón S (2011) La movilidad sustentable en Quito: una visión desde los más vulnerables [Tesis de maestría, Facultad Latinoamericana de Ciencias Sociales]. www.flacsoandes.edu.ec

Gudynas E (2010) Desarrollo sostenible: una guía básica de conceptos y tendencias hacia otra economía. Otra Economía, IV(6):43–66. www.riless.org/otraeconomia

Gutiérrez A (2010) Movilidad, transporte y acceso: una renovación aplicada al ordenamiento territorial. Scripta Nova. Revista Electrónica de Geografía y Ciencias Sociales, 14. https://revistes.ub.edu/index.php/ScriptaNova/article/view/1736/1712

Gutiérrez A (2012)¿Qué es la movilidad? Elementos para (re) construir las definiciones básicas del campo del transporte. Bitacora 21(2):61–74. http://www.redalyc.org/articulo.oa?id=74826255011
Hammershøj LG (2017) Diagnosis of the times. Handbook of Anticipation. Springer International Publishing, pp 1–20. https://doi.org/10.1007/978-3-319-31737-3_64-1
Hernández D (2017) Transporte público, bienestar y desigualdad: cobertura y capacidad de pago en la ciudad de Montevideo. Revista De La CEPAL 122:165–184
Hernández D, Falavigna C (2016) Inmovilidad en la ciudad: el perfil del ciudadano "inmóvil" en Córdoba y Montevideo. Implicancias metodológicas y sustantivas. Universidad Católica Del Uruguay, pp 1–12.
Hidalgo R (2022) La trama del Metro. Patrones de transformación en torno a las estaciones del Metro de Santiago (Chile). Dearq 32:46–58. https://doi.org/10.18389/dearq32.2022.06
Jirón P (2017) Presentación. Movilidad Urbana y Género: experiencias latinoamericanas. Revista Transporte y Territorio, 16:1–8. https://doi.org/10.34096/rtt.i16.3600
Keivani R (2010) A review of the main challenges to urban sustainability. Int J Urban Sustain Dev 1(1–2):5–16. https://doi.org/10.1080/19463131003704213
Lazo A (2008) Transporte, movilidad y exclusión. El caso de Trasantiago en Chile. Scripta Nova, XII(270):741–798. https://www.ub.edu/geocrit/sn/sn-270/sn-270-45.htm
Levy A (1999) Urban morphology and the problem of the modern urban fabric: some questions for research. Urban Morphol 3(2):79–85
Lively G (2017) Anticipation and Narratology. Handbook of anticipation (1st ed., vol 1). Springer International Publishing, pp 1–20
Lupano JA, Sánchez RJ (2009) Políticas de movilidad urbana e infraestructura urbana de transporte. Documentos de Proyectos, 230(LC/W.230-P/E):65
Maldonado D (2018) Influencia del transporte y la movilidad en el desarrollo urbano de Quito, construcción del Metro de Quito, influencia urbana de la estación Jipijapa en el sector [Tesis de maestría]. Pontificia Universidad Católica del Ecuador
Manheim M (1979) Fundamentals of Transportation Systems Analysis (1st ed., vol 1). MIT Press, Cambridge, Mass
MDMQ (2015) *Municipio del distrito metropolitano de quito secretaría de movilidad visión estratégica de la movilidad*
MDMQ (2020). Plan de Acción de Cambio Climático de Quito
MDMQ (2021) Plan Metropolitano de Desarrollo y ordenamiento territorial
Metro de Medellín (2023) Relacionamiento con la comunidad. Cultura Metro. https://www.metrodemedellin.gov.co/cultura-metro/relacionamiento-con-la-comunidad/?hs_preview=WqETznOG-145687407978
Metro (2015) Metro de Santiago—Memoria Anual 2015. Memoria Anual. https://www.metro.cl/minisitio/memoria2015/ma-cap-1.php
Metro (2022) Nuestra História—Gobierno Corporativo—Metro de Santiago. Metro de Santiago. https://www.metro.cl/gobierno-corporativo/historia
Miller R, Poli R, Rossel P (2013) The discipline of anticipation: exploring key issues 1:4. http://www.costa22.dk/
Miralles-Guash C (2002) Ciudad y Transporte: El binomio Imperfecto (1st ed., vol 1). Editorial Ariel
Miralles-Guasch C, Cebollada À (2009) Movilidad Cotidiana y Sostenibilidad, una interpretación desde la geografía humana. Boletín de La A.G.E. 50:193–216
Montezuma R (2003) Ciudad y Transporte: La movilidad Urbana. In: Balbo M, Jordán R, Simioni D (eds.), La ciudad inclusiva (1st ed., vol 1). Naciones Unidas, CEPAL, p 313
Monzón A (2005) Gestión del transporte metropolitano. In: Rojas E, Cuadrado-Roura J, Fernández JM (eds.), Gobernar las metrópolis (1st ed., vol 1). Banco Interamericano de Desarrollo, pp 409–464
Moreno D (2020) La Política de movilidad en el Distrito metropolitano de Quito: El caso del Metro en Calderón [Tesis de pregrado]. UDLA

Olarte SH (2021) Which mobility variables are related to urban poverty? The case of Santiago De Chile And Quito. Archit City Environ 16(47). https://doi.org/10.5821/ace.16.47.9510
Oleas D (2019) ¿Cómo inciden los intereses y las instituciones en el cambio de agenda de la política de movilidad urbana? [Tesis de doctorado]. Facultad Latinoamericana de Ciencias Sociales
Oleas D, Albornoz MB (2015) La bicicleta y la transformación del espacio público en Quito (2003–2014). Letras Verdes 19:24–44. https://doi.org/10.17141/letrasverdes.19.2016.1817
Ortiz C (2022) Exclusión social e inmovilidad. Un análisis al diseño de la primera línea del Metro de Quito [Tesis de maestría]. Facultad Latinoamericana de Ciencias Sociales
Ortuzar S (1983) Santiago's Metro. Transport 113–116
Paz-Gómez D, Santelices M (2020) (In) capacidades de política en tiempos de Covid-19: comprendiendo las respuestas económicas de Colombia y Ecuador. Análisis Político, 100(1):72–91. https://doi.org/10.15446/anpol.v33n100.93362
Pinto V, Córdova M, Bell D (2023) The right to the city and contemporary housing policy in Ecuador. Hous Stud. https://doi.org/10.1080/02673037.2023.2180491
Plan V (2016) Los activistas que protestan contra las excavaciones del metro. https://Www.Planv.Com.Ec/Historias/Urbano/Activistas-Que-Protestan-Contra-Excavaciones-Del-Metro-Quito
PMM (2009) Plan Maestro de Movilidad para el Distrito Metropolitano de Quito 2009–2025
Poli R (2017) Introducing anticipation. In: Handbook of anticipation. Springer International Publishing, pp 1–14. https://doi.org/10.1007/978-3-319-31737-3_1-1
Pozueta J (2000) Movilidad y planeamiento sostenible : hacia una consideración inteligente del transporte y la movilidad en el planeamiento y en el diseño urbano (1st ed., vol 1). Instituto Juan de Herrera. https://books.google.com/books/about/Movilidad_y_planeamiento_sostenible.html?hl=es&id=GkmuAAAACAAJ
Proaño G (2012) Cultura "ciclera" en Quito, políticas de movilidad: estudio de caso ciclópolis y al sur en bici [Tesis de maestría]. Universidad Andina Simón Bolivar
Quezada Larriva A, Orellana D, Guerrero Balarezo ML, García JA, Cárdenas Villenas G, Osorio Guerrero P (2023) Impact of Quito's first Metro line on the accessibility to urban opportunities. J Transp Geogr 108. https://doi.org/10.1016/j.jtrangeo.2023.103548
Quintero J (2011) Inventarios Viales y Categorización de la Red Vial en Estudios de Ingeniería de Tránsito y Transporte. Revista De Ingeniería 20(30):65–77
Quintero-González J-R (2017) Del concepto de ingeniería de tránsito al de movilidad urbana sostenible. Ambiente y Desarrollo 21(40):57–72. https://doi.org/10.11144/JAVERIANA.AYD21-40.CITM
Rivas ME, Suárez-Alemán A, Serebrisky T (2019) Políticas de Transporte Urbano. In: BID. Banco Interamericano de Desarrollo, p 52
Rodriguez DA, Vergel-Tovar CE (2013) Sistemas de Transporte Publico Masivo Tipo BRT (Bus Rapid Transit) y Desarrollo Urbano en America Latina. Land Lines 16–24
Rodriguez D, Vergel-Tovar CE (2018) Urban development around bus rapid transit stops in seven cities in Latin America. J Urban 11(2):175–201. https://doi.org/10.1080/17549175.2017.1372507
Rodríguez Vignoli J (2008) Movilidad cotidiana, desigualdad social y segregación residencial en cuatro metrópolis de América Latina. Revista Eure, XXXIV(103):49–71
Salar Khan M (2016) Metro de Quito: better or bitter? Analysis of getting around Quito. PODIUM 29:93–105
Sánchez R (2021) Genealogía y criminalización de una subcultura urbana caso Metro de Quito, 2018 [Tesis de maestría]. Universidad Andina Simón Bolivar
Sheller M, Urry J (2006) The new mobilities paradigm. Environ Plan A 38(2):207–226. https://doi.org/10.1068/a37268
Silva J (2010) Movilidad urbana: análisis de los fallos de mercado ocasionados en el Distrito Metropolitano de Quito. Periodo 2009 [Tesis de pregrado]. Pontificia Universidad Católica del Ecuador
Vasconcellos EA (2015) Transporte urbano y movilidad (1st ed., vol 1). Universidad de San martín

Vergel-Tovar E (2019) ¿Es la densidad suficiente? Análisis de la relación entre la densidad poblacional y la demanda de pasajeros en los sistemas de transporte público masivo tipo BRT en Curitiba, Quito y Bogotá. Medio Ambiente y Urbanización 90(1):53–82 (30). IIED-América Latina

Winchester L (2006) Desafíos para el desarrollo sostenible de las ciudades en América Latina y El Caribe. Revista Eure XXXII:7–25

Zahariadis N (2007) The Multiple streams framework: structure, limitations, prospects. In: Sabatier PA (ed) Theories of the policy process. Westviwe Press, Boulder, pp 65–92

Chapter 5
"*Se Va a Caer*" (Patriarchy Will Fall): Care at a Possible Intersection Between Future Studies and Feminism

Daniela Losiggio and María Pozzio

"Y ahora que estamos juntas,
y ahora que sí nos ven,
Abajo el patriarcado
Se va a caer
Se va a caer;
Arriba el feminismo
Que va a vencer
Que va a vencer"
Popular and feminist song (Argentina 2018–2020)

Abstract In this chapter, the authors propose a dialogue between future studies and feminism. The feminist slogan "*Se va a caer*" (Patriarchy will fall) helps them think about this movement in terms of future narratives and the relationship between utopia-dystopia and uchronia in the context of threats and the rise of the ultra-right. Then, they propose some theoretical genealogies that allow them to conceptualize the notion of care and consider how practical proposals inspired by this concept can be regarded as anticipation.

Keywords Care · Future · Feminism · Patriarchy

D. Losiggio · M. Pozzio (✉)
CONICET, Universidad Nacional Arturo Jauretche, Buenos Aires, Argentina
e-mail: mpozzio@unaj.edu.ar

D. Losiggio
e-mail: dlosiggio@unaj.edu.ar

S. Gonzalez-Arellano and B. Gandlgruber (eds.), *Cities as Anticipatory Systems*, The Urban Book Series, https://doi.org/10.1007/978-3-031-78162-9_5

5.1 Introduction

In *Cuando el feminismo era mala palabra* (Tarducci et al. 2019), the authors describe the origins and actions of feminist demands during the years before and after the dictatorship in Argentina (1976–1983): at that time, these were isolated actions with low participation and, in general terms, promoted by urban and middle-class women. Since then and until today, on the 40th anniversary of democracy, feminisms have become plural and massive, as well as a significant actor in the political arena (we talk today about feminisms in the plural and, with it, we include popular feminisms, transfeminisms, indigenous women's movements, and lesbian movements, among others).

In the case of Argentina specifically, feminisms occupied the streets, starting with the first Ni Una Menos against femicide in 2015 and, later, with the *marea verde* (green tide), the name given to the massive vigils that accompanied the parliamentary debate on abortion in 2018 and 2020. The massification of feminisms constitutes a predominantly urban phenomenon, and this leads us to think about the city not as a given space—subsequently "occupied" by citizenship that fits its precepts—but as a system of networks that enable a "complex collective behavior" (see Chap. 2 Gonzalez-Arellano).

On the one hand, feminisms challenge the normative aspects of public infrastructure. Parks, public buildings, streets, and bus stops constitute power devices that regulate people circulation (Harvey 2007). Parallel, the glitter, the ornaments, the masks, the costumes, the colors, and the carnival—all of them feminist symbols in Argentina—call the sovereign exercise and manage to damage the ritualized and masculinized ways of inhabiting the city (Losiggio 2020; Losiggio and Taccetta 2022).

On the other hand, the massification of feminism expresses that cities have the capacity for self-organization, not necessarily arising from a control center, even when they can bend political wills and become institutionalized (Gonzalez-Arellano 2022). This phenomenon has attracted worldwide attention and required a fundamental component: social networks. Social networks extended the urban space throughout the country, generating a digital continuity in the city. Women and diversities in peri-urban or rural spaces participated in the demonstrations through the viralization of images and selfies showing outfits that responded to different slogans and hashtags.

Besides, this massification has also influenced public interest, agenda-setting, and, consequently, public policy. That is an unprecedented phenomenon. For the first time, public policies—including those aimed at public works and urban planning—incorporate the problems traditionally understood as private or intimate (sexual and reproductive health, gender identity, domestic violence).

In short, feminisms stresses the connotations of the city as the place of the public masculine (as opposed to the private feminine), showing that the city has the capacity to organize itself to exceed its precepts and devices. Finally, they manage to become institutionalized in different public policies. The song that became the score of the massive feminist demonstrations in Argentina in 2018, and says *Ahora que sí nos ven (…) abajo el patriarcado, se va a caer* (Now that they do see us, down with the patriarchy, it will fall), can be understood as the synthesis of the different anticipatory strategies of Argentine feminisms of the period 2015–2018; building a city in the present where that visibility is not only possible but desirable and safe; a more egalitarian city that materialize in actual campaigns against street harassment, murals that remember the victims of femicides, red benches to denounce violence against women, feminist taxi services to guarantee safe mobility, ad hoc daycare centers. In brief, as feminist urban planning organizations propose, each place of the patriarchal city must be transformed into a place of resistance by mapping, signaling, and sustaining organization networks. At the same time, the city must be the city of the young and the city of the old, your city and the global city.

In this article, we are especially interested in analyzing feminist anticipatory strategies around care. If we learned any lesson from the pandemic, it is that care activities are the basis for the sustainability of life. The idea of care represents an inversion of patriarchal and capitalist ways of understanding social existence: from the margins to the center, reproduction is an essential activity: housing, mobility, security, health, care of the land, food sovereignty, education over-exploitation, extractivism, pollution, and accumulation.

In times of the re-emergence of neoliberal economic policy proposals and the rise of new rights, statistics show us that it is still women (especially trans, indigenous, and rural women) and LGBTIQ + people who are affected the most by successive economic crises. Women and LGBTIQ + people continue to be those who mostly do domestic labor and those who embody the lowest-paid branches of employment in the market. Expressed in multiple proposals with significant impact on the urban scale and the world of work, the demand aims to develop public policies that make visible ("now that they do see us," says the song), de-privatize, and de-feminize care. Policies include creating care and parenting spaces, incorporating care issues—based on play—in school curricula, and paternity leave. The projection of public policies that can destroy the classic sexual division of labor—one of the main supports of patriarchy—can then be thought of as a form of anticipation toward a more egalitarian world.

Thus, throughout this article, we propose to analyze some notions of the future and anticipation that currently appear in feminist fiction and are much more effective than other genres in describing the threats that fall on women and diversities in the current context. Then, we will recover the genealogy of the sexual division of labor question to show the actual problem around care (some statistics and diagnoses). Finally, we will describe care public policies as forms of political anticipation and propose community care developed by women from popular neighborhoods in large cities as an example of this anticipation.

5.2 Feminism, Threats, and Dystopias

As Geneviève Liveley (2017) and María Mercedes Patrouilleau (2020) have stated, discourses and narratives provide a privileged setting for studying futures. Our motto, "The patriarchy will fall," can be understood through this lens. Now, if the score of the so-called green tide connotes optimism—the achievements of recent years in terms of rights support this emotion—it is also true that we write in the imminence of a global scenario with proposals that threaten to plebiscite rights, withdraw affirmative policies of parities and quotas, and defund sexual and reproductive health that can be victorious. So, what about the future?

The idea that the patriarchy will fall does not dictate the need for that event. The phrase constitutes an expression of desire founded on signs of the present, reliable information, scientific diagnoses, and collective networks alternative to the oppressive and extractivist ones. It aims to reopen the possibilities of the future and in some way expresses what anthropologist Arjun Appadurai (2015) calls the "politics of possibility": anticipation is not describing a destiny or reflecting a causality, but instead manifesting the possibility of doing something in the present to transform the future that appears obvious. That implies intervening in the temporal. Literature can do this by inventing a particular fictional hyperbole that cannot be accused of being tremendous because it belongs to the literary genre.

Utopias, uchronies, and dystopias bring us closer to critical concepts for the anticipatory exercise. Let us see one of our favorite dystopias, *The Handmaid's Tale*, Margaret Atwood's novel, transposed into a television series by Bruce Miller. Which concept of the future does this novel offer?

The term dystopia refers to fictions that outline a totalitarian society through a pernicious inversion of utopia, where our ideals are distorted and become nightmares. The dystopian narrative contains a critical and counter-hegemonic view of our contemporary society, implicitly including a warning. Margaret Atwood's novel describes the society of Gilead, shaken by mass infertility that causes a return to conservative religiosity. A caste of theocratic military and political leaders overthrows the democratic government and imposes a dictatorship that enslaves fertile women, prohibits women's public participation, and criminalizes gender diversity. For us, Gilead constitutes a warning: the phenomenon of generalized sterility is a metaphor for the control over our reproduction that we people with pregnant capacity have gained since the feminist struggle (the starting point of the so-called second wave). Indeed, radical theories about the origin of patriarchy revolve around this hypothesis: the will of men to control the reproductive power of women.

Furthermore, if Shulamith Firestone's utopia consisted of a dictatorship of the reproductive class (people with gestational capacity) and the retention of the means of reproduction, *The Handmaid's Tale* is presented as its opposite mirror. It plays with the idea of a reactionary reappropriation of that power by the patriarchy. Reading and circulating this text in a scenario like the current one allows us to diagnose the present critically and becomes a rational element of anticipation: indeed, we are threatened.

What are those threats? Several analysts have pointed out that the sexist and misogynistic discourse of the new rights in Latin America has the support of some sectors of the Pentecostal and Evangelical Churches (Bárcenas Barajas 2022), as well as Catholic organizations such as the Opus Dei (Vaggione 2005). In the debate on the right to abortion, these groups and their ideas gained visibility with a counter-symbol, the light blue scarves, and currently find expression in the political group of the Argentinian president, Javier Milei (Incaminato 2023). Within an ultra-liberal, free-market, anti-statist, denialist, and reactionary agenda, but also in its public interventions, his platform embodies a bellicose anti-feminist discourse with additives of the obscurantism typical of the radical right. The possibility of an electoral victory and a government by this political force openly threatens the lives of women and LGBTIQ + people. It refers to Atwood's dystopia as a latent horizon and reveals the necessary relationships between failed utopias, dystopias, and catastrophes (Engélibert and Guidée 2015). In the context of threat, dystopian fictions serve to make us understand the dangers of the future. The wish song "The patriarchy will fall" must not lose its meaning, but now it must be evoked with its disturbing reflection. If the essential thing about dystopia is to reflect a world that is distanced from the real one, but that serves to reveal it, these dystopian fictions reveal that patriarchy is not going to fall meekly without first giving furious kicks of reaction. In the context of our country, this reaction is embodied in the anti-feminist proposals of the new "libertarian" extreme right.

The Senegalese thinker Fellwine Sarr (2019) has been pointing out how the hypertrophy of the present is linked to two intentions of neoliberalism: to install the idea of no alternative (a future without promises) and the omnipresence of narratives where the only thinkable horizon is a posthuman future. Against this hypertrophy, he proposes "re-opening the future." The feminist slogan that predicts a future where "the patriarchy is going to fall" implies naming desire, leaving a time without promises, and investing the present and the future with meaning.

The idea of re-opening the future involves thinking about the possible challenges and future scenarios for women and the agendas of feminism (Hurley et al. 2008). In some way, the advances of the different feminisms of the world support this notion that "the future is feminist," but that, at the same time, depends on our actions. We are now interested in considering how we understand the care agenda proposed by feminisms in Argentina as a way of anticipation.

5.3 Care as Anticipation

5.3.1 *The Narrative of the Sexual Division of Labor*

However, what do we understand by care? To refer to the notion of care and understand its anticipatory possibilities, it is first necessary to analyze the Marxian narrative on the sexual division of labor and observe its consequences for the political struggles from the twentieth century to the present.

In the early writings of Karl Marx and Friedrich Engels and, eminently, in the great *Capital* (1867), it appears recurrently the idea of a natural division of labor, immemorial but still current, determined by sex and age, physical strength, and the different "roles in the sexual act" (Marx and Engels 2014, p. 21). With the development of the capitalist mode of production, they maintain, an increasingly specialized social division of labor emerged (famously described by Adam Smith before Marx). However, this social division does not replace the natural one. Women continue to be responsible for non-productive work (reproductive work). Then, in the factories, women and children work according to their physical conditions; in a word, less qualified jobs (Marx 2001).

Later, in *The Origin of the Family, Private Property and the State* (1884), Friedrich Engels tried to refine his understanding of the devaluation of female work and women in general. Its starting point is the thesis of Lewis H. Morgan and Johann J. Bachofen on archaic society and matriarchy, which were very much fashionable then and have been quite refuted today, especially in the field of anthropology. Based on these ideas, Engels maintains that there were extended families in "barbarian" kinship systems, parenting was shared, and monogamy did not exist. In a word, these were "communist homes" (2017: 60). In such contexts, descent could only be established through the maternal line (the only one genuinely verifiable), which contributed to granting women a robust social status.

With the invention of private property, men sought to guarantee better control over their offspring, appropriating them for the family economy. Maternal rights were then abolished, and women became "instruments of reproduction," those in charge of gestating, giving birth, caring for, and raising children; they were expelled from unproductive work (2017: 72). In this way, female oppression and the feminization of specific tasks were considered derivatives of the invention of private property. Consequently, Engels believed that with the communist revolution and the abolition of private property, female exploitation would disappear. Care, education, feeding, and cleaning would no longer constitute unpaid work.

Engels' theses were, to a certain extent, welcomed by feminists within communism, and for a long time, both demands overlapped. Although there have always been objections to the text, the most famous and fundamental one, dedicated to historical materialism, appears in the third chapter of *The Second Sex* by Simone de Beauvoir. Beauvoir maintains that Engels's narrative confirms the truths of historical materialism and leaves unexplained the singular character of the oppression it sought to criticize. The exploitation of women beyond the economic sphere (which is undeniable) must be read in its specificity. It is, above all, sexual: "In vain does rationalist materialism try to ignore the dramatic character of sexuality" (Beauvoir 2015, p. 121).

Feminist theory begins with an elementary premise: reproduction constitutes a historical fact inseparable from production. This proposal has been projected since the end of the 1960s. That year, the Argentine Isabel Larguía and the American John Dumoulin (2019), both committed to the Cuban Revolution, wrote in Havana a controversial and critical article that attacked the new Cuban regulations on the prohibition of the inclusion of women in certain kinds of employment. Larguía and Dumoulin (2019) invite us to think that the divorce between reproduction and production constitutes an ideological operation within Marxist theory and praxis. In the famous "The Unhappy Marriage between Feminism and Marxism" (Hartmann, 1981), Heidi Hartmann agrees with these ideas: if genders are distinguished, it is not because one group is capable of reproducing and the other is not. It is precisely the other way around: genders are the result of the need for the sexual division of labor, of one group alienating the other, both from material resources and from their sexuality.

Nancy Folbre and Ann Ferguson introduce another exciting idea in an unjustly forgotten text. The authors consider that female exploitation extends far beyond the domestic sphere. They refer to a "sexual-affective production," a notion through which it is possible to make the exploitation of women intelligible in a broad sense, transcend mere "procreation and raising children," and observe under the same lens "the provision of care, affection, and sexual satisfaction" at the social level (Ferguson and Folbre 1981: 314).

In neoliberalism and post-Fordism, social reproduction becomes a productive field for capital, which is converted into a source of profit. For this reason, we speak today of a "care crisis" (Kussy et al. 2023). Neoliberal economic policies damage what they should exploit but never destroy: the ability to feed, shelter, educate, and heal—the basis not only of capitalism but of existence itself—is put at risk for the entire population (Fraser et al. 2019).

These authors allow us to understand that for gender equality to exist economically, merely incorporating women into the labor market is insufficient. For true gender equality (but also class and racial equality) to exist, it is necessary to deconstruct patriarchal imaginaries, among which is the idea that women are the "natural" people in charge of care and "emotional" tasks and, on the other hand, these tasks must be understood as fundamental for the sustainability of life.

We live today in a systemic socio-ecological crisis directly linked to neoliberal capitalism and the damage that human beings themselves generate as a means of survival. This historical moment is called the Anthropocene, where humanity has become the cause of existence; no area of the environment has been affected by technology. In this context, feminism, together with environmentalism and communitarianism, has been proposing that it is necessary to project a new culture, called in Latin America "Buen Vivir" (Sumac Qamaña, in Aymara, and Sumak Kawsay, in Kichwa) that places in the center the caring, not only of other humans but also of the environment that allows the sustainability of life.

An inversion of the dominant narrative is thus proposed. What was previously understood as "natural reproduction" (outside sophisticated production, intellect, and culture) must be placed in the center as the most essential.

5.3.2 *Feminist Economies and Care Ethics*

Today, we call "care" the organization of material and symbolic resources that reproduce biological and social sustainability. They include self-care, direct care of other people (the interpersonal activity of care), the provision of the preconditions in which care is carried out (sanitation, food preparation), and the management of care (coordination of schedules, supervision of the housework). Feminized subjectivities have historically carried out these poorly paid or unpaid tasks and are also marked according to class and race (Tronto 1993, p. 112). The feminization of these tasks is neither natural nor casual: it has historically contributed to social non-recognition, invisibilization, and the symbolic—and therefore economic—devaluation of care jobs and those who carry them out. Care is thus either romanticized (in the figure of the mother) or vilified (the "dirty jobs" of cleaning or grooming elderly or sick people) but rarely recognized as valuable and value-producing work (Rodríguez Enríquez and Partenio 2020).

According to the CEPAL repository of information on time use in Latin America and the Caribbean (CEPAL, 2019a, b, 2021), women dedicate more than twice as much time to caregiving as men on our continent. The overload of unpaid work constitutes a barrier to participation in the remunerative labor market: women are the most informally employed, they suffer vertical segregation in public and private hierarchical organizations, and they receive much lower salaries than men (which is known as the gender pay gap). Furthermore, in the labor market, they occupy the branches of employment associated with tasks understood as "feminine" (such as domestic service, health, teaching, and social assistance). In feminist economics, these branches are called the "care sector." The countries with the highest concentration of women in this paid sector are Argentina (42.8%), Uruguay (38.4%), Chile (34.9%), and Brazil (33.7%).

This diagnosis allows us to observe more precisely women's greater economic dependence and relative impoverishment (OIT 2019). Authors such as Gago and Cavallero (2019) have shown that the recent financialization of the economy is mainly harmful to women, who are the ones who go into debt the most to support households.

During the COVID-19 pandemic, women became the primary caregivers and were responsible for the people who made up the new "risk population." With the pandemic, the care burden for women in the domestic sphere increased, and the inequality gap between women (which includes the cis, trans, urban, migrant, and native population) and men increased, too (MinCyt and MMGyD 2020). In the unprecedented exceptionality caused by the global pandemic, the word "care" acquired a relevance that cannot be understood as rhetoric. The pandemic highlighted that no decent life is possible without care, so it opened the possibility of problematizing it from an economic, ethical, and political perspective. It is necessary to deprive, de-feminize, and address the universality of care as an immediate policy goal. State institutions could consider equity strategies for their workers, incorporating care as a human right. Putting it this way means proposing a prospective intervention; it suggests lines of action in the present to outline a future that we understand is necessary.

5.4 Caring Cities

In Argentina, as we have already mentioned, women continue to be the prominent managers of household care and activities, as well as those in charge of "replacing collective infrastructure [for which the State takes no responsibility] in conditions of extreme precariousness" (soup kitchens, voluntary activities in schools, promoters of support mechanisms in situations of violence, health, and problematic consumption, etc.) (Gago and Cavallero 2019: 142).

In the popular neighborhoods of Argentina, women oversee the soup kitchens; if there is no work or the salary is not enough, they organize themselves. Under a roof, in their houses, back yards, or the streets, a pot, a fire, and whatever falls inside: noodles, vegetables, little bones, and meat, whatever they can find. The soup kitchens could serve as an indicator of the economic situation of the most impoverished sectors of Argentina. Every block has a pot, a soup kitchen, and a "glass of milk" because families have problems solving the most basic: food and the reproduction of life.

As we have been suggesting, the sexual division of labor is the basis of the feminization of care and its invisibility. Well, here, in the phenomenon of soup kitchens managed by women, we have an example of this and, at the same time, a sample of how care is becoming politicized: women take care; they resolve the reproduction of households, but not in the privacy of their homes, not in isolation, but with organization and visibility. Together and in the streets.

Liliana Albertazzi (2017) suggests that the micro-genesis of anticipation is the learning of living organisms from those experiences that have allowed them to survive. Well, in the peripheral neighborhoods of the big cities of Argentina, these learnings are historical: every time there was a crisis, women were in charge of "coming out to offer a pot." If, in the multiple actions developed in large cities, it is possible to discern anticipatory properties, do the experiences of the popular soup kitchens and what women do in the peripheral neighborhoods contain an idea of how to make a city that cares? How do you develop a care system's germ from a learned, palliative, reactive action?

Lately, the State has begun to develop institutional strategies for the deprivatization and defeminization of care. The thesis that the State is the origin of the sexual division of labor and, therefore, cannot be its remedy (Kussy et al. 2023) is not useful for understanding what is happening in countries like Argentina for several reasons. On the one hand, it responds to a societal perspective of the State, which observes it as a reflection of the market's interests. On the other hand, it assumes that it is a unitary and homogeneous actor, losing sight of the different logic of its agencies and the different rationalities that can manage it. In Argentina, in recent years, there have been exciting experiences developed by the Dirección Nacional de Economía, Igualdad y Género (National Direction of Economy, Equality and Gender), in the Ministerio de Desarrollo Social (Ministry of Social Development), and in some provincial ministries, such as that of Women and Gender Policies of Buenos Aires Province, to revert gender inequality in terms of wage gaps, and horizontal and vertical segregation. One of them, and probably the most important, is the inauguration of Early Childhood Spaces and Community Care Spaces around the national territory.

As part of the same context and political culture—where state capacities are built from social demands and state agencies are the subject of interpellation and demand to guarantee new rights, the urban social movements of Argentina have developed the project salary law for community cooks. Moreover, from these experiences, the State learned timid but relevant lessons.

In short, between what women do in the peripheral neighborhoods—the popular soup kitchens—the institutionalized proposals of the Ministries—national and provincial—and the salary project for community cooks, we can see several visions of the future that overlap and interact. As already stated in this book, "Futures emerge from a diversity of visions that are continually updated with the arrival and consolidation of events in the present, new information, or changes in values. However, amid this diversity, concrete futures are institutionalized, identifiable" (González Arellano, see Chap. 2 in this book). In these proposals, future scenarios coexist and complement each other as ways of imagining and promoting a shared space in cities. As they become institutionalized, they may converge with more hegemonic images of the future, such as those proposed by the UN Sustainable Development Goals. In the dialectic between the subaltern and the hegemonic, the capacities of these different actors develop. The big question is how it will be possible, in an adverse and threatening context, to project a more inclusive vision of a city.

5.5 Final Reflections

Throughout this article, we tried to put concepts of feminism into dialogue with future studies. We prefigure the present as a threat, and we have shown a counterpoint within the kind of desire that is expressed in the words "The patriarchy will fall." This slogan has taken over the streets of Argentina and other urban regions of the southern cone of Latin America. We have reviewed the theoretical contributions that allow us to discuss the sexual division of labor as a material structure that sustains patriarchy and to review the principles of other care ethics. We wanted this theoretical dialogue to allow us to understand the proposal for a comprehensive care policy as a system of anticipation. If these are action devices, intervention proposals that encourage future scenarios, undoubtedly the policies and institutions that implement measures of this type not only encourage an idea of the future (that idea that reopened feminism in a moment of hypertrophy of the present and the end of emancipatory narratives) but instead push it, in material terms, also in a socio-affective and symbolic way.

The proposed dialogue also sheds light on the different scenarios and threats that attack the women's and LTBGI + movements: on acquired rights, that is, on the present, but also, and above all, on the future. In this sense, the political proposal that places care at the center must not only be understood as an anticipation of these threats but also as a way of appropriating and giving concrete and utopian meaning to the desired, imagined times, to the intimate way in which past and present are linked, and above all, to the time to come. Every time a community cook in a soup kitchen prepares food for a child, she reproduces his existence. She has faith in the future of that child, her community, and the generations to come while, at the same time, updating a collective form of care that is an ethical message and a supply for State policies. That is, it establishes a priority and a form of action to be recognized and financed by the Comprehensive Care System for which it is fighting and which must consider the salary of those who care, in this case, cooking.

Every time a community cook prepares and delivers food, she makes that soup kitchen a space of care, a piece of home in the neighborhood. With that daily act, she citizens that corner, by definition private and feminine, which is the kitchen (and from where comes the Greek notion of *oikos*, hearth, the house, the fire that must be kept burning). That means the anticipation is carried out and takes shape in a social and, in this case, urban space.

If the city is a space for promoting new futures, these visions of shared spaces, of cities that extend the notion of care to the public sphere, can serve as inspiration. The literacy future works propose how the unimagined—via technological development, imagination, and science fiction—enters the agenda of urban policies (Gonzalez-Arellano 2014). Here, we tried to expand that possibility to understand how collective intelligence is developed not in the imagined nor the unimagined but in the unseen, the historically invisible.

This book has invited us to think of cities as anticipatory systems; then, the dialogue we have carried out should allow us to imagine and anticipate future cities, cities that put care at the center. A city that cares is a home. A city that, like the invisible cities written by Italo Calvino, is all cities. Furthermore, stop being invisible.

References

Albertazzi L (2017) Microgenesis of anticipation: windowing the present. In: Poli (eds) Handbook of anticipation. Springer International Publishing. https://doi.org/10.1007/978-3-319-31737-3_7-1

Appadurai A (2015) El futuro como hecho cultural: ensayo sobre la condición global. Fondo de Cultura Económica, Buenos Aires

Bárcenas Barajas K (coord) (2022) Movimientos antigénero en América Latina. Cartografías del neoconservadurismo. Ciudad de México, UNAM-IIS

Beauvoir de S (2015) El segundo sexo. Cátedra, Valencia

Engélibert JP, Guidée R (2015) Actualité de l'utopie. Utopie et catastrophe: revers et renaissances de l'utopie (xvie-xxie siècle) [en línea]. Rennes: Presses universitaires de Rennes, 2015 (generado el 15 septembre 2023). Disponible en Internet: <http://books.openedition.org/pur/177786>. ISBN: 9782753591592. https://doi.org/10.4000/books.pur.177786

Engels F (2017) El origen de la familia, la propiedad privada y el Estado. Akal, Madrid

Ferguson A, Folbre N (1981) The unhappy marriage of patriarchy and capitalism. In: Sargent L (ed) Women and revolution. Black Rose Books, Montreal, pp 313–338

Fraser N, Bhattacharya T, Arruzza C (2019) Manifiesto de un feminismo para el 99%. Herder, Madrid

Gago V, Cavallero L (2019) Una lectura feminista de la deuda. Vivas libres y desendeudadas nos queremos. Fundación Rosa Luxemburgo, Buenos Aires

González-Arellano S (2022) The future scenarios of cities: an analysis of their institutional construction. Int J Urban Reg Res 46(5):865–873. https://doi.org/10.1111/1468-2427.13103

Hartmann H (1981) The unhappy marriage of marxism and feminism. Towards a more progressive union. In: Sargent L (ed) Women and revolution. Black Rose Books, Montreal, pp 1–42

Harvey D (2007) Urbanismo y desigualdad social. Siglo XXI, Buenos Aires

Hurley K et al (2008) Future studies and feminism. Conversation. Futures 40. Elsevier, pp 388.407. https://doi.org/10.1016/j.futures.2007.08.004

Incaminato N (2023) El género en disputa: el voto libertario y el post-feminismo de las Pibas de Milei. 25 de junio de 2023, Sección Opinión en Diario.ar, disponible en: https://www.eldiarioar.com/opinion/genero-disputa-voto-libertario-post-feminismo-pibas-milei_129_10323102.html

Kussy A, Palomera D, Silver D (2023) The caring city? A critical reflection on barcelona's municipal experiments in care and the commons. Urban Studies 60:2036–2053

Larguía I, Dumoulin J (2019) Hacia una concepción científica de la emancipación de la mujer. Desde la Cuba revolucionaria: feminismo y marxismo en la obra de Isabel Larguía y John Dumoulin. comps. Mabel Bellucci y Emmanuel Theumer. Buenos Aires: CLACSO, pp 117–120

Liveley G (2017) Anticipation and narratology. In: Poli (ed) Handbook of anticipation. Springer International Publishing. https://doi.org/10.1007/978-3-319-31737-3_7-1

Losiggio D (2020) Universal y afectiva. La esfera pública en el pensamiento político feminista. Las Torres De Lucca 17:139–165

Losiggio D, Taccetta N (2022) The virtual museum and female joy in Argentina: an exploration of the revolt. Intérmedialités. Históire et théorie des arts, des lettres et des techniques, p 38

Marx K (2001) El Capital. Fondo de Cultura Económica, México

Marx K, Engels F (2014) La ideología alemana. Akal, Madrid

Patrouilleau MM (2020) Epistemología y crítica de la prospectiva y los estudios del futuro. Una visión desde América Latina. In: Patrouilleau M, Albarracín Deker J (coords.) (2022). Prospectiva y estudios del futuro: Epistemologías y experiencias en América Latina. CIDES-UMSA, La Paz. En Memoria Académica. Disponible en: https:www.memoria.fahce.unlp.edu.ar/libros/pm.5423/pm.5423.pdf

Rodríguez Enríquez C, Partenio F (2020) Sostenibilidad de la vida desde la perspectiva de la Economía Feminista. Madreselva, Buenos Aires

Sarr F (2019) Reouvrir les futurs. In: Achille M, Felwine S (dir.) Les ateliers de la pensée. Politique des temps: imaginer les devenirs africains. Dakar, Phillipe Rey eds

Tarducci M, Trebisacce C, Grammático K (2019) Cuando el feminismo era mala palabra. Algunas experiencias del feminismo porteño. Espacio editorial, Buenos Aires

Tronto J (1993) Moral boundaries. Routledge, Nueva York y Londres

Vaggione JM (2005) Entre reactivos y disidentes. Desandando las fronteras entre lo religioso y lo secular. Defensa de los derechos sexuales en contextos fundamentalistas, ed. Católicas por el Derecho a Decidir. CDD-Córdoba, IGLHRC, Buenos Aires

Informs

CEPAL (2019a) La autonomía de las mujeres en escenarios económicos cambiantes
CEPAL (2019b) Oportunidades y desafíos para la autonomía de las mujeres en el futuro escenario del trabajo
CEPAL (2021) Repositorio de información sobre uso del tiempo de América Latina y el Caribe
MinCyt/CONICET, MMGyD (2020) Diagnóstico de la situación de las mujeres rurales y urbanas, y disidencias en el contexto de COVID-19. Disponible en: https://www.conicet.gov.ar/wp-content/uploads/resumen_ejecutivo_mujeres_y_covid_-_mincyt-conicet_-_mingen.pdf
OIT (2019) Panorama laboral en América Latina

Chapter 6
Inclusive Urban Health Systems and Universal Well-Being: Mirage or a Real Future?

Bikramaditya K. Choudhary

Abstract The survival of the human race refers to the living humans dependent on their health and health systems, including preventive and curative measures. The contemporary shift toward urban settlement as the place of residence makes it imperative to understand the intricacies of urban spaces and their liveability. Cities as a healthy system and economy as a system of cities are often considered synonymous, though they are quite a contra in neoliberal regimes. The people are left to take care of their health in large cities, including occupational health and trauma cases, partly due to the economic processes that make the city. The idea of an "anticipatory city" is rightly posed as it holds the future. It is essential to point out that many individuals and institutions have answers for the future without knowing the problems; they have a vision for an unknown destination. This paper will conceptualize important questions for the larger masses destined to live in oblivion as their voices have been silenced in the loud noise of the economic successes of existing and emerging "great powers." Based on the discursive contextualization, I conclude that the core challenge for the future city is the tacit fortification of inequality on different counts and the unabashed defense of increasing consumption.

Keywords Diseases · Preventive health · Neoliberal economy

6.1 Introduction

It has been established more than once that there is no relationship between collective human achievement and the common good. On the one hand, cities are the real future of humanity in terms of preferred forms of settlement; at the same time, cities are termed "nightmares" or "dystopias" (Ameel 2016; Latham and Hicks 2014; Sulzbach 2013). The desirable utopia or contra to it can be conceptualized in multiple ways, including the cultural amalgamation of multiple stands of social practices, the cultural

B. K. Choudhary (✉)
CSRD/SSS, Jawaharlal Nehru University, New Delhi, India
e-mail: bkcjnu@gmail.com

S. Gonzalez-Arellano and B. Gandlgruber (eds.), *Cities as Anticipatory Systems*, The Urban Book Series, https://doi.org/10.1007/978-3-031-78162-9_6

conflicts on the same basis, the higher levels of living standards and contra to the same, and so on. Cities are a certain kind of spatial fix, a set of spatiality produced filled with ideologies, actions, conflicts, and negotiations.

Soja (2000: 6) argues that the processes of producing spatiality or "making geographies" begin with the body, the construction, and the performance of the self, the human subject, as a distinctively spatial entity involved in a complex relationship with our surroundings. On the one hand, our actions and thoughts shape the spaces around us, but at the same time, the larger collectively or socially produced spaces and places within which we live also shape our actions and thoughts. Production and reproduction of spaces shape places as they get identified, creating class- and class-based place-making. Class-based place-making, in return, creates an exclusionary cityscape with restrictions and the formation of "gated communities" and "ghettos" simultaneously within the same city. Cities in the contemporary world are utopia and dystopia at the same time, based on who is in charge of a particular space. The questions of power and marginalization also become factors in considering cities or parts of cities as either category. The advent, persistence, and continuance of the current pandemic established once again that cities can survive or perish together.

Countries like India, except during the second peak, considered COVID-19 a "disease of the city" as major mortality was reported from the large cities and remained concentrated in the large cities as it has a "long tail." It is important to note that India still (until the last census) has most people living in rural areas. During the pandemic's peak, millions of urban workers returned to villages to save themselves from the wrath of the disease and consequent "lockdown." The sufferers of the pandemic are mostly poor, as most poor households became dependent on government support to fulfill their basic needs, whereas, during the same period, large business corporations reported unexpectedly faster growth in their wealth and income. For those who view cities through a dystopian lens, the ongoing COVID-19 pandemic, with people fleeing urban areas, is another stark illustration. It raises the question: Can we envision and plan our cities more optimally to address such challenges? Can we anticipate the future? Can such pandemics be better managed? Can the sufferers of this pandemic (mainly poor and migrant workers) not be the sufferers of the next pandemic? The questions bring another set of questions, including vulnerability to disaster, inequality in terms of being affected, the nature and extent of resilience, and the possibility of a "fractured future."

We know that the distribution of health risks is uneven, as is the access to healthcare facilities. The existence and persistence of such inequalities can be seen globally when comparing life expectancy at birth in Japan (83.7 years) and Sierra Leon (50.1 years). Similar inequalities across gender and economic quintiles are visible nationally in rich and poor countries. There is a 20-year gap in male life expectancy between Glasgow's richest and poorest areas (Haining 2021). India is not an exception, and in any study, including in everyday newspaper reporting, one can come across glaring inequalities between men and women or rich and poor (Asaria et al. 2019). On top of that, intersectionalities across caste, gender, and economic categories make things complex. What is expected of the health policy of a democratic welfare government is a clear articulation to contain existing inequalities to move toward a healthier, more inclusive society.

More often than not, the issue related to health care raises the question of equity, especially in poorer countries. The history of human civilization is also the history of fighting diseases successfully and unsuccessfully. From Plague to Ebola, the relative success of human efforts has been varied across geographies at certain times. A partial explanation for such varied success in dealing with the "health issues" draws our attention to states' varying nature and priorities. The role of state and public policy has entered the debate since the existence of the "nation-state," especially the articulation of the "welfare state." The spatial dimension is critical in developing societies where healthcare institutions are concentrated in large urban areas and major diseases are primarily concentrated in rural areas or urban slums. Socio-cultural hierarchies and the varying purchasing power of individual households add to the complexity of health care and disease prevalence in these societies. The prevalence of disease also has roots in differentiated vaccination and varying vulnerabilities across the layers and strata of the population. It is, therefore, pertinent to discuss the nature of economic opportunities and differentiated vulnerabilities rooted in the socioeconomic conditions of individuals and families. The following sections of the paper will discuss the urban healthcare system, followed by healthcare facilities in Delhi. The last section will discuss the outlined conditions impacting cities and their future.

6.2 Inequality, Informalities, and Vulnerabilities in Delhi-NCR

Urban spaces are known for visible and invisible concentration and deprivation. Some thrive in cities at the cost of millions in countries like India. The metropolitan cities of India are known for their vibrant business environment, sprawling shopping plazas, international standard superspecialist health facilities, grand venues for business meetings, and so on. At the same time, these cities are also known for their vast homeless population, large slums, poor sanitation, unliveable environmental conditions, excessive pollution, scarcity of water, unsafe city transport, and so on. The rising income and galloping income gap move almost in a linear manner. Informality is the key constituent of these cities, whether in planning, the job market, land tenure, or housing. With that being acknowledged, urban regions have been a preferred destination for improving the standard of living for millions of underprivileged people in emerging countries. This is especially important because the villages that used to be their traditional homelands face economic stagnation.

The deteriorating condition of people living in rural areas leads to a heavy influx of migrants to urban areas. The growing inequality among the people and the growing inequality in the profit-making capabilities of urban processes have promoted urban informality through sub-contracting. A significant trend in growing inequality can be seen through the profit-making processes of different sectors. The high-profit industries dominate major urban economies. They have increased the price of space and land in the big cities (Sassen 2015). Low-income people can hardly afford the space, and hence, we witness high-density slums like Dharavi in Mumbai and scores of homeless people across Indian cities.

The informal settlement is described as a built environment that is either a slum, illegal or legal tenure (Wekesa and Styen 2011), or squatters of different kinds. Three interrelated characteristics, viz., physical, social, and legal, define the informal settlement (Wekesa and Styen 2011). Due to being an illegal settlement, the physical characteristics of such a settlement include the lack of adequate infrastructure like roads, electricity, water supply, toilets, schools, health centers, and markets. The people living in informal settlements mainly belong to a low-income group or are daily wage earners working in different unorganized sectors of the economy. The residents of these settlements lack ownership of the land on which they built their abode. The land people squat on is mostly government, public, or marginal land parcels like railway tracks, marshy lands, and more.

These settlements lack basic amenities and have improper housing, leading to vector-borne disease and making the residents vulnerable to communicable diseases. In developing countries, the proliferation of informal settlements results from many actors and processes, including speculative urban land markets. On the one hand, the speculative land market allows real estate actors to own the city while making houses out of reach for millions of families (Bhagia and Bose 2023; Sassen 2015). Private entities own more buildings, and the conventional building of houses has become beyond the reach of the urban poor (Wekesa and Styen 2011). The proliferation of informal settlements is witnessed due to market failure and public policy failure under globalization. It is well documented that, citing a lack of funds, the local government and municipality have abandoned providing proper infrastructure and basic amenities to these settlements. The land cost is so high that it is difficult for the average individual to build houses, which leads to a proliferation of slums and squatters.

In the twenty-first century, urban growth is increasing in the developing world, but city-related theories remain confined to the developed world. The future of urban was thought to lie in Third World cities like Mumbai and Rio de Janeiro (Messey 2001; Roy 2005); urban transformations in the developing world remained particularly associated with squatters. It is seen as a generalized mode of metropolitan urbanization. The idea of distributive justice in urban planning remained in writing and scholarship (Roy 2005) at best, and inequality and injustices continue to perpetuate on people unhindered (Srivastava 2015; Choudhary 2019; Saroj et al. 2020).

Informal hypergrowth of the city is considered (Roy 2005) as one of the causes of growing informality; however, the classical study of Engels established that the same forces that generate squatter in one place will produce it elsewhere unless the process of urbanization that is based on accumulation is not changed (Choudhary 2005). In India, most of the urbanization results from urban informality, with no provision of adequate infrastructure and a weak municipal governance structure. Most of these settlements are developed along the major transportation routes within the metropolitan boundaries. Informality is a different type of market where affordability accrues through the absence of formal planning and regulation. These are complex phenomena of legality and illegality where squatter settlements grow through land invasion (Roy 2005). Informality indicates to whom this belongs and who has the right to a city (Harvey 2003). The right to include and exclude plays in a manner that is influenced and questioned diabolically (Bloom 2004). Delhi accounts for 14.6% of the population living in informal settlements (Census of India 2011). A severe issue in Delhi is the lack of affordable housing for the lower and middle classes. Slums and Unplanned Housing: In 2014, the Delhi government assessed slums and JJ colonies, estimating that 0.33 million households, or around 1.7 million people, lived in these settlements, making up roughly 10% of Delhi's total population. The lack of housing, together with the miserable living conditions of the homeless, the abundance of slums, including unauthorized colonies and Jhuggi-Jhopri (squatter) settlements, the constant influx of people into the city, etc., force people to live in substandard conditions with little or no access to essential services and inadequate housing (Economic Survey Report 2022–23). Informality in housing and land ownership means frequent displacement within the city. Delhi has witnessed impoverishment due to displacement in many phases. The city witnessed demolition before the Asian Games in the early 1980s; in the 1990s, rapid urban construction witnessed large-scale displacement; and in 2010, the largest displacement was witnessed in Delhi during the Common Wealth Games. There were slum policies adopted or introduced with some conditions for their eviction in the 1990s, but the judicial intervention undermined the policy to a large extent (Bhan 2009). The evicted often lose their claims, as revealed in studies such as the Almitra Patel Case (2005) and the Okhla Factory Owners Association Case (2002), highlighting the state's failure to fulfill its obligation to provide alternative resettlement for evicted families in these situations (Dupont 2000; Ghertner 2014).

6.3 Access to Basic Services and HealthScape

Urban water supply systems are vital in ensuring access to safe drinking water for millions in different cities. The cities of the global south are known for having multiple waterscapes. In this broad primary circuit, urban local bodies supply water, and people living in the rest of the city spaces are left to fend for themselves. At times, water vendors, both formal and informal, make water available to them. The issues of water supply, access to clean water, and distribution systems in the cities have become central to discussions on improving access to safe water for millions in the Global South, like India (Sarkar and Choudhary 2020). The inadequacy of water supply in Indian cities, especially across informal settlements with precarious tenure status, such as slums and unauthorized colonies, is often highlighted (Dutta and Chaudhuri 2020; Ranganathan 2014; Björkman 2015). Conditions of water inadequacy across planned neighborhoods for the poor (resettlement and rehabilitation colonies) have a similar fate, as either they depend on vendors or, sometimes, on water ATMs in cities like Delhi (Sarkar and Choudhary 2020). The gap between water demand and supply continues to exist despite some augmentation efforts by the local bodies to supply water.

Delhi's National Capital Territory (NCT) has always had difficulties dealing with water, as the city has hardly access to groundwater, often 40 m below the surface. Further, the Delhi metropolitan limits contain the most polluted stretch of the Yamuna River. Even during monsoons, urban flooding in the metropolis, even with low rainfall, is commonly witnessed as faulty drainage and unplanned construction in the *laldora* (village boundary) area. Ensuring that citizens have access to safe, adequate, and affordable portable drinkable water and hygienic sanitation has to be a fundamental public service the government must ensure for the well-being of citizens. As per the Master Plan of Delhi 2021 formulated by DDA, the proposed water requirement is 80 gallons per capita per day (GPCD), of which 50 GPCD are for domestic uses. Domestic uses comprise 30 GPCD portable drinking water and 20 GPCD non-portable. The demand for water in Delhi based on the norm is 60 GPCD by 2021 is 1260 MGD to meet the requirement of the population of about 21 million people, whereas the installed capacity of water plants has increased from 810 MGD in 2009 to 943 MGD in 2022. There is still a more than 300 MGD gap, and the poor face the brunt.

New Delhi's sprawling suburb Gurugram, developed as a city to be built and serviced by private real estate developers, exemplifies the necessity of considering public transport and road safety as a public health concern. With a sparse public transport network, the city is home to approximately 2 million people and is served by a paltry network of 20 routes and 142 buses. This leaves large swathes of the city, at best, underserviced and, at worst, unserved. The Rapid Metro Gurgaon, another privately conceived and built project, serves the commercial and new sectors of the city and is notorious for its poor network spread of 12.85 km and only 11 stations. In the absence of any public transport network worth its name, private cars are the preferred mode of transportation, while the urban poor are mainly at the mercy of shared private autos and cabs ferrying people throughout the city and beyond. The Statistical Abstract of Haryana (2014–15) reported an average of 323 cars per 1000 people in Gurugram compared to 88 cars per 1000 people in Delhi (Economic Survey of Delhi 2014–15). The city's lack of public transport networks burdens the urban poor in more ways than one. A significant portion of income is spent on commuting to and from work for the urban poor, which could otherwise be spent on addressing other basic needs. At the same time, commuters on foot or bicycle are also at higher risk of road accident injuries, incurring out-of-pocket expenditure on medical expenses. Gurugram's planned expansion of the road network and underdevelopment of public transport has only ensured an ever-increasing demand for cars, which feeds into the demand for even wider roads and rising levels of air pollutants throughout the year. Temperature inversion in the winter months causes heavy smog in Gurugram, affecting visibility for drivers and commuters alike, leading to frequent accidents and greater exposure to harmful air pollutants for people who walk or bicycle as a means of commute. While the city administration has been focused on developing a non-motorized network plan and bike-sharing facilities, it barely addresses the core issue of negligible public transport with poor feeder services, which could significantly impact public health. It is further established that most real-time pollution sources contributing to Delhi's particle pollution are due to vehicles (CSE 2021). The transportation sector contributes around half of the pollution from solely local sources in Delhi if the total is determined by considering the sources' local contributions alone, excluding those from outside sources.

Without a robust public transport system, city dwellers use individual vehicles to commute; they even use the mass transit system for last-mile connectivity. The identified vulnerability of people due to poor water supply, exposure to dangerous transport, and resultant pollution, along with the absence of sewage connectivity, pose changes to the city's future. About 16 percent of people have no access to safe drinking water, and 27 percent have no access to toilet facilities, as per the 2011 census in India (Census of India 2011). Our earlier study (Das and Choudhary 2024) validated the presence of open defecation and defecation alongside open sewage in Delhi; similar conditions persist in other large cities in India and the Global South. Untreated excreta contaminating groundwater often leads to the consumption of unsafe drinking water, which in turn results in the occurrence of diarrhea, dysentery, parasitic worm infection, and many other water-borne diseases (Ministry of Drinking Water and Sanitation 2011–12; Chaplin 2011; Ortiz-Correa et al. 2016; Patil et al. 2014; Saroj et al. 2020). This leads to a loss of working days, resulting in substantial economic losses (Ministry of Drinking Water and Sanitation 2011–12; Ortiz-Correa et al. 2016). The WASH levels (Water, Sanitation, and Hygiene) remain problematic for a sustainable city and inclusive society in urban India. To develop efficient policy interventions, there is a need for recognition of the existing comprehensive studies related to inequality in WASH, along with its levels and trends (Goli et al. 2011; MoHUPA 2013; Saroj et al. 2020; Sarkar and Choudhary 2020).

6.4 Urban Healthcare System

The prevalence of disease coexists with human existence. The differentiated prevalence of communicable and non-communicable diseases, along with their estimation, has made it possible to relate them to different types of risk and facilitate the analysis of social epidemiology.

Poor preventive measures regarding public health provisions, including clean water and sanitation, mean a wider prevalence of communicable diseases. The disease once appeared needs to be treated to bring back the "healthy human." The concept of health as a fundamental human right for all people, irrespective of "…race, religion, political belief, economic or social condition," was advocated in the WHO constitution (1946).

The novelty of this definition is that, for the first time, it emphasizes "complete physical, mental, and social well-being," contrasting with the previously narrow focus on the "absence of disease or infirmity." This is the most popular and widely accepted definition of health. Last (1988), however, argued this as a "poor operational definition" and presented his definition with measurable attributes for which indicators can be identified and relative vulnerabilities can be mapped for different strata of people. Health is "…a state characterized by anatomic integrity, the ability to perform personally valued family work and community roles, the ability to deal with physical, biological, and social stress, a feeling of well-being, and freedom from the risk to untimely death" (Last 1988). The role performance conception comes close to this definition; however, assessing health in this particular conception becomes impossible when we look at the statistics available from different urban authorities. What is significant to note is that the condition of health in the cities of the global south can still be estimated only based on reported morbidity during large-scale surveys like NFHS (National Family Health Survey), DHS (District Health Survey), or registered morbidity, i.e., hospital-based data.

Health policies worldwide, including India, claim to be directed toward achieving universal health coverage, though the idea of universal and public remains contested. Achieving universal health care requires significant investment in the health infrastructure. However, when we look at the investment, it is less than 1 percent of the gross domestic product (GDP), which is a significant reason for the inadequate health infrastructure in India. On the other hand, due to inadequate public healthcare facilities, people use both formal and nonformal private healthcare facilities, resulting in high out-of-pocket expenditures. Such expenditures push the poor into debt traps and remain indebted to nonformal moneylenders (Choudhary 2010). A health system has to have more than one component, and its efficiency depends upon the integrated functioning of these components to achieve better health care for all. However, in articulation, "all" seems to be inclusive.

In contrast, in operation, "all" is exclusive to those who can afford the ever-increasing high-cost lifestyle and high-end private healthcare services that incur heavy "out-of-pocket" expenditure. A well-functioning health system working in harmony is built on having trained and motivated health workers, a well-maintained infrastructure, and a reliable supply of medicines and technologies backed by adequate funding, robust health plans, and evidence-based policies (Balasubramaniam et al. 2018). It is important to look into the investment pattern in public health care to achieve the ideal "health for all" mission. The rhetoric of "health for all" and the actual implementation of "universal health coverage" seem to be operating on a different scale, as there is a later need for targeted investment in preventive services, including water and sanitation services and accessible public curative health care services. Studies on healthcare expenditure among the urban poor highlight various socioeconomic inequalities in urban areas (Sharma et al. 2020). The social production of health inequality, as shaped by neoliberalism, has to be understood in this historical context of reshaping the capitalist order, primarily based on an unfettered market with the motives of greed and avarice. This divides societies into very rich and very poor groups. A paradox of coexistence of great wealth and great poverty is the reality of the current era, especially in countries like India, which are proliferating. This offends the notion of a just and equal society. Under the pressure of wealthy nations and transnational corporations, the neoliberal reforms implemented by international financial institutions like the International Monetary Fund, the World Bank, and the World Trade Organisation have produced adverse health effects.

The advent of these economic measures in the late 1970s and early 1980s coincided, in fact, with reduced worldwide gains in both economic growth and life expectancy. The question of why health inequalities exist has led to considerable empirical work and much theorizing. The empirical evidence that has been mobilized and studied is vast and goes from survey data to a range of other variables as disparate as crime statistics and voting behavior. Integrating these survey data with psychosocial theories has emerged as one of the most popular and dominant methodological approaches. Indeed, during the 1990s and 2000s, the "psychosocial hypothesis" became widely accepted as conventional wisdom. Wilkinson (2003) and other researchers in Europe and the United States built an original research program on social inequalities in health (Kawachi 2006). These inequalities persist unabated globally, and India is no exception.

6.5 Healthcare Facilities in Delhi

In Delhi, various healthcare facilities are available, both public and private. The Government of Delhi claims to provide accessible and affordable healthcare facilities to the Delhi population and the migratory and floating population. It has been providing primary, secondary, and tertiary healthcare facilities. Certain areas in Delhi do not have access to primary healthcare services within their reach, and these areas mainly comprise the poor and vulnerable population. There has been a wide gap in the demand and supply of healthcare services, evidenced by the long queues in the public healthcare system reported in the media during the COVID-19 health crisis. To fill the gaps in services, the Government of Delhi came up with a four-tier system of healthcare, in which the Aam Aadmi Mohalla Clinics provide primary healthcare.

The neighborhood clinics (aka Mohalla Clinics in Delhi) have been designed to offer accessible, high-quality primary healthcare services within the communities right at their doorstep (Lahariya 2016, 2020; Agrawal et al. 2020). The healthcare crisis is far from over with these initiatives, as private hospitals have gotten land on a subsidized rate with commitments to provide care to the poor; however, they often fail to stand up to their commitments. Research on healthcare facilities for Economically Weaker Sections (EWS) in Delhi offers evidence regarding the percentage of bed occupancy for EWS beds. The bed occupancy data for 2012–13 to 2015–16 shows that the bed occupancy of 41% of the hospitals was below 10%, while only two hospitals had more than 30% occupancy by EWS (Shukla 2018). Even in the Public–Private Partnership (PPP) model, private agencies continue to profit without considering providing healthcare services to the EWS. Many hospitals have been accused of refusing EWS patients, delaying surgery schedules, and charging fees for several procedures. Often, on the flimsy technical ground of an income certificate being "older than six months," treatment is denied. It has been reported that patients have to pay for certain services whenever treatment is provided. Delhi exhibits socioeconomic segregation, and the health-seeking behavior of low-income households differs from that of middle- and high-income households. The public healthcare system, designed to offer healthcare to the less privileged, is scarcely utilized by them for various reasons. The low-income households are forced to use private healthcare facilities, which incur high "out-of-pocket expenditure." Their limited financial capacity burdens them, frequently leading them into a debt trap. It is pertinent that the government should focus on an equitable provision of healthcare to all sections of society, with particular emphasis on primary healthcare services (Gupta and Sengupta 2013). However, the existing reality is that the state has introduced an insurance-based health provisioning system where the poor face multiple discrimination, which sometimes includes paying for treatments that are not required. The land has been provided to various hospitals by different authorities of the Delhi government on lease or at subsidized rates, which came with the clause of serving the EWS patients at no cost.

6.6 Discussion: Future City as Health City

It is known to us that one of the major achievements of human civilizations is the improvement in individual health. Improving life expectancy at birth, reducing maternal and child mortality, and continuing efforts to reduce pain for people at different stages of life have been major achievements. However, the poorer sections of society still spend a sizable proportion of their income on health as catastrophic or out-of-pocket expenditures. Accompanying this is the emergence of what is referred to as the health divide and increasing inequalities in social determinants of health. The social determinants of health, also considered "solid fact" (Wilkinson Marmot 2003), include unemployment, early life, work, food, addiction, transport, and so on. These determinants are further based on access to education, environmental quality, income, preventive activities, and others. Various factors play a role in shaping the health status of individuals and communities in an urban environment. These include the quality of the physical and social environment, stress levels, educational opportunities, the local economy, and urban congestion (Leeuw 2011). The connection between social conditions and health is close to the extent that "the magnitude of health inequalities indicates the impact of social and economic inequalities on people's lives" (Marmot and Allen 2014).

Access to resources, institutional arrangements to address the inability of specific individuals and groups to access resources, and the opportunities to earn a livelihood, which primarily determine income, are dimensions of vulnerabilities that make or break individuals' internal and external resistance to illness. Whether we include indicators on which data is available like socio-demographic variables or proxy variables that indicate individuals' capabilities, persistent inter-urban and intra-urban inequalities are a matter of concern for the future of the city and future cities, which I discuss later in this section. Social vulnerabilities also get expressed in terms of social capital, including social networks and connections, which allow or hinder access to limited public health services in large cities in third-world countries. Two types of indicators, namely the socioeconomic vulnerabilities index and built environmental index, can be used to estimate the level of vulnerability of the masses in large cities, and the intervention in minimizing these vulnerabilities would make us move toward a better and healthier city. The built environmental index (BEI) may include distance to the nearest hospital, population and housing density, age of building stock, average age of sewers and water pipelines, quality municipal roads, exit routes in case of emergency, and so on (Zhong et al. 2022). However, the major challenge in cities in the global south is to have reliable data on these aspects.

Further, the housing stock would matter for those living in the houses, even if they are dilapidated. There is a sizable proportion of people who are either homeless or living as squatters, and without legal rights, the possibility of accessing healthcare becomes more difficult. In such a scenario, universal health coverage is the answer if we plan for a healthy city in the future. The NRHM (National Rural Health Mission), ASHA (Accredited Social Health Activists), and ANM (Auxiliary Nurse) were introduced in 2007 onwards to address the health needs of the masses in India, including vaccination. Delhi introduced the Mohalla Clinic and Mobile Clinic to address the everyday health needs of people at the neighborhood scale. A decentralized healthcare system was introduced in Delhi in 2015 as a form of the neighbor clinics (Mohalla clinics) at the level of primary healthcare centers. However, these clinics are yet to be available across the city and are far from achieving the desired results. More and more people of all social classes, genders, and ages must utilize the services available in these clinics. A specialized form of these clinics for women, i.e., Mahila Mohalla Clinics, is still in its nascent stage and needs to start functioning on a large scale. These specialized clinics are aimed at providing maternal and child healthcare services. These clinics are supposed to be at every K; if this happens, people's everyday health needs can be addressed. If a good network of Mohalla Clinics is established, it definitely has the potential to change the face of Delhi's healthcare system and help reduce the patient burden on the tertiary care public hospitals. Apart from these clinics is a network of primary healthcare centers and Delhi's mother and child care homes. The government needs to take cognizance of this fact and integrate the existing system with the Mohalla Clinics network. The services in these primary healthcare centers should be made more versatile.

Mobile health clinics are an important component of urban healthcare, and their availability and services should be broadened. The health documents of the patients can be stored digitally and need to be accessible at all levels of healthcare facilities so that there is data on ailments, and health planning can be put in place for the future. In absence of the reliable data on registered and reported morbidity, health planning remained only a mirage. Further, if the documents are stored virtually, they can be accessed anytime at any facility level. Full use of digital advancements should be made to make the system integrated. However, the question of privacy and consent adds complexity. Both the central and the state governments claim to offer health insurance. However, in reality, the actual number of patients who benefit from these schemes is questionable. Moreover, insurance-based health services, in contrast to

UHC, are based on collective risk and cater to non-health services, including financial management. There are diseases not covered under the ongoing insurance schemes, like "Arogya Kosh" and "Arogya Nidhi," besides the RSBY and Ayushman Bharat. Universal Health Coverage essentially holds the future of the city significantly. My idea of an anticipatory system that will save the citizens and residents in large cities would be based on universal health care. The experience of East European countries, where UHC is in place compared to out-of-pocket healthcare systems, tells us that it is better for saving lives during a pandemic. Kerala's experience in India has a similar outcome, especially with lower mortality despite high prevalence. The infected persons could be treated and put under quarantine when needed.

Universal Health Coverage (UHC), as defined by the World Health Organization, encompasses the assurance that all people have access to adequate and high-quality health services, including prevention, promotion, treatment, rehabilitation, and palliation, while also ensuring that the utilization of these services does not subject the user to financial hardship (WHO 2023). Draft National Health Policy 2015 acknowledged that the population growth rate has significantly declined and GDP has improved, but health inequity has reversed. The situational analysis leading to National Health Policy 2017 in the opening line reads,

> India is now the world's third-largest economy in terms of its Gross National Income (in PPP terms) with the potential to develop more equitably. As never before, it possesses a sophisticated arsenal of interventions, technologies, and knowledge required to provide health care to its people. However, the inequities in health outcomes continue to remain.

The inherent contradiction in the government's claims persists. Despite asserting the status of the fastest-growing economy and acknowledging existing inequities, governmental investment in India to address glaringly unequal health outcomes based on caste, income, geography, and religion is noticeably lacking. The healthcare needs of the people are changing, and there is a continuous need for the healthcare system to adapt itself. High expectations for a clear roadmap to minimize disparities in access to healthcare services remain unaddressed. The city's future depends on expectation; in other words, we know expectation is the future even though "false consciousness" remains the tool of power to manipulate the masses, and their expectations get shaped in different directions.

Additionally, India was expected to transition toward a "right-based approach to health." However, the flat refusal to entertain the possibility of articulating health as a right, despite soon becoming the third-largest economy, is disconcerting. The existence of health and well-being inequalities across the cityscape is used to clamor for improving living conditions for diverse groups, especially the most vulnerable. Following the broad debate of different SDGs, it is almost agreeable, at least in the literature, that health and sustainability objectives are to be part of urban planning. Pandemic recovery strategies encouraged "temporary urbanism" (Andres et al. 2021) to be considered a solution. They argue that temporary urbanism offers better living conditions and overall well-being and brings alternatives to foster resilience and coping. Temporary urbanism is an anticipatory system; if the temporariness addresses the multiple challenges, it would be a boon for the city's future. The introduction of temporariness in planning and provisioning would, however, be detrimental to the citizens of future cities. The temporary nature of it allows those in power to use it to their advantage, and when there is a possibility of the poor gaining, change the structure. Moreover, the argument is based on the premise that the situation of urban living, that is, the modern city, is given, and we need to develop resilience in the event of any health emergency or otherwise. I wish to argue otherwise, based on the premise that the city of contemporary times needs to be altered fundamentally, at least in two ways: "reduce consumption" and "provide universal health care."

6.7 Conclusions

The world has witnessed over the decade that "fundamentalism," supremacist claims, and dividing the voters are the "universal" means to reach power. In return, power is used to subvert the future by indulging in three precise forms of exploitation: labor, environment, and natural resources. The question is: What future are we looking for? If there exists a "we" or if the world is divided between "us" and "them." On the one hand, hyper-consumerism is the order of the day with technology and innovative measures to spend time and energy enhancing what Singh and Hickel (Singh and Hickel 2023) call capitalogenic diseases and on the other degenerative disease of the poor old persons and diseases of deficiency scar the masses in the same cityscape. The vital choice that humanity has to make is to either engage in a "universalized" echo for the capitalogenic disease or to eradicate vulnerabilities that put millions at risk of being sick. The cities of the future need to vision the long-term intentions, purpose, and aim that will ameliorate the glaring inequities in access to basic services and access to healthcare, meaning, in brief, to provide universal provisioning of basic needs, including water and sanitation services along with ensuring universal healthcare services to everyone.

References

Ameel L (2016) Cities utopian, dystopian, and apocalyptic. The Palgrave Handbook of Literature and the City, pp 785–800

Andres L, Bryson JR, Moawad P (2021) Temporary urbanism as policy alternative to enhance health and well-being in the post-pandemic city. Curr Environ Health Rep 8:167–176

Asaria M, Mazumdar S, Chowdhury S et al (2019) Socioeconomic inequality in life expectancy in India. BMJ Glob Health 4:e001445. https://doi.org/10.1136/bmjgh-2019-001445

Balasubramaniam P, Rao N, Sharma G, Maraganty P, Sathyanarayana TN, Lakew D, Karua F, Bloom G, Ebata A, Gatellier K (2018) Innovations for universal health coverage: a south-south collaboration to transform health systems in Africa and India, innovations for universal health coverage collaboration. Public Health Foundation of India, Delhi

Bhagia M, Bose M (2023) Who owns the city? Neoliberal urbanism and land purchases in Gurgaon, India. Urban Studies. https://doi.org/10.1177/00420980231184784

Bhan G (2009) This is no longer the city I once knew. evictions, the urban poor and the right to the city in millennial Delhi. Environ Urban 21(1):127–142
Björkman L (2015) Pipe politics, contested waters: embedded infrastructures of millennial Mumbai. Duke University Press
Choudhary BK (2005) Waste picking a marginalized activity in urban space. Urban India 25(2):93–116
Choudhary BK (2010) Political economy of waste-pickers: their health and well-being. Trans Inst Indian Geogr 32(1):63–80
Das R, Choudhary BK (2024) Cultural compulsion vis-à-vis poor sanitation in cities: curious case of open defecation in the capital city of Delhi. Indian J Soc Work (forthcoming)
Dupont V (2000) Mobility patterns and economic strategies of houseless people in Old Delhi. Delhi: Urban Space and Human Destinies 99–124
Dutta P, Chaudhuri C (2020) Role of the private sector in escalating medical inflation: evidence from 75th round of NSS data. Econ Pol Wkly 55(22):18–21
Ghertner DA (2014) India's urban revolution: geographies of displacement beyond gentrification. Environ Plan A 46(7):1554–1571. https://doi.org/10.1068/a46288
Goli S, Arokiasamy P, Chattopadhayay A (2011) Living and health conditions of selected cities in India: setting priorities for the National Urban Health Mission. Cities 28(5):461–469
Gupta I, Sengupta P (2013) Health seeking behaviour in urban Delhi: an exploratory study. World Health Population 3(2). https://doi.org/10.12927/whp.17580
Haining K (2021) Mind the gap: Glasgow life expectancy divide. https://the-gist.org/2021/10/mind-the-gap-glasgows-life-expectancy-divide/
Harvey D (2003) The right to the city. Int J Urban Reg Res 27(4). https://doi.org/10.1111/j.0309-1317.2003.00492.x
Lahariya C (2016) Delhi's mohalla clinics: maximising potential. Econ Polit Wkly 15–17
Lahariya C (2020) Access, utilization, perceived quality and satisfaction with health services at Mohalla (Community) Clinics of Delhi, India. J Family Med Prim Care 9:5872–5880
Last JM (1988) Dictionary of epidemiology, 2nd edn. Oxford University Press, New York and Oxford
Latham R, Hicks J (2014) Urban dystopias. The Cambridge companion to the city in literature, pp 163–174. https://doi.org/10.1017/cco9781139235617.014
Leeuw de E (2011) Do healthy cities work? A logic of method for assessing impact and outcome of healthy cities. J Urban Health: Bull New York Acad Med 89(2). https://doi.org/10.10007/s11524-011-9617-y
Ranganathan M (2014) 'Mafias' in the waterscape: urban informality and everyday public authority in Bangalore. Water Alternatives 7(1)
Roy A (2005) Urban informality: toward an epistemology of planning. J Am Plann Assoc 71(2):147–158
Sarkar UD, Choudhary BK (2020) Reconfiguring urban waterscape: water kiosks in Delhi as a new governance model. J Water Sanitation Hyg Dev 10(4):996–1011
Saroj SK, Goli S, Rana MJ, Choudhary BK (2020) Availability, accessibility, and inequalities of water, sanitation, and hygiene (WASH) services in Indian metro cities. Sustain Cities Soc 54:101878
Sassen S (2015) Who own our city and why this urban takeover shall concern us all. https://www.theguardian.com/cities/2015/nov/24/who-owns-our-cities-and-why-this-urban-takeover-should-concern-us-all
Sharma A, Singh AK et al (2020) Out-of-pocket expenditure on healthcare among the urban poor in India. Econ Pol Wkly 55(32):49–55
Shukla S (2018) EWS beds in Delhi: private profiteering on public sufferings. Econ Pol Wkly 53(18):53–59
Singh G, Hickel (2023) Capitalogenic disease: social determinants in focus. BMJ Global Health
Soja EW (2000) Postmetropolis: critical studies of cities and regions. Blackwell

Sulzbach C (2013) From urban nightmares to dream cities: revealing the apocalyptic cityscape. Constructions of space V-place, space and identity in the ancient mediterranean world
Wekesa BW, Styen G (2011) A review of physical and socio-economic characteristics and intervention approaches of informal settlements. Habitat Int 35(2):238–245
WHO (1946) Constitution of the World Health Organization. Geneva
WHO (2023) Universal Health Coverage. Universal health coverage (UHC) (who.int)
Wilkinson RG, Marmot M (eds) (2003) Social determinants of health: the solid facts. World Health Organization
Zhong J, Liu W, Niu B, Lin X, Deng Y (2022) Role of built environments on physical activity and health promotion: a review and policy insights. Front Public Health 10:950348. https://doi.org/10.3389/fpubh.2022.950348

Chapter 7
Anticipating the Future of Iranian Cities: High-Tech Versus Nature-Based Solutions

Mohammadjavad Mahdavinejad, Farzaneh Gharaati, and Mamdouh Ibrahim Zaky Mohamed Galil

Abstract One of the most critical questions about our future is, "What will our cities be like in 2050?". The answer to this question shows our definition of development, humanity, nature, and what we expect from life. The ever-growing world population will intensify the climate change challenge, global warming, etc. Therefore, the research aims to anticipate the possible urban future scenarios and to outline the most prescriptible scenarios for the global south. The research methodology is based on the primary literature of the Discipline of Anticipation (DoA) by Riel Miller, Roberto Poli, and Pierre Rossel for future studies. It was conducted by co-design discussion involving a focus group discussion (FGD) methodology in three phases for expectations and hopes, reframing, and rethinking for flourishing ideas based on future literacy (FL). It was implemented with an interdisciplinary approach ranging from visionary and utopian futures to participatory, critical, or integral futurism by 2050. The conclusion of the research promoted nature-based solutions, socio-economic empowerment, carbon neutrality, high-performance architecture, against irrational ambitious capital-intensive developments. It is to focus on measuring happiness instead of incomes, circular economy instead of short-term gains, nature-based value creation instead of conventional mercantilism, for the MENA region and the global south.

Keywords Discipline of anticipation (DoA) · Global south · Sustainable and green future · Future generations · Low-carbon cities · Carbon neutrality · High-performance architecture · MENA region

---

M. Mahdavinejad (✉) · M. I. Z. M. Galil
College of Engineering and Architecture, University of Nizwa, Nizwa, Oman
e-mail: mahdavinejad@unizwa.edu.om

M. I. Z. M. Galil
e-mail: mamdouh@unizwa.edu.om

F. Gharaati
Department of Architecture, Tarbiat Modares University, Tehran, Iran
e-mail: farzaneh.gharaati@modares.ac.ir

S. Gonzalez-Arellano and B. Gandlgruber (eds.), *Cities as Anticipatory Systems*, The Urban Book Series, https://doi.org/10.1007/978-3-031-78162-9_7

7.1 Introduction

One of the most crucial questions about our future is, "What will our cities look like in 2050?" This simple question relates to our understanding of our current actions and relationship with nature, which affects our future. On the other hand, the rise in the average surface temperature of the Earth has resulted in adverse consequences linked to climate change on a global scale (Saaty and De Paola 2017). Notably, urbanization is the main factor behind this problem (Khan and Zaman 2018). According to the Department of Economic and Social Affairs of the United Nations Secretariat, the world's population is predicted to increase from 7.6 billion in 2017 to 8.6 billion by 2030 and 9.8 billion by 2050 (DESA 2022). Moreover, it is anticipated that by 2050, approximately 70 percent, or around 6.4 billion people, will live in urban areas (Moriarty and Honnery 2015; Kotzen 2021). While urbanization does offer benefits, such as opportunities for escaping poverty and the potential for long-term sustainability (Khan and Zaman 2018), it also comes with negative consequences. These include pollution, increased resource and energy consumption, and the proliferation of landfills (Saaty and De Paola 2017). As a result, the international community has established various goals to reduce half of global CO_2 emissions, as outlined in the 2016 Paris Agreement (Uwasu et al. 2020). As mentioned, cities are considered the essential drivers of these challenges and are acknowledged as complex and dynamic organisms (Asprone and Manfredi 2015). They offer opportunities and challenges at the same time. This is while, conventional urban planning practices have resulted in excessive greenhouse gas emissions. Moreover, the rise of new needs, globalization, demographic shifts, technological advancements, epidemics, privacy concerns, and data security issues demand the creation of more comprehensive and integrated urban models (Visone 2020) through careful foresight and future visioning in planning and management (Saaty and De Paola 2017). These measures are essential to ensure the ability to adapt to the risks and uncertainties of the future (Likitswat 2019).

Iran, as a country in the Global South, with its millennia-old historical cities, is no exception in this field. While numerous Iranian cities still adhere to traditional planning and models, contemporary Iranian cities have encountered significant shortcomings. Due to these shortcomings, adopting a futuristic vision for sustainable future cities instead of having a nostalgic perspective of the past is imperative. This means a balanced approach that adopts traditional and modern models while reasonably assessing potential future developments is necessary (Sharifi and Murayama 2013).

While Iran's population in 2023 stands at approximately 80 million people, demographic projections for Iran until 2050 anticipate a population of roughly 100 million individuals (PopulationPyramid 2023). Iran is ranked among the top 10 countries regarding carbon emissions, with 75% of its population residing in urban areas. Notably, between 1990 and 2019, Iran's energy consumption increased by 5–8%, a rate approximately five times higher than the global average. A significant portion of this energy consumption occurs in the capital city of Tehran, which has transformed over the past two centuries into one of the world's largest cities, with approximately 9 million residents. Furthermore, buildings, especially in urban areas, contribute significantly to energy consumption, exceeding the global average by 2.5–4 times, with 70% of this consumption stemming from public buildings, which waste up to 60% of their energy (PROVIDAS 2023). Given the demographic growth, cultural changes, and the production of greenhouse gases, particularly in metropolises like Tehran, conducting a thorough study is essential to formulate plans for a greener and more sustainable urban future. As there has not been a comprehensive urban regeneration plan developed for the future of Iranian cities, anticipating and creating scenarios for these cities' green and sustainable future of these cities becomes even more critical. Therefore, this study seeks to anticipate future urban scenarios and outline the most recommended scenarios specifically for the global south, especially Iran.

7.2 Background and Context

According to Wikipedia and UNIDO (2016), a sustainable city, also known as an eco-city or green city, is designed to consider its social, economic, and environmental impacts, often referred to as the three pillars. A sustainable city is a platform for low-carbon industrialization, turning today's cities into livable economic powerhouses with clean energy innovations. Indeed, it aims to provide a resilient habitat for the present generation without compromising the ability of future generations to have the same experience. Sustainable cities are dedicated to minimizing required energy, water, and food inputs and drastically reducing waste, output of heat, and air pollution—CO_2, methane, and water pollution. The significance of this endeavor stems from the fact that cities significantly influence the lifestyles of future generations (Likitswat 2019; Nakagawa 2020).

However, research findings indicate that among the three dimensions—social, economic, and environmental—environmental indicators have received less attention in proposed future city frameworks (Khan and Zaman 2018). The importance of these indicators cannot be overstated, as neglecting them could lead to cities becoming unlivable due to the impacts of climate change, necessitating the construction of entirely new cities (Moriarty and Honnery 2015). Drawing from the insights of authoritative articles and a comprehensive review of ideas and concepts, it is evident that creating sustainable cities for the future demands a multidisciplinary approach and comprehensive management through co-design (Riffat et al. 2016).

Over the years, various terms and labels have been employed to envision "ideal" and "future" cities. In 1516, Thomas More introduced the concept of Utopia, which depicted an idealized, fictitious island nation characterized by a lack of private property and a high degree of religious tolerance (Lang 2007). Utopia is a term that conceptually relates to future cities and is usually interpreted as an ideal city or society. Since both strive for more favorable and better living situations through innovation, utopia concepts can inspire future cities. Both serve the areas of technology, environmental sustainability, harmony with nature, justice, and social equality (Bellagamba 2016).

Different names and designations have been assigned to alternative visions of future cities over time:

- Age-Friendly City
- Compact City
- Creative City
- Circular City
- Eco-City
- Eco-Smart City
- Garden City
- Global City
- Green City
- Hanging City
- Livable City
- Low-to-Zero Carbon City
- Regenerative City
- Resilient City
- Sharing City
- Smart City
- Vertical City
- Zero Waste City

For instance, a *compact city* represents a densely populated urban area with reduced parking spaces and a strong emphasis on public transportation. Typically, compact cities feature limited population density and abundant green spaces and recreational areas, which can contribute to a more sustainable lifestyle for future generations (Saaty and De Paola 2017). Conversely, *Vertical Cities* constitute a network of interconnected, eco-friendly, self-sustaining towers that extend up to a mile into the sky, reaching up to one kilometer or a mile (Kotzen 2021). The concept of a *Hanging City* is conceived to preserve natural landscapes and prioritizes an ecosystem-based design, along with achieving energy self-sufficiency. Meanwhile, *Green Cities* are designed by harnessing natural forms and resources to construct future residences (Saaty and De Paola 2017). Furthermore, several popular ideas, such as Circular Cities, Smart Cities, Eco-Cities, and Eco-Smart Cities, have gained traction in the discourse on future urban development.

Circular cities represent a novel and practical approach to addressing the intricate and pressing challenges associated with sustainable urban development, arising in response to the dynamics of urbanization. The notion of a "circular city" is derived from the principles of the circular economy model. Circular city strategies align closely with the objectives outlined in the Sustainable Development Goals (SDGs), particularly emphasizing SDG11 and SDG12 (Gravagnuolo et al. 2019).

Ecology, as defined, involves the scientific examination of the interplay between organisms and their surroundings. Consequently, one can infer that ecological cities represent intricate and manufactured ecosystems with a strong synergy between society, economy, culture, and nature. These *ecological cities* are fundamentally concerned with recycling materials, achieving efficient energy utilization, implementing regulatory mechanisms, maximizing economic advantages, fostering societal harmony, and advancing the coexistence of humans and nature (Li and Yang 2015).

Smart city originally emerged as a simple reference to delivering energy and technology solutions to pressing sustainability issues, such as climate change and resource scarcity (Shah 2023). Over time, this concept has evolved and acquired diverse interpretations (Visone 2020). It includes comprehensive plans encompassing various urban infrastructures and functions, including services, commerce, transportation, communication, water, and energy (Saaty and De Paola 2017; Javed et al. 2022).

In contemporary discussions, the most favored concept is that of an *eco-smart city*—a city that focuses on economic, social, and spatial aspects to prioritize environmental protection equally. It emphasizes preserving the environment, conserving natural resources, limiting carbon dioxide emissions, reducing water consumption, and utilizing renewable resources. This concept emerges as a response to numerous factors, such as climate change, the vulnerability of the global economic system, urbanization, the digital transformation of society, higher expectations of life quality, demographic changes, and a growing public awareness of the need to protect the environment for the benefit of current and future generations. It represents the future's form—a shift from an industrial to a post-industrial civilization (Cheng and Xie 2011) (Fig. 7.1).

Fig. 7.1 A vision of an eco-smart city (Joachim 2018)

7.3 Methodology

Future thinking is an age-old human phenomenon. However, modern futures research advances far beyond that, and the systematic exploration of alternative futures is defined as foresight instead of prediction or prophesizing. Modern futures research was established in the late 1940s as a scientific field (Heinonen et al. 2017). All attempts to "know the future" mean thinking and anticipating that the future is equally present in all phenomena, conscious or unconscious, physical or ideal. The complete prediction of change in our world is practically and theoretically unattainable and nothing more than a virtual existence (Pillkahn 2008).

Regarding prediction or anticipation, "a predictive system is a system that contains a predictive model of itself and its environment, which allows this system to change its state at one moment by the model's predictions about a later moment." Practically speaking, the problem is that the data and models used to anticipate the future are incomplete and need to be supplemented. Theoretically, the challenge of anticipation is that our world is in some sense "creative" and that innovation happens. Many forecasting exercises are based on a modeling approach tested using past data (Poli 2017). Another fundamental challenge when imagining the future is the absence of future generations in the process, which causes intergenerational conflicts (Hara et al. 2021; Uwasu et al. 2020; Timilsina et al. 2020). That **is while creating** a series of failures, such as global warming and biodiversity loss, that will affect future generations (Saijo 2020). Notably, the future is unpredictable, and we can only create images and ideas of events we will face. Also, there is not one future but multiple possibilities. Furthermore, the future can be influenced based on decisions (Heinonen et al. 2017; Pillkahn 2008). Moreover, as we move further from the present, the number of future possibilities increases because various other factors influence what will happen (Pillkahn 2008) (Fig. 7.2) (Many attractive scientific researches have been carried about the Scenario Cone, see: Gall et al. 2022).

Before conducting anticipation research, defining the specific knowledge area and choosing the approach strategies is crucial. Additionally, it is logical to gather some references or real-world examples to provide context. Evaluating anticipations poses a challenge because the future does not come with concrete data, making it hard to determine the success of predictions. Even if predictions eventually align with reality, there is often no real-time assessment method (Miller et al. 2013). We can turn to Futures Literacy (FL) to tackle these challenges. Like learning a language, FL equips us with the skills to navigate the future effectively. It is like learning to read and write in a new language of possibilities (Miller 2018). FL involves understanding "what to know," "when to know," and "why it is important to know." One of its strengths is its ability to distinguish between different ways of using future knowledge (Miller et al. 2013).

Anticipation has three general levels:

- The first level (forecasting) is part of the correct prediction of future studies. Its models tend to consider a very short or very long time horizon.

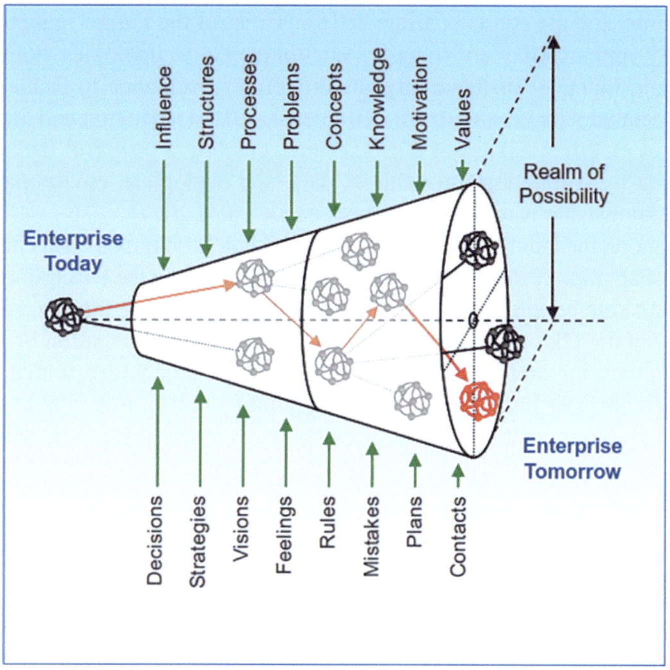

Fig. 7.2 The scenario cone: the realm of possibility (Pillkahn 2008)

- The second level (foresight) includes most traditional futures studies. It is not predictive, so the main output of most foresight exercises is the production of several possible futures. Foresight is often qualitative and includes and even focuses on discontinuities. As practiced, foresight usually shows a limited acceptance of complexity.
- The third level (anticipation) is based on the results obtained from forecast and foresight models, and the goal is to implement them in decision-making and actions.

Anticipation exhibits many temporal patterns, from micro anticipations embedded in perception to usually longer forms of social anticipation, ranging from seconds to years and decades (Poli 2017).

During the startup phase, several critical questions appear: "Who should be invited to take part?", "how should they be extended invitations?" and "Where should the process take place?" (Miller 2018). The challenge here is that participants often need to shift their perceptions of the future to shape a shared vision collectively. This typically involves collaborative work within small groups. Through this process, participants often appreciate that the future comprises diverse aspects, both pessimistic and optimistic. It is important to note that the choices regarding participants and settings depend on factors such as the specific group, the topic, available time, and the context (Miller 2018). "Cities of the Future" adopts an interdisciplinary approach that encompasses environmental technologies, holistic urban development, financial sustainability, and effective governance to tackle the challenges associated with urban growth (Riffat et al. 2016; Harrington and Stryja 2021; Heinonen et al. 2017; Likitswat 2019). Hence, creating a powerful meta-narrative of the future by combining information, data, and ecological, environmental, and individual knowledge is necessary (Saaty and De Paola 2017).

Furthermore, the magic concept of a "shared vision" strengthens real commitment instead of adaptation (Pillkahn 2008). Therefore, in applying the Discipline of Anticipation (DoA) methodology in future urban studies, developing future urban visions for regions in the Global South is imperative. This needs the creation of scenarios, specifically focusing on the emergence of green cities in the future, with a particular emphasis on cities in Iran. To put this strategy into practice, it is vital to recognize the challenges based on the circumstances, try to overcome them, and offer answers for each.

7.4 Vision for Urban Future of the Global South

Global South countries are primarily situated in the Southern Hemisphere, including regions such as Asia (excluding Japan, Taiwan, Singapore, South Korea, and Macau), Central America, South America, Mexico, Africa, and the Middle East (excluding Israel) (Cruz 2015). In the future, cities in the Global South are anticipated to face rapid urbanization while concurrently experiencing rising temperatures due to the impacts of climate change and urban development (Bardhan et al. 2020). Many articles propose the implementation of "smart city" technologies as a remedy for urban challenges in the Global South (Borkowska and Osborne 2018; Visone 2020). However, it is essential to recognize that relying solely on these technologies will not be sufficient for shaping and achieving a green and sustainable future. This issue requires careful management and the importance of continuous study for planning future cities. In addition to exploring the ideas and concepts, there is a need to consider solutions based on nature-based solutions approaches and advanced technology.

In urban development, future cities share a common aspiration, regardless of the labels attributed to them: to furnish their citizens with the most optimal and desirable urban experiences (Khan and Zaman 2018). Furthermore, involving local communities and civil society in making decisions about urban matters is essential. It is crucial to listen to the opinions of all people during the urban development process (Saaty and De Paola 2017). Our investigation into the concept of ideal cities has led us to identify two conceptual paradigms: high-technology and tall buildings approaches and nature-based solutions.

7.4.1 High-Technology and Tall Buildings

Tall buildings will be essential in future cities because of the world's rapidly increasing urban population and the migration of people from rural-to-urban areas due to growing security concerns. Tall buildings have critical features like preserving agricultural land and offering multi-level connectivity, height, and multifunctional and biophilic content (Ali and Al-Kodmany 2012). They stand out for their ability to connect multiple towers with skywalks instead of isolated single towers. While most people imagine the vertical city in the air, it is conceivable that these cities could easily be extruded underground as well, especially since some functions do not require natural light (Fig. 7.3). This concept of the "vertical city" creates a network of three-dimensional green spaces throughout the city, unlike traditional cities with separate parks and gardens, the biophilic vertical city creates a network of three-dimensional green spaces throughout the city (Kotzen 2021). Technological advancements make all these innovations possible (Riffat et al. 2016).

7.4.2 Nature-Based Solutions

Nature-based solutions (NbS) emerged in the late 2000s, initially introduced by the World Bank and IUCN (2009), primarily emphasizing the significance of biodiversity preservation and mitigating climate change. The European Commission has recently defined NbS as "actions inspired, supported, or imitated from nature" (Kabisch et al. 2017). These actions involve the creation of urban green spaces, such as parks and street trees, which can help alleviate high temperatures in cities and regulate air and water flow. Architectural solutions, including green roofs and wall installations, can also reduce temperatures and conserve energy (Kabisch et al. 2017). NbS aims to develop sustainable urbanization to improve economic growth, enhance environmental conditions, restore degraded ecosystems, adapt to climate change, implement carbon reduction measures, improve environmental risk management, and promote multifunctionality (Atanasova et al. 2021).

The concept of NbS is intricately linked to other ideas such as sustainability, resilience, ecosystem services, human–environment interaction, reinforcement, reconstruction, co-creation, and co-design of urban green networks, as well as green/blue infrastructure and biomimicry concepts. Compared to conventional approaches, NbS presents a more efficient and cost-effective development strategy (Lafortezza et al. 2018; Dushkova and Haase 2020). Concepts like ecosystem-based adaptation (EbA) (focused on providing services that support human adaptation to climate change), green infrastructure (GI) (aimed at supporting native species), and ecosystem services (ESS) (aimed at facilitating people's benefits from ecosystems) are closely intertwined with the NbS concept within scientific literature (Kabisch et al. 2017). A notable characteristic of NbS lies in its broad nature and scale, ranging from expanding forested areas to absorbing gaseous pollutants and spanning from micro to

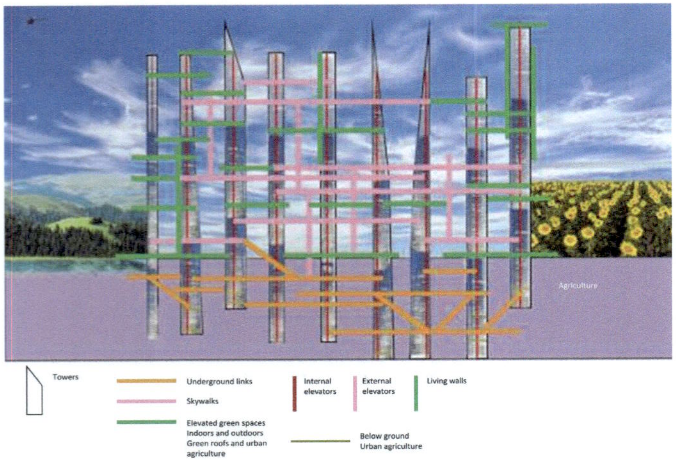

Fig. 7.3 Left: concept of the biophilic vertical city, right: Mexico city's proposed 75-storey underground buildings in the future (Kotzen 2021)

macro scales. This encourages collaborative approaches for co-design, co-creation, and co-management, often called "co-co-co" (Kabisch et al. 2017; Lafortezza et al. 2018). The concept of a circular economy, which is also connected to nature-based solutions, holds significant importance. Circular economy (CE) is a system that aids cities in accomplishing a transition from managing their resources in a linear to a circular manner. At its core, fundamentally, the primary challenges of circular economy (CE) are resource conservation and reuse, waste reduction, and resource utilization cycle establishment (Atanasova et al. 2021; Langergraber et al. 2021). Key aspects of CE include the recovery and preservation of the water cycle (including the management of rainwater), water treatment, and waste treatment, alongside efforts in recycling and reusing resources. Simultaneously, CE strongly emphasizes enhancing energy efficiency and harnessing energy recovery methods (Atanasova et al. 2021). Furthermore, it highlights the importance of building reuse, epitomized by the concept of Zero Energy Buildings (nZEB) (Riffat et al. 2016). Embracing vertical greening systems and green roofs, promoting food and biomass production, and facilitating the creation of (public) green spaces also rank among CE's fundamental principles (Langergraber et al. 2021; Heinonen et al. 2017). These principles necessitate a holistic, cross-sectoral, and interdisciplinary approach (Atanasova et al. 2021).

7.5 Scenarios for the Green Future of Iranian Cities

7.5.1 Traditional Cities of Iran

Historical urban planning in Iran can be categorized into two primary periods: pre-Islamic and post-Islamic. Our understanding of pre-Islamic cities primarily relies on archaeological findings, and as a result, access to direct documentation about the social life in these cities is scarce. Consequently, figuring out specific urban patterns for this period is challenging. In the post-Islamic era, a distinctive transformation occurred in urban planning as new urban forms emerged. The combination of pre-Islamic and Islamic concepts gave these post-Islamic cities a unique identity that distinguished them from their pre-Islamic counterparts. To gain a deeper understanding of cities in Iran, it is essential to familiarize with the main urban elements and their functions:

- *Mosques*: After the Arab invasion, structural changes were made in cities to meet Islamic needs. The mosque was one of the main elements of a square connected with other elements.
- *Bazaar*: Is originally developed to perform commercial and production functions and usually has a linear shape.
- *Meydan*: It was created in the pre-Islam era.

- *Mahala*: Traditional Iranian cities were divided into several residential neighborhoods called Mahala. People with similar ethnic and religious backgrounds were gathered together in the same neighborhood. Narrow winding alleys connected the center of the neighborhood with the market on one side and ended in a dead end on the other. The alleys had a hierarchy and became narrower away from the *Bazaar*. These winding passages were covered intermittently to provide greater security against horse raiders if the walls were breached.
- *Zorkhaneh* (house of power): It was created in the pre-Islam era. They were not only places for physical exercise but have been centers of social and political activity for centuries.
- *Hammam* (public baths): They were health and social places where people go to bathe, massage, shave, and talk, and they can be found in all the traditional cities of the Islamic world.
- *Persian garden*: It is an exceptional and even unique evidence of cultural traditions and has become a prominent feature in private residences, palaces, public buildings, and collections related to charitable or religious institutions, such as tombs (Sharifi and Murayama 2013).

Therefore, the reasons for the geometric formation of traditional Iranian cities can be summarized as climate, agriculture, and religion. In terms of climate, for example, the streets of Tehran are arranged in the main directions, and the houses face north to take advantage of the summer breezes from the Alborz mountains. From a religious point of view, because Muslims must face Mecca when they pray, Iranian mosques, the main element of Iranian cities after Islam, are rectangular with open courtyards, and the Qibla wall must face southwest and toward Mecca. The streets are often parallel to the mosques, and the grid system may have been developed along the streets around the mosques perpendicularly and toward the southwest (Bonine 1979).

7.5.2 Contemporary Cities in Iran

Contemporary cities in Iran have experienced significant transformations over time. While the initial steps toward modernization began during the period of Naser al-Din Shah in the 1870s, with his visit to Paris serving as a catalyst, the peak of these transformations occurred in the early 1920s when Iranian cities witnessed substantial structural changes. Unfortunately, urban renewal principles were often applied without considering historical growth patterns or preserving local cultural values. The introduction of automobiles marked a significant moment in the urban landscape, leading to street systems designed primarily to accommodate motorized transportation needs. The Approval of the Law on Municipalities in 1930 and the Street Widening Law of 1933 played crucial roles in this transformation, as straight and axial street networks began to disrupt the traditional urban fabrics.

Consequently, the cohesive urban structure started to erode, causing the fragmentation of key elements of the traditional city. Over time, these elements lost their functionality and efficiency, necessitating a redefinition. Historically, urban routes were meticulously designed to ensure pedestrians' comfort and give precedence to environmental considerations. However, in contemporary developments, streetscapes have shifted their orientation toward cars, often neglecting pedestrian safety. Traditional cities were adept at creating outdoor spaces that provided relative respite from adverse weather conditions. Still, new developments often disregard these principles, resulting in uniform dimensions across the country, regardless of the local context.

Additionally, there is insufficient consideration of hierarchy in the new street patterns, diminishing the traditional order from public to private spaces. As activities are segregated in modern zoning systems, residents spend less time in their neighborhoods, impacting social interactions and reducing the sense of belonging to these areas. Beyond physical changes, increased rural-to-urban migration and the Iran-Iraq war further complicated the urban landscape. The war forced residents from war-torn regions to migrate to other cities in search of land to build their homes under the Urban Land Law of 1987. These multifaceted factors have collectively shaped the contemporary cities of Iran, highlighting the need for a holistic understanding of their development and transformation (Sharifi and Murayama 2013) (Fig. 7.4.) Hence, exploring city streets is crucial for shaping future city concepts, especially in Iran, where significant street changes followed modernization. As Boudreau (2022) argues, studying streets can reveal hidden ways to influence the future.

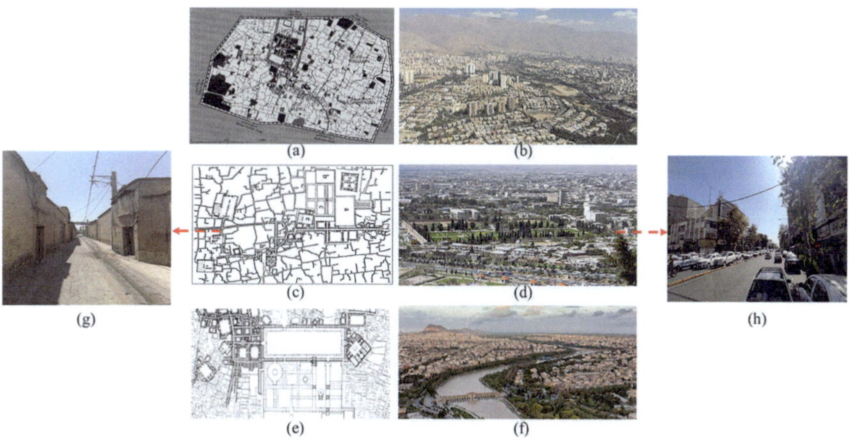

Fig. 7.4 **a**, **c**, **e** Historical context of Iranian cities: Tehran, Shiraz, and Isfahan, respectively, **b**, **d**, **f** The current view of the Iranian cities: Tehran, Shiraz, and Isfahan, respectively (Jamei et al. 2021, Shirazi and Falahat 2015), **g** Alleys and historical districts of Shiraz city, **h** The current streets of Shiraz city (authors 2023)

7.5.3 Presenting Scenarios

Through an examination of ideal city concepts, we have outlined two distinct scenarios that envision the future cities of Iran in 2050. These scenarios are integral to the evolution of urban landscapes, offering unique solutions to meet the challenges of sustainable urbanization. We present these scenarios as two compelling paths for urban development in the coming decades, focusing on the most populous cities in Iran, including Tehran, Isfahan, Shiraz, Mashhad, Tabriz, and Karaj. The finer details of these scenarios will be left to urban planners' and experts' expertise and contributions to avoid any orientation. With the DoA method, we are operating on a distinct time frame. We decided on the year 2050, around thirty years from now.

Scenario 1: Technology and Tall Buildings

The first scenario centers around integrating technology and tall buildings, resulting in intelligent and vertical cities. This vision hinges on making intelligent various combinations of urban life, people, the environment, and lifestyles (Visone 2020). Additionally, it entails the construction of interconnected tall buildings that redefine urban living. This approach stimulates sustainability through innovative technologies and vertical urbanization.

Scenario 2: Nature-Based Solutions

The second scenario promotes nature-based solutions, combining the principles of an eco-city, a green city, and a zero-carbon city. This vision considers city regeneration, ecosystem restoration, and the sustainable utilization of resources and energy (Lafortezza et al. 2018). It incorporates green/blue infrastructure into contemporary urban planning (Gearey et al. 2019).

These two scenarios have been chosen based on their capacity to enhance sustainable urbanization and address critical urban challenges. Both nature- and technology-based solutions contribute significantly to the following aspects:

- *Ecosystem Restoration*: The scenarios support ecological health by making restoring damaged ecosystems and vital services easier. This is crucial for densely populated Iranian cities like Tehran, which face pollution challenges.
- *Climate Change Adaptation and Mitigation*: These scenarios integrate climate change strategies by reimagining gray infrastructure as part of green /blue infrastructure. Because it considers the potential of Iranian cities to adopt green/blue infrastructure, leveraging their advantageous locations.
- *Risk Management and Resilience*: They enhance risk management and urban resilience by offering multifunctional nature-based designs, reducing pollution, protecting biodiversity, mitigating heat stress, and enhancing water retention (Dushkova and Haase 2020). Most Iranian cities like Tehran rank among the cities with significant carbon emissions.
- *Microclimate Regulation*: They regulate microclimates, prevent floods, purify water, and promote sustainable food production (Langergraber et al. 2020). Given

their access to rainwater during various seasons, Iranian cities possess substantial potential for harnessing microclimates to obtain clean water.

- *Circular Economy*: These scenarios, especially Nbs, encourage long-term material design, maintenance, repair, reuse, reconstruction, renovation, and recycling, reducing waste and minimizing environmental impact (Langergraber et al. 2020). This is especially relevant for many Iranian cities with rich historical heritage, as reusing resources aligns with the circular economy concept. Additionally, buildings that release greenhouse gasses and carbon emissions may be able to mitigate their effects through circular renovation.
- *Job Creation:* These scenarios generate new employment opportunities that align better with the needs of urban populations (May et al. 2022). Given the substantial population, job creation holds significant importance.

7.5.4 Results

A series of expert meetings, spanning three sessions of approximately one and a half hours each, were held to deliberate and strategize on the future cities of Iran in 2050. The experts were generally chosen based on their experience and knowledge background in the context of Iranian cities. The objective was to apply their knowledge and insights to imagine the urban landscape of Iran's future based on considered scenarios and the four distinct images that depict various trajectories:

1. Continuous Growth: Illustrating a future that aligns with conventional expectations.
2. Transformation: Envisioning a future that exceeds expectations, striving for substantial improvement.
3. Collapse: Picturing a future worse than expected, marked by challenges and failures.
4. Protective/Disciplined Society: Imagining a future that unexpectedly departs from expectations (Cruz 2015; Boyko et al. 2012).

Eight scenarios appeared after determining the four trajectories and considering them with the two basic scenarios. During the initial meeting, experts discussed and decided that the most promising approaches for shaping Iran's future cities are the "Transformation" and "Protective/Disciplined Society," which correspond to numbers 2 and 4, respectively. This decision was made with the common goal of collaborating toward improving Iran's future. Nevertheless, it was acknowledged that substantial challenges and obstacles must be anticipated and addressed to realize these scenarios successfully.

During the following discussions, a crucial realization occurred when specialists acknowledged that depending only on natural or technological solutions was not enough. They concluded that the cities of Iran in the future of 2050 would require a well-balanced combination of all strategies and scenarios. However, they chose to place the scenarios on the "Transformation" mode and make judgments in this mode because the "Protective/Disciplined Society" mode is not very certain in Iran until 2050, based on accurate data.

In the third session, the experts were asked to imagine themselves as the Imaginary Future Generation (IFG) facing an intergeneration dilemma (Hiromitsu et al. 2022). It was marked by challenging and controversial discussions as experts collaboratively presented various solutions and strategies tailored to the unique characteristics of Iran's future cities in 2050. These deliberations centered predominantly on densely populated cities, including Tehran, Isfahan, Shiraz, Tabriz, Mashhad, and Karaj.

Through these rigorous and insightful expert meetings, a comprehensive framework for the future cities of Iran in 2050 began to take shape, emphasizing transformation according to a harmonious blend of nature-based and technology-driven solutions.

We also faced several practical challenges: (1) creating scenarios; (2) selecting participants; (3) forecasting uncertainties; and (4) estimation of effective variables until 2050. In order to address the first challenge, the statistical data of Iran in the last ten years and articles on predicting Iran's future were studied. However, we refrained from providing specific statistics and data to the participants as we opposed providing them with any orientation. We tried to select urban development and planning experts for the second challenge because this type of meeting is the first of its kind in Iran. In order to evaluate the third challenge, we consulted other experts on the concepts. The fourth challenge was the hardest one. The research results show that climate change and global warming are the most obvious factors affecting Iran's future. However, this method can be combined with other quantitative methods under the condition of completing and solving the challenges in future research.

7.5.5 Discussion

The experts' conceptualizations of the scenarios to represent Iran's future cities in 2050 are provided in this section. Besides, a comparison with the findings of our studies is made. Thus, challenges about the scenarios will be discussed depending on the methodology.

Scenario 1: Technology and Tall Buildings with Transformation Mode

Urban areas in the global south, especially Iran, grow gradually rather than all at once. In this scenario, the conceptual framework of sustainable cities in all social, economic, and environmental dimensions is used to imagine the sustainability and intelligence of cities. Technology is essential in every aspect; ideas and solutions can only be implemented with a fundamental shift (Wickenberg et al. 2021). The historical context, dominating beliefs, customs, and past behaviors reflect the identity of Iran's traditional cities, such as Tehran, Isfahan, and Shiraz. As a result, there is a high potential for value creation through technological integration in historical contexts and legacy. Furthermore, it is possible to apply traditional patterns in various contexts and cities, focusing on the social aspects of evolving and developing technologies.

High-rise structures, a new generation of buildings, referred to as "green skyscrapers/earth-scrapers" or "zero energy buildings," are becoming more energy efficient and capable of producing as much energy as they consume. Furthermore, vertical farms have a bright future. The digital revolution and technological advancements in construction materials are expected to influence the architectural style and functions of buildings in Iranian cities (Mahdavinejad et al. 2024). This strategy needs the participation of stakeholders, designers, and policymakers in collaborative processes that will produce a city model that intelligently manages its resources, encourages the development of innovative urban transportation systems, and strives for economic sustainability and energy efficiency. On the other hand, quality of life considerations in this scenario go beyond innovative information and communication technologies and reach urban sustainability, which is achieved through optimizing resources, increasing energy efficiency in buildings, and reducing land use. Finally, it follows the precise and comprehensive mechanisms of a smart city (Wickenberg et al. 2021) following intelligent optimization methods (Fakhr et al. 2023; Shirzadnia et al. 2023; Najafi et al. 2023; Goharian et al. 2023).

The challenge we may face in this scenario is the construction of tall buildings; while they symbolize modernization and progress, they create significant demands on infrastructure and transportation systems. In addition, they can have a profound effect on the historical contexts and skyline of the city. However, a potential future problem is an unhealthy competition between designers and developers who are only focused on achieving greater prestige, potentially ignoring citizens' real needs and preferences. Furthermore, there is a danger that local culture, ecology, and contextual relevance may be overlooked in the pursuit of architectural grandeur. This idea is probably because the experts are well aware of the events of the development of contemporary cities in the 1900s for Iranian traditional cities, and they hope to prevent it from happening again.

Scenario 2: Nature-Based Solutions with Transformation Mode

According to experts, since most of Iran's future challenges are environmental, climatic, and rising temperatures, presenting this scenario is very important for Iran. To illustrate this scenario, the experts highlighted some examples of passive methods of renewable energy usage in traditional Iranian cities (Dezfuli et al. 2023; Saadatjoo et al. 2023). Using traditional badgirs (wind deflectors), Asbad (windmills), shading in narrow alleys, and geometric proportions of elements (Dezfuli et al. 2024; Latifi et al. 2023), especially in hot and dry cities are a few examples of these techniques. Using these traditional patterns not only helps to conserve energy and improve energy efficiency in buildings and cities but also contributes to value creation. These views are most likely based on the expert's knowledge in the field of Iranian cities, especially traditional ones. Like traditional cities, the residents' biological, social, and perceptual characteristics and prevailing weather conditions must be considered in contemporary cities. These factors play a significant role in evaluating the happiness of citizens. When adopting nature-based solutions, aesthetic appeal and values are essential to citizens and require collective governance. This approach can seamlessly integrate into various urban programs (Frantzeskaki 2019).

Another idea that experts mentioned was the circular economy, which leads to clean production methods by reusing scraps and waste. Iranian cities have a high potential for this idea because the historical context and traditional patterns are highly flexible (Cavaleiro de Ferreira and Fuso-Nerini 2019; Gharaati et al. 2023). This transformational change can promote sustainable growth, stimulate economic development, and reduce the adverse effects of climate change while minimizing the consumption of raw materials (Lakatos et al. 2021). This is probably because experts seek to introduce innovative forms of value creation and break away from traditional ways of value creation. They aspire to harness the boundless potential of creativity and innovation to cultivate unique forms of value that resonate with the evolving dynamics of Iran's urban landscape.

Challenges and Considerations

The proposed scenarios, although promising, come with a set of major challenges that must be addressed holistically. The most critical challenges are legal deficiencies, lack of capacities, and public awareness in Iran. Both technology-based and nature-based approaches face these common challenges. Experts also mentioned the concern about over-reliance on technology. While technological advances offer significant benefits, over-reliance on them can raise questions about the flexibility and sustainability of these cities.

On the other hand, when it comes to Iranian cities, there is a need for cultural preparation of Iranian cities to accept the proposed scenarios. In addition, considering factors such as climate, economy, and the unique historical context of many Iranian cities, special solutions must be devised. Another challenge is the relationship between tall buildings and considerations related to the aesthetics, culture, and overall texture of Iranian cities. Determining how to address this issue is critical because implementing and planning these solutions requires careful study. Finally, applying advanced technology approaches in Iranian cities creates a challenge because it requires significant structural and fundamental changes. In addition, a significant challenge is the potential for high real estate prices that usually appear very quickly in Iranian cities. As Iranian cities embrace intelligent and sustainable features, property values may increase, potentially making housing unaffordable for certain population segments.

In addition, the complex communication between Iranian intelligent cities and their non-smart counterparts creates challenges. Ensuring the integration and seamless cooperation between these two types of cities requires careful planning and implementation. High electricity costs are another concern, especially in the context of high-tech urban infrastructure. Imagine a city intricately connected and controlled through satellite and Internet networks, highlighting the advantages and disadvantages of such connectivity. While offering greater security and convenience, it raises legitimate concerns about personal space and privacy. Maintaining privacy in Iranian cities is a significant concern, which makes it an important assignment. On the other hand, the potential vulnerability of the entire system to the Internet, such as disruption, is a significant concern. Furthermore, the advent of new technologies may inadvertently lead to a decrease in physical activity in individuals, potentially leading to long-term health problems such as increased rates of disease and obesity (Shichiyakh et al. 2016).

Therefore, instead of relying on conventional measures such as income and short-term economic gains, our research addresses a new paradigm that places city dwellers' happiness and well-being at its core. Experts emphasize the new urban vision, which makes the happiness and well-being of Iranian city dwellers its primary goal. They support a shift toward comprehensive urban well-being and emphasize metrics that measure the city dwellers' quality of life, satisfaction, and happiness rather than income. This change also emphasizes deviating from the conventional pursuit of immediate financial gains. Therefore, instead of income, experts suggest a circular economy, instead of short-term profit, value creation in new forms instead of traditional forms.

7.6 Conclusion

This study aims to provide new strategic concepts and scenarios for the future cities of the global south in 2050, with a focus on Iran. To achieve this, a series of meetings were conducted with five experts with deep knowledge of traditional Iranian cities based on the Discipline of Anticipation (DoA) and Future Design (FD). A range of alternatives can be described for Iran's future cities in 2050 through scenario-based literature. This makes it possible to compare future differences directly. However, the outcome of the experts' discussion brought two scenarios: technology and tall buildings and nature-based solutions with the transformation mode in 2050.

According to the findings, considering both scenarios simultaneously is more beneficial than doing so. Furthermore, the quality of life and the demands of future generations are intrinsically related to their happiness. Given that both scenarios have a high chance of being welcomed in Iranian cities, to overcome environmental challenges, details, and solutions of the scenarios should be designed under Iran's traditional and cultural values and the design principles of Iran's historical cities. It is also necessary to use each city's specific characteristics, such as hierarchy and street orientation, to consider future climate changes.

In characterizing the capacities to anticipate futures in Iran, two scenarios show comprehensive potential. In the first scenario, Iran has promising developments in technology and infrastructure, particularly in the construction of tall buildings. There is also a rising focus on intelligent and green buildings in academic areas. This gives us hope for a futuristic vision toward technological innovation and sustainable development. Socio-culturally, Iranian society is strongly inclined to embrace technological adoption and future visioning. Economically, solid Internet security networks and developing intelligent tools by expanding knowledge-based companies will support further research in future research efforts. The second scenario shows latent opportunities along with challenges. Despite facing economic vulnerabilities, Iran has unrealized potential in environmental and historical fields. The existence of laboratories and inactive farmland provides a suitable field for experiments and efforts to protect biodiversity. Meanwhile, abandoned historical and heritage sites provide revitalization and adaptive reuse opportunities. In the socio-cultural aspect, the educational infrastructure in universities provides a platform for cultivating futuristic thinking.

Meanwhile, the socio-cultural dimension's narrative is more prevalent. Because it appears that social trends—becoming increasingly evident—affect how Iranians perceive their future prospects. Although this was an exercise in the context of Iran, it is necessary to provide detailed scenarios for other contexts and gather expert comments through DoA and FD workshops such as IFG. Then, analyze the trends and predictions and make more accurate plans for the future of the cities.

Preparing Iranian cities culturally can be possible by accepting the presented scenarios' society. This method can be made more inclusive by using participatory approaches, particularly those that involve citizens, enhance collaboration between researchers and policymakers, acknowledge cultural values, traditions, and beliefs, forecast their evolution, and incorporate them into scenario planning.

It is essential to mention that to strengthen the process of futures thinking, it is possible to increase awareness through educational programs, public understanding, and cooperation between government organizations, networks, and academic institutions. It is critical to incorporate many viewpoints, listen to everyone in the community, and embrace the most innovative voices and concepts.

References

Ali MM, Al-Kodmany K (2012) Tall buildings and urban habitat of the 21st century: a global perspective. Buildings 2(4):384–423

Asprone D, Manfredi G (2015) Linking disaster resilience and urban sustainability: a global approach for future cities. Disasters 39(s1):s96–s111

Atanasova N, Castellar JA, Pineda-Martos R, Nika CE, Katsou E, Istenič D, Langergraber G et al (2021) Nature-based solutions and circularity in cities. Circ Econ Sustain 1(1):319–332

Bardhan R, Debnath R, Gama J, Vijay U (2020) REST framework: a modelling approach towards cooling energy stress mitigation plans for future cities in warming Global South. Sustain Cities Soc 61:102315

Bellagamba U (2016) From ideal to future cities: science fiction as an extension of utopia. Philos Technol 29:79–96

Bonine ME (1979) The morphogenesis of Iranian cities. Ann Assoc Am Geogr 69(2):208–224

Borkowska K, Osborne M (2018) Locating the fourth helix: rethinking the role of civil society in developing smart learning cities. Int Rev Educ 64(3):355–372

Boudreau J-A (2022) City of repair: practicing the future in Mexico City. Int J Urban Reg Res 46(5):851–864. https://doi.org/10.1111/1468-2427.13102

Boyko CT, Gaterell MR, Barber AR, Brown J, Bryson JR, Butler D, Rogers CD (2012) Benchmarking sustainability in cities: the role of indicators and future scenarios. Glob Environ Change 22(1): 245–254

Cavaleiro de Ferreira A, Fuso-Nerini F (2019) A framework for implementing and tracking circular economy in cities: the case of Porto. Sustainability 11(6):1813

Cheng S, Xie S (2011) Making eco-smart city in the future. In: 2011 International conference on consumer electronics, communications and networks (CECNet), pp 1380–1383, IEEE

Claudio Providas (2023) UNDP Resident Representative in Iran, Iran tackles its cities' carbon emissions, 21 Oct 2021. https://www.undp.org/blog/iran-tackles-its-cities-carbon-emissions

Cruz SO (2015) Alternative futures of global governance: scenarios and perspectives from the Global South. Foresight 17(2):125–142
DESA (2022) World population prospects 2022. Summary of Results, New York
Dezfuli RR, Bazazzadeh H, Taban M, Mahdavinejad M (2023) Optimizing stack ventilation in low and medium-rise residential buildings in hot and semi-humid climate. Case Stud Therm Eng 52:103555
Dezfuli RR, Mehrakizadeh M, Najar BSA, Bazazzadeh H, Mahdavinejad M (2024) Geometric investigation of entrance proportions of houses from the Qajar to the beginning of the early Pahlavi in Dezful City (1789–1979). Front Archit Res 13(1):57–78
Dushkova D, Haase D (2020) Not simply green: nature-based solutions as a concept and practical approach for sustainability studies and planning agendas in cities. Land 9(1):19
Fakhr BV, Mahdavinejad M, Rahbar M, Dabaj B (2023) Design optimization of the skylight for daylighting and energy performance using NSGA-II. J Daylighting 10(1):72–86
Frantzeskaki N (2019) Seven lessons for planning nature-based solutions in cities. Environ Sci Policy 93:101–111
Gall T, Vallet F, Yannou B (2022) How to visualise futures studies concepts: revision of the futures cone. Futures 143:103024
Gearey M, Robertson L, Anderson J, Barros P, Cracknell D (2019) Re-naturing the city for health and wellbeing: green/blue urban spaces as sites of renewal and contestation. Plann Cities Nat Theor Strateg Methods 153–168
Gharaati F, Mahdavinejad M, Nadolny A, Bazazzadeh H (2023) Sustainable assessment of built heritage adaptive reuse practice: Iranian industrial heritage in the light of international charters. Historic Environ Policy Pract 14(4):498–532
Goharian A, Daneshjoo K, Shaeri J, Mahdavinejad M, Yeganeh M (2023) A designerly approach to daylight efficiency of central light-well; combining manual with NSGA-II algorithm optimization. Energy 276:127402
Gravagnuolo A, Angrisano M, Fusco Girard L (2019) Circular economy strategies in eight historic port cities: criteria and indicators towards a circular city assessment framework. Sustainability 11(13):3512
Hara K, Kitakaji Y, Sugino H, Yoshioka R, Takeda H, Hizen Y, Saijo T (2021) Effects of experiencing the role of imaginary future generations in decision-making: a case study of participatory deliberation in a Japanese town. Sustain Sci 16:1001–1016. https://doi.org/10.1007/s11625-021-00918-x
Harrington AR, Stryja M (2021) Eco smart city method of promoting environmental sustainability and sustainable development in an industrialised and digitalized society. In: Internet and new technologies law. Nomos Verlagsgesellschaft mbH & Co. KG, pp 275–294
Heinonen S, Parkkinen M, Karjalainen J, Ruotsalainen J (2017) Energizing peer-to-peer urban futures—challenges for urban governance. Proc Eng 198:267–282
Hiromitsu T, Kitakaji Y, Hara K, Saijo T (2022) What do people say when they become "future people"?—positioning imaginary future generations (IFGs) in general rules for good decision-making. Sustainability 10, 13(12):6631. https://doi.org/10.3390/su13126631
Jamei E, Ahmadi K, Chau HW, Seyedmahmoudian M, Horan B, Stojcevski A (2021) Urban design and walkability: lessons learned from Iranian traditional cities. Sustainability 13(10):5731
Javed AR, Shahzad F, ur Rehman S, Zikria YB, Razzak I, Jalil Z, Xu G (2022) Future smart cities: requirements, emerging technologies, applications, challenges, and future aspects. Cities 129, 103794
Joachim M (2018) Post carbon city-state, rezoned circular economy. http://www.terreform.org/projects_urban-new-york-city-state.html
Kabisch N, Korn H, Stadler J, Bonn A (2017) Nature-based solutions to climate change adaptation in urban areas: Linkages between science, policy and practice. Springer Nature
Khan S, Zaman AU (2018) Future cities: conceptualizing the future based on a critical examination of existing notions of cities. Cities 72:217–225

Kotzen B (2021) Future cities: speculation on the case for vertical biophilic cities. Ecocycles 7(2):59–80
Lafortezza R, Chen J, Van Den Bosch CK, Randrup TB (2018) Nature-based solutions for resilient landscapes and cities. Environ Res 165:431–441
Lakatos ES, Yong G, Szilagyi A, Clinci DS, Georgescu L, Iticescu C, Cioca LI (2021) Conceptualizing core aspects on circular economy in cities. Sustainability 13(14):7549
Lang SB (2007) Predicting or inventing the future? Common grounds in architecture and strategic marketing. Predicting Future 499–506
Langergraber G, Pucher B, Simperler L, Kisser J, Katsou E, Buehler D, Atanasova N et al (2020) Implementing nature-based solutions for creating a resourceful circular city. Blue-Green Systems 2(1):173–185
Langergraber G, Castellar JA, Pucher B, Baganz GF, Milosevic D, Andreucci MB, Atanasova N et al (2021) A framework for addressing circularity challenges in cities with nature-based solutions. Water 13(17):2355
Latifi M, Mahdavinejad MJ, do Vale CP, Iannaccone G (2023) Asymmetric symmetry: an analytical framework of hidden mathematics in Isfahan's valuable Qajar houses; a video-ecology approach
Li J. Yang T (eds) (2015) China's eco-city construction. In: Chapter 1. Ecological crisis, eco-civilization, and eco-cities. Springer
Likitswat F (2019) Future cities: new generation's visions of sustainability concepts and models. Future Cities Environ 5:9–9
Mahdavinejad M, Bazazzadeh H, Mehrvarz F, Berardi U, Nasr T, Pourbagher S, Hoseinzadeh S (2024) The impact of facade geometry on visual comfort and energy consumption in an office building in different climates. Energy Rep 11:1–17
May C, Gandlgruber B, D'Costa A, Botelho AJJ, Balestro M (2022) Brazil and India transitioning to a knowledge economy. In: 34th annual meeting. SASE. https://sase.confex.com/sase/2022/meetingapp.cgi/Session/6024
Miller R, Poli R, Rossel P (2013) The discipline of anticipation: exploring key issues. fumee.org
Miller R (2018) Futures literacy laboratories (FLL) in practice: an overview of key design and implementation issues. Transforming Future 95–109
Moriarty P, Honnery D (2015) Future cities in a warming world. Futures 66:45–53
Najafi Q, Gorji-Mahlabani Y, Goharian A, Mahdavinejad M (2023) A novel design-based optimization method for building by sensitivity analysis. J Solar Energy Res 8(2):1446–1458
Nakagawa Y (2020) Taking a future generation's perspective as a facilitator of insight problem-solving: sustainable water supply management. Sustainability 12(3):1000
Pillkahn U (2008) Using trends and scenarios as tools for strategy development: shaping the future of your enterprise. John Wiley & Sons
Poli R (2017) Introducing anticipation. In: Handbook of anticipation. Springer, pp 1–14
PopulationPyramid, Iran (Islamic Republic of), population in 2050. 10 Feb 2023. https://www.populationpyramid.net/iran-islamic-republic-of/2050/
Riffat S, Powell R, Aydin D (2016) Future cities and environmental sustainability. Future Cities Environ 2(1):1–23
Saadatjoo P, Badamchizadeh P, Mahdavinejad M (2023) Towards the new generation of courtyard buildings as a healthy living concept for the post-pandemic era. Sustain Cities Soc 104726
Saaty TL, De Paola P (2017) Rethinking design and urban planning for the cities of the future. Buildings 7(3):76
Saijo T (2020) Future design: bequeathing sustainable natural environments and sustainable societies to future generations. Sustainability 12(16):6467
Shah H (2023) Beyond smart: How ICT is enabling sustainable cities of the future. Sustainability 15(16):12381
Sharifi A, Murayama A (2013) Changes in the traditional urban form and the social sustainability of contemporary cities: a case study of Iranian cities. Habitat Int 38:126–134

Shichiyakh RA, Klyuchnikov DA, Balashova SP, Novoselov SN, Novosyolova NN (2016) Smart city as the basic construct of the socio-economic development of territories. Int J Econ Financ Issues 6(1):157–162

Shirazi MR, Falahat S (2015) Urban oasis, towards a prototype for sustainable urbanity: the case of Tehran. Int J Soc Syst Sci 7(4):334–351

Shirzadnia Z, Goharian A, Mahdavinejad M (2023) Designerly approach to skylight configuration based on daylight performance: toward a novel optimization process. Energy Buildings 286:112970

Timilsina RR, Nakagawa Y, Kotani K (2020) Exploring the possibility of linking and incorporating future design in backcasting and scenario planning. Sustainability 12(23):9907

UNIDO (United Nations Industrial Development Organization) (2016) UNIDO sustainable cities initiative, hubs of innovation, low carbon, industrialization and climate action, Austria

Uwasu M, Kishita Y, Hara K, Nomaguchi Y (2020) Citizen-participatory scenario design methodology with future design approach: a case study of visioning of a low-carbon society in Suita City, Japan. Sustainability 12(11):4746

Visone S (2020) Eco-Smart governance in the transformation processes of future cities. Masdar city: a model for sustainable cities. Sustain Mediterr Constr Land Cul Res Technol 207–211

Wickenberg B, McCormick K, Olsson JA (2021) Advancing the implementation of nature-based solutions in cities: a review of frameworks. Environ Sci Policy 125:44–53

Part IV
Building Anticipatory Systems in Urban Sectors in Mexico

Chapter 8
Imagining Asbestos-Free Futures—Participatory Mapping as Elements to Anticipate Transitions Based on Hope

Nora A. Morales Zaragoza and Laura E. Quiroz Rosas

Abstract This article reflects on findings about co-creating visions of futures around an asbestos-exposed community in the periphery of Mexico City. It highlights the importance of addressing complex problem approaches from the Transition Design approach in the context of social-ecological systems. We emphasize the need to generate alternative scenarios from the locals that challenge traditional normative urban futures. Our methodological approach was based on interviews and prospective participatory mapping workshops with community inhabitants, who generated visions that reflect their anticipatory assumptions and social dynamics around the problem, allowing them to understand the situation in the present better and to use the future as an asset to promote different levels of interventions considering the spatial dimension. The concept of hope worked as a force reinforced by popular communities' narratives that impulse action toward imagined desirable visions. Finally, we reflect on the need to incorporate spatio-temporal analysis of desires and hopes with the community's inhabitants and other actors to generate comprehensive territorial interventions that contemplate urban-environmental regeneration in broader time frames.

Keywords Socio-ecological systems · Asbestos exposure · Transition design · Participatory cartography · Hope and anticipation of futures

---

N. A. Morales Zaragoza (✉)
Department of Design Theory and Processes, Universidad Autónoma Metropolitana-Cuajimalpa, Mexico City, Mexico
e-mail: nmorales@cua.uam.mx

L. E. Quiroz Rosas
Department of Social Sciences, Universidad Autónoma Metropolitana-Cuajimalpa, Mexico City, Mexico
e-mail: lquiroz@cua.uam.mx

S. Gonzalez-Arellano and B. Gandlgruber (eds.), *Cities as Anticipatory Systems*, The Urban Book Series, https://doi.org/10.1007/978-3-031-78162-9_8

8.1 Introduction

This article aims to communicate findings related to the development of the capacity to imagine futures on an empirical case of participatory action research involving an asbestos-exposed community in the periphery of Mexico City. From a socio-ecological systems approach, the collective creation of visions from the future within a diversity of actors is crucial to understand complex problems and catalyze change.

Our methodology involved previous interviews among stakeholders and experts from various sectors and a participatory mapping workshop with community inhabitants where they imagined alternative desirable futures related to the problem of asbestos. The first phase of the project, resulted in four preliminary scenarios that confronted views among stakeholders, specially exposing their own anticipatory assumptions and ideas about progress, human development, personal fulfillment, among other aspects. The main confrontation we observed signaled two main perspectives of knowledge, a more deterministic one, that organized human agency toward a more feasible future following the logic of the industrial city, (usually coming from experts and researchers) and a hopeful one, that envisioned a full recovery of health and wellness pointed toward an urban regeneration environment (from the local community).

Our hypothesis considers that the incorporation of the spatio-temporal dimension and the desires and hope from the people participating in the workshop, generated two novel future scenarios reflecting emerging anticipatory assumptions and revealed the social dynamics surrounding the problem, thus allowing a better understanding of the present situation to reconcile the paradigm of anticipating the future and promoting intervention in the present.

The first section of this chapter explains the concept of socio-ecological systems (SES) and Transition Design (TD) approaches, which address the need to generate gradual rather than radical changes in the behavior of society and its relationship with the environment, considering long-term visions toward urban transformations and sustainability. Then we explain the concepts of "transition" and "transformation" which have been central to a variety of current interdisciplinary discourses including some stated in this book, that are concerned on how change manifests and the need to understand dynamic interactions among diverse individuals and communities (see González-Arellano in Chap. 2).

We take both approaches for the action research project of asbestos exposure within an interdisciplinary group of researchers and local participants in order to obtain insight for transformative spaces by creating novel visions of the future and explore how these narratives could create emergent opportunities for change.

The second part reflects upon the current global problem of the asbestos industry, and the impact that it has caused to humanity at an urban scale. We identify some initiatives that the European Commission Act is pursuing to better protect people and the environment from asbestos and ensure an asbestos-free future, which reveals some conscious use of the future and the explicit application of anticipation connecting to future studies. Then we present our case study at San Pedro Barrientos' (SPB) community in Mexico City, and how the integration of prospective participatory mapping techniques worked to generate future scenarios.

In the third part, we analyze the two emerging scenarios from the perspective of hope and its relation to anticipation and the construction of future literacy capacities in the community, explaining how they connect with the notion of urban regeneration (Manzini and Jégou 2003). We also reflect upon anticipatory processes emerging and the type of logic of action and transformative capacity that participants had from the SPB case.

Finally, we conclude that more exploratory approaches involving visual socio-spatial methods are needed to co-create future visions with a diverse group of actors in the context of socio-ecological systems, which could contribute to creating awareness and scale-up actions toward complex global environmental problems.

8.2 Socio-ecological Systems and Transition

The global crises and problems caused by the global environmental change that characterizes the Anthropocene (Steffen et al. 2011) have stimulated interest in science, policy, and society for change from a systems approach (Folke 2006).

This new geological era presents essential environmental and social challenges, such as the depletion of natural resources and increasing inequality and social vulnerability, which require research and innovative approaches to understand the dynamic interactions between ecosystems and human societies (Becker 2012). Socio-ecological systems (SES) approach refers to complex adaptive systems that require a holistic understanding and an interdisciplinary approach to study and address their challenges (Preiser et al. 2018).

Despite some academic efforts to integrate complex adaptive systems into SES research, such as cities, there has been limited reflection to identify relevant methods and practical approaches to study SES, specially about intentionally guiding transition toward more sustainable infrastructures and economies. Given this lack of clarity, a common thread is needed to address the threats we face in an equitable, fair, and sustainable manner that recognizes the need to *transition* and *transform ourselves* in order to create fundamentally different social-ecological interactions at all levels that reduce inequality and social vulnerability (Gunderson and Holling 2012; Walker et al. 2004; Folke et al. 2010; O'Brien 2022).

The terms "*transition*" and "*transformation*" have emerged in political and scientific discourses, which have been used to refer to radical, large-scale, nonlinear, and structural changes in complex adaptive systems to generate a sustainable society (Feola 2015; Patterson et al. 2017). Both concepts provide nuanced perspectives on how to describe, interpret, and support social change. Their differences are due, on the one hand, to their etymological origins and, on the other hand, to the use they have been given by different research communities interested in *transition* or *transformation* (Hölscher et al. 2018).

Transition has been used mainly to analyze changes in specific social subsystems (energy, mobility, cities), maintaining a systemic view between technical, environmental, and institutional interactions (Loorbach et al. 2017). *Transformation,* on the other hand, is more commonly applied to refer to large-scale changes, with short and long-term actions that can be global, national, or local and involve the interaction of the components of the socio-ecological system, integrating concepts such as resilience and planetary boundaries (Folke et al. 2010). In the following (Table 8.1), some premises of coincidences and differences between both concepts are presented, emphasizing the need to apply integral approaches for their effective implementation at different scales in the system, from normative or regulatory futures, usually taken by formal agreements between governments, plans, or laws that require compliance with certain behavioral patterns as in the case of the Global Goals for Sustainable Development (González-Arellano 2022). As well as local and even individual initiatives within existing conditions that opposed imposed processes and perspectives.

In addition to the operational differences outlined above, this paper used methodologies drawn from both concepts to expose the case study: transformational spheres (O'Brien and Sygna 2013), social cartography (Diez Tetamanti and Chanampa 2016) and Transition Design (Irwin et al. 2015).

Table 8.1 Comparison between applications of terms of "*transition*" and "*transformation*" concepts adapted from Hölscher et al. (2018)

Scale of change	Transition	Transformation
System approach	Complex adaptive systems	
	Social, institutional, and technological change in social subsystems	Processes of large-scale social change (global, regional, local)
Dynamics and processes	Identifies patterns and mechanisms of complex systems, including dependency, emergence, and thresholds	
	Describes how nonlinear change occurs, focusing on the dialectic between limits and opportunities	Attempts to explain emerging patterns of change and the impact of outcomes
Regulations	Change is questionable and may be desirable or undesirable	
	Results focus on the change from unsustainable to sustainable system status	The results focus on creating safe and fair operating spaces, avoiding undesirable changes in the system
Agency and Governance	Multi-stakeholder process that enables innovation, learning, collaboration, and knowledge integration	
	Development of disruptive interventions to support sustainability transitions	Respond to the implications of change (e.g., risks, vulnerabilities); individual motivations and values that support transformations

8.3 Transition Design and Future Studies

The TD framework developed by Irwin et al. (2015, 2020) recognizes that we are living in "transitional times" and takes as its central premise the need to catalyze social transitions toward more sustainable and just futures, with the understanding that design has a key role to play in facilitating the performance of these transitions. It is based on a heuristic structure around four interrelated areas with some unique features. This design approach is linked to a growing awareness of a systemic problem and analysis with multi-stakeholder involvement and a clear orientation toward longer time horizons explicitly informed by visions of sustainable futures (Escobar 2016).

After the understanding of the problem has been explored with different participants, the phase of co-creation common preferred futures follows by using *backcasting,* a methodology used in the future studies, developed by Karl Dreborg (1996), that focuses on imagining and planning for desirable and sustainable futures, and then working backward to identify the actions needed to achieve those futures. Unlike forecasting, which focuses on predicting the future based on past and present trends, backcasting begins with a vision of a preferred future and then analyzes how to get to that future from the present.

The collective generation of those visions in long-term futures forces the stakeholders to identify the more problematic structures in shaping their pathway to navigate transition. Some articles explain the adaptation of these techniques in various projects such as agroecological transition (Morales Zaragoza et al. 2021), biocultural change of cancer (Valadez-Blanco et al. 2020), water resilience (Morales and Barraza 2020), and sustainable tourism in Latin American contexts.

In the Asbestos case, while four scenarios (described ahead) were already developed following the DT approach, we emphasize the use of the "Prospective Social Mapping Workshop," conducted only with local community participants, which generated two novel future scenarios, and their emerging anticipatory assumptions that unveiled social dynamics of the local community surrounding the problem, this allowed to better understand the problem by integrating a more holistic view of the factors involved and promote different scales of interventions for the project.

The next sections we stress the use of this participatory mapping role in the construction of future visions using local participants' own imagination and suggest something that could be added to the DT model. We would add a complementary analysis from a socio-ecological system perspective by incorporating the notion of hope and well-being as anticipatory forces in relation to a vision of regenerative and the eco-sanitary urban future. We draw this notion, from future studies literature and particularly the work of anthropologist Rebecca Bryant, who incorporates an affective and emotional dimension to the concept of anticipation and hope, drawing a distinction to think about it relation to the past; she suggests that a thick connection gives us insight into what she calls *teleologies* of action.

8.4 The Global Problem of the Asbestos Industry and Its Impact on Health and the Environment

Asbestos has always been a highly dangerous, cancer-causing substance, the World Health Organization considers it a hazardous and deadly material affecting thousands of people worldwide (World Health Organization 2015). Despite being confirmed as a highly dangerous carcinogenic substance and banned among several countries, asbestos fibers are still present in the urban environment and are responsible for many preventable deaths around the world (European Commission 2022) as they have been used for decades, mainly in the construction industry. Ideas of globalization and development have had important effects in supporting commercial strategies of the multinational asbestos industries, introducing them to emerging economies. Even governments have been the leading promoters of this material due to the demand for economical and heat-resistant building materials for the development of cities despite these well-known dangers. This practice has inevitably been used worldwide in many public and residential buildings (Wallis et al. 2020).

There is epidemiological evidence confirming that all types of asbestos and its fibers are implicated in the causes and development of various diseases (such as mesothelioma, lung cancer, asbestosis, pleural thickening, and pulmonary plaque) and premature deaths. Another surprising aspect is that 30 years after the asbestos boom in the 1970s, the true extent of this hidden danger is exposed (Wallis et al. 2020).

Although many countries have achieved a total ban on the use and manufacture of asbestos, unfortunately, the history of regulation shows that efforts are insufficient, the processes are slow, and sometimes corrupt to mask the risk control of the material itself with a false promise (Hadley and Rennell 2013, p. 36). In that sense, the commercialization and use of this material have prevailed over human, environmental, and labor rights. Despite advances in the regulation and prohibition of the use and manufacture of asbestos products in the countries of the Global North (Canada, the United States, the European Union, and Japan), the countries of the Global South in Latin America, Asia, and Africa, including Mexico, continue to lag in this area.

Taking the case of the European Union, even though all forms of asbestos have been banned since 2005, it is still present in older buildings and represents a threat to health when asbestos-containing materials are altered, and fibers are released and inhaled, for example, during renovations. Therefore, the European Commission (EC) for the Prevention and Control of Cancer[1] (European Commission 2022) has generated a program in 2005 that seeks to modify labor policies to protect the environment and people from asbestos exposure and it is considering how to ensure an asbestos-free future. It addresses three advocacy priorities in its objectives:

- The search for improved diagnosis and treatment of diseases caused by asbestos.
- Through the detection and safe removal of installed asbestos.

[1] The commission comprises member states of the European Union EU and workers and employers from the asbestos' industry.

- To treat waste containing the material.

8.5 Environmental Issues and Future Studies

These types of environmental issues initiatives concern a particular interest in futures studies, which, according to Miller (2018), leads us to explore the concept of "futures literacy" (FL) as a tool for understand collective anticipatory processes and systems that project actions within the imminent and rapid urbanization of the world, which primarily can be used to innovate in the form of urban organization and foster decision-making within a wide diversity of problems and risks.

Understanding the nature of the future of environmental, human health, and toxicity is related to the development of the capacity to imagine actions to catalyze change today. The example of this European plan demonstrates how simply thinking about what might happen with the asbestos installed in buildings and ask ourselves how it will directly affect the way society will live in the future can lead to developing anticipatory skills and improving the design of collaborative processes (Miller 2018, p. 8) this complex system, can support the capacity building of people affected by a problem to "break free" (Nussbaum 2013), and transform their reality.

The main rationale of "Futures Literacy Framework" (FLF) emerges as a foundational theory for exploring how people perform anticipatory awareness activities and provides an analytical ground for mapping these activities in search of solutions to global problems that no longer work with current systems and leads us to explore new ways of approaching what we want to change about the dominant regime. The three aspects of the free-asbestos program proposal show a clear line of planning and preparation as an instrument to get to a future goal, but how about new kinds of futures for different purposes. Is there a way to look for more alternatives?

8.6 Hope and Anticipation in the Construction of Futures

In the book "*The Anthropology of the Future*," Rebecca Bryant and Daniel Knight propose an innovative approach based on six orientations or domains of the future (Anticipation, Expectation, Speculation, Potential, Hope, and Destiny) which are interrelated and show ways the future may orient the present, and shape the way people experience and think about the future and how these can affect the present (Bryant and Knight 2019). Each orientation can be analyzed from social discourses and practices in different societies.

We picked anticipation and hope from Bryant's anthropological perspective, because she examines how humans imagine, construct, and relate to the future, both individually and collectively, suggesting that anticipation and hope are not only mental states, but also social practices rooted in specific cultural contexts.

Both orientations are fundamental components of the human condition, shaping our actions and perceptions of the world around us. Bryant theory explores how these notions influence various aspects of life, such as urban planning, survival strategies, politics, and religion. She provides a valuable theoretical basis for understanding how we imagine and construct the future.

For Bryant, hope is a dynamic force that drives action and imagination toward the future, shaped by specific social and cultural contexts and with important ethical and political implications (Bryant and Knight 2019). This orientation holds on to the belief that the future will be better, as it is shaped by people's present experiences and the possibility of returning to specific standards or ways of life that motivate them to act in the present.

Anticipation is seen as a form of future orientation that drives people's action and imagination. This orientation to the future is seen as a fundamental human capacity that involves preparing for the future, considering different possibilities and facing the uncertainty and anxiety that accompany the expectation of the unknown and involves these key features:

1. Preparation and planning: Anticipation involves the ability to anticipate future events and prepare for them. This may include planning concrete actions to achieve goals or avoid risks.
2. Expectations and possibilities: Anticipation involves having expectations about what might happen in the future and considering different possibilities and scenarios. This ability to imagine alternative futures is fundamental to anticipation.
3. Uncertainty and anxiety: Anticipation can be accompanied by uncertainty and anxiety about what the future holds. Bryant recognizes that anticipation can generate both positive and negative emotions, depending on individual circumstances and expectations.
4. Cultural and social context: Like hope, anticipation is shaped by the social and cultural context in which it is found. The ways in which people anticipate the future may be influenced by cultural norms, social expectations, and past experiences.

On the other hand, a characteristic of hope is the collective sense of experiencing a critical juncture in which something is happening, usually accompanied by a time of previous austerity or promises that have not been fulfilled. Hope begins as something lost with some restorative or prideful power taken from the general public or nation (Bryant and Knight 2019). In that sense, it can be compared to our case since the perception of having lost one's living environment due to the intervention of the industry specifically related to asbestos contamination, but extrapolated to the current presence of other industries such as cement or the vast warehouses and storage buildings from the speeches of the workshop participants or informants interviewed from the community, denotes a sense of helplessness that plunges them into a state of discouragement as they have no possibility of recovery but at the same time the promise that they can recover the lost standards as citizens or the lost desire to reconfigure themselves as a community.

8.6.1 Hope as a "Latent Moment" of the Future

Taking it further, Ernst Bloch's relates hope to a kind of **anticipatory consciousness** in his book "The Hope Principle" (2007)[2] where he explores utopian futures and describes a moment "yet to be realized," which could also be explained as a kind of "exaggerated anticipation perpetrated by a kind of desirable imaginary reinforced by popular stories, fairy tales, fiction, theater, and other mass media." Bloch, based on Schatzki's notion of "practice-arrangement bundles" that create their time–space structures, tells us that these imaginary (produced, consumed, and shared) functions as a kind of "affective temporal spaces" or *teleoaffects* that influence our future actions. The "*teleoaffections*" refer to the set of accepted images that combine with the technologies of the imagination to produce a kind of "latent moment" that attracts us to "something" indeterminate (Bloch 1995, p. 18).

For Bloch, hope is a future orientation that functions as a bridge that connects the gap between potential and actuality, and its value lies in contemplating the future by allowing us to think about possibilities; it asks us to move away from the focus on the present state of social relations and focus on what they may become.

Both Bloch, Bryant and Knight (2019) relate the domains of hope and potential by stating, "The immanence of not-yet-realized potential becomes a resource of hope," detailing that when a potential has not been realized or is frustrated, there is the possibility of the emergence of hope, providing the "momentum" of mobilization and momentum toward actualizing the future.

One quality of hope worth mentioning is the notion of "perpetual motion," seen as a potential tendency that propels toward something; hope arises in the gap between the potential and the actual and can guide us toward the future even if it is perceived as an insignificant force, it can be built as an unstoppable force. What may begin as a gathering of collective desires can be transformed into seeds of hope that seek to materialize the form of what does not yet exist.

In this sense, we can affirm that the scenarios created by the local participants reflect a sense of collective hope as a community, starting from the moments of daily life emerges an uncertain view of "of what has not yet been conquered, nor has it been completely won."

It is important to reiterate that the reinforcement of the historical potential of the past is a critical aspect of improving the mobilization capacity of the collective. However, the low credibility of the participatory methodologies, the constant time pressure of the project, as well as the lack of openness of the research actors to include aspects that involve the imagination and creativity of the people as drivers of citizen action, and the presence of an overtly political orientation of the project, led to the conclusion of the other social mapping workshops with the community.

If hope is nourished by indeterminacy (or ambiguity, as the case may be), it is also rooted in potentiality and the possibility that it concludes with the impossibility of realization. Therefore, some anthropologists, such as Bloch (2007), suggest that it is

[2] The philosopher's original work was written in 1938 during his exile in the United States of America and later reworked after his return to Germany in 1947.

essential to study it alongside disappointment, a way in which potential is partially actualized.

According to Bloch, the first level of disappointment is that the "here and now" is close to the present. The raw experience that brings us from the drifting dream to another state: that of immediate proximity, where the recognition of reality is more complex and obscure than the "desirable image."

Disappointment is the recognition that the actualization never considers the potential or the paradox of what can happen but the realization "that the dream seemed firmer or, in any case, brighter than its realization" (Bloch 2007).

What builds hope is precisely the idea that ways of life do not necessarily coincide with the existing reality but the surplus potential of future orientation, which provides the motor or movement toward that indeterminate "something."

In his book *The Methods of Hope* (2004), Miyazaki argues that hope is an indeterminate orientation to the future and a method of living. People displaced from their land in 1880 in Fiji continue to seek compensation from the government despite continuous rebuttals and failures. Drawing on ethnographic theory and taking up Bloch's concept of indeterminacy, he criticizes the Greek-derived assumption that humanity is captured by the notion of history as a theological course of truth about the essence of life.

The future does not unfold according to the planned or imminent teleology.[3] The future will always surprise and stimulate us. Indeterminacy is at the heart of the theory of hope as a method.

8.7 The Case of Asbestos Exposure in San Pedro Barrientos

The participatory action project that frames this research work, entitled: "Transdisciplinary approach to the epidemic of pleural malignant mesothelioma due to asbestos exposure in the community of San Pedro Barrientos (SPB)" is funded by the Mexican National Council of Science and Technology and aims to influence polluting processes policies of toxic damage as well as address the socio-environmental impacts associated with sources of natural and anthropogenic origin.

San Pedro Barrientos is located in the north part of the metropolitan area, the municipality of Tlalnepantla de Baz, one of the most industrialized areas in Mexico City. In the early 1940s, one of the leading Asbestos Mexican factories settled in the neighborhood, with the promise of economic development and well-being of their inhabitants (mainly family workers), after 58 years of the final factory departure in 1998. The place was left as a polluted environment and a poor and socially fragmented population of less than 6000 inhabitants survivors of exposed workers families and

[3] In the context of Rebecca Bryant's work, "teleology" refers to the study of ultimate purposes or goals in nature, society, and history; it involves the understanding of how people imagine and strive to achieve certain goals or end states in the future.

an increase in pulmonary malignant mesothelioma cases associated with asbestos exposure (Gopar-Nieto et al. 2016).

Although the project was approved in February 2022, the researchers began to organize and develop a previous diagnosis with experts, community inhabitants, and other stakeholders a year before, in remote mode, due to the confinement of the COVID-19 pandemic that started in Mexico in March 2020. The initial stages of research (following DT methodology), involved academic actors, institutions, activists, and community leaders understanding the complexity of the asbestos problem in context of ecological health from a multilevel perspective. The participants explored historical events that impacted the community and region during the years, such as statistics about the pleural malignant mesothelioma cases of the exposed population, continued with the identification of historical landmarks and community practices and activities that reflect their regular lifestyles until today. After some research based on interviews with experts and secondary reviews that put into context the dimension of this toxic impact, some co-creation workshop sessions were held, in which the interdisciplinary group came up with four preferred desirable future scenarios, which we describe next:

Scenario 1 "Recovery of green spaces and isolation of the endangered zone." This scenario situates San Pedro Barrientos community in a medium-term future, a space where green areas are recovered with trees and plants among residential areas, the areas exposed to contaminated dust have been identified and isolated by a thick wall from the people. The inhabitants have improved risk management measures due to well-informed programs invested in screening and early diagnoses held by the government and citizens. Local employers and businesses are organized in a commission responsible for risk management and protecting inhabitants against exposure.

Scenario 2 "(S.O.S) A call for help to the urban authorities." This scenario captures a tone of urgency within the problem of contamination and asbestos exposure, covering a short-term vision. It is an urgent call for "rescue actions" from the authorities for the victims of asbestos exposure, similar to situations of natural disasters (flood, fire, or earthquake), where people affected are quickly "extracted" from the site and relocated to safeguard temporal areas. This scenario also contemplates support for victims of asbestos-related diseases.

Scenario 3 "Total ban of asbestos in Mexico." This scenario, also situated in a short to medium temporal span, shows a Court government legislative congress scene where Mexico has banned all uses and products of asbestos. This regulatory-oriented future does not necessarily contemplate detailed implementation law issues, such as the cessation of importation and use or the agreements of "controlled use" or "asbestos substitute materials" still being used in the construction industry. It does not consider funding available to support health prevention or removal of asbestos materials in buildings and other urban infrastructure.

Scenario 4. For a future free of environmental pollution. This scenario envisions a long-term future where the inhabitants of San Pedro can walk freely through the streets and engage in sports and recreational activities without worrying about air pollution or the environment. The former factory and industry site is now a center for asbestos research.

These scenarios caused some concerns and discomfort among the scientific and academic researchers of the project, because they were considered improbable and not feasible for achieving solution initiatives from the locals, some expressed their concerns and irresponsibility to create false hopes or expectations for the inhabitants. This reaction was in part natural due to the lack of understanding of the backcasting methodology and our limited experience to deal with experimentation and imagination processes to generate new innovations and visions of the future.

According to Riel Miller (2018), the above situation is a common aspect that arises when people "use futures," and confronts two types of anticipatory systems; one that he calls **anticipation for the future** (AfF), which refers to a type of plan or desired future to which people bet based on the probability of happening (which refers to the researchers assumption based on their current information and knowledge perceived as improbable to realize). The other type is **anticipation for emergence (AfE)** which contemplates emergence as a kind of "non-future" or alternative option to the dominant perspective, here the important thing is not having a goal to reach or achieve plan or preparation, but to be able to imagine something beyond-now, to create a disposable construct, which need not be limited by probability or desirability. As Miller puts it, (AfE) is about a strange future that allows us to feel and make sense of present aspects with some novelty that will always be overshadowed by anticipation for futures.

At this point of the project, we wanted to pursue other ways to incorporate the local community in the generation of desirable scenarios so we conducted a "Prospective Social Mapping Workshop" following Diez Tetamanti and Chanampa work on social mapping (2016).

8.7.1 "Social Cartography" Workshop with the Inhabitants

Participate social cartography is a collection of methods, attitudes, and beliefs that allow the people of a community to express and analyze the conditions of the reality in which they live through the collective production of maps, with the objective of planning, taking actions, monitoring, and evaluating their own results in a territory (Chambers 2006). The use of this tool allows, among other things, to involve and empower the community in decision-making, in addition to raising awareness about a conflict. This mapping method also adheres to the proposal Deleuze (2017) who argues that subject and object are not divorced, and the communion of hands amalgamates in the process of shared and mobile experience. This process of experience places both the researcher and the object of research on a common plane where the experience is the process that unifies them in the dialogue and exchange of information, by which both the research subject and the researched subject share the experience.

Fig. 8.1 Three participants in the prospective social mapping workshop in San Pedro Barrientos

The prospective social mapping workshop[4] was held on September 22, 2022, at San Pedro's Public High-school (Fig. 8.1) to complement the future visions generated in the research project and produce a collective map that communicates the experiences, activities, and perceptions around the daily spaces of the inhabitants of the community, encouraging reflection on the legacy of the past, current concerns, and needs for future generations.

The workshop was attended by seven participants from affected families in the area. We began by giving a guided presentation of the activity's objectives in which we introduced the research initiative, the local demand for the project, and the need to coordinate other intervention events in the territory.

Three key questions were posed: How do people live in SPB today? How did they live in the past? What are the main concerns of the community members? What challenges do they face, and what desirable futures do they aspire to?

[4] Diez Tetamanti and Chanampa conceive social cartography as a productive process that involves a succession of events and phases around the conjunction of collective representations and transformations from the production of new spatial senses (Diez Tetamanti and Chanampa 2016).

8.7.2 Workshop Structure

Stage 1. Base map Construction. Location of prominent spaces in the community (public spaces, streets, schools, and factory areas).
Stage 2. Spaces of the past to be intervened (existing objects, current conflicts, social practices, and social relations).
Stage 3. The Envision future. The imagination of a desired future, incorporating new objects, conflicts, and practices, and looking for relations between objects and society.

8.7.3 Workshop Key Findings

The workshop fostered an atmosphere of dialogue and information exchange between participants and researchers, resulting in three maps (Fig. 8.2) that gave feedback for two novel future scenarios and five key findings for the project:

K1. Political segmentation. Inhabitants used to join political parties because they hoped to influence their community and create a reputation of leadership; however, after corruption and deceit, they became divided and tried to stay away from politics; leaders of the colony looked out for their benefit.
K2. Decrease in recreational and convivial places. Recreational areas in the neighborhood, such as soccer fields and sports fields (which were promoted by the factory in the past), have been privatized, generating an economic barrier to the organization of collective sports activities such as soccer. Today, it is easier and cheaper to organize activities outside the colony. We highlight the reflection of one participant: "Industry and development have caused a series of damages to our community; they have taken away recreational areas, where we used to walk on Sundays or play sports, the places where we used to sit and live together, are now poisoned places."
K3. Localized contamination. Most of the workshop participants identify the contaminated areas near the location of the old asbestos factory, and others, such as those near the roads, speak of the disappearance of entire families due to asbestos-related diseases. They also mention a change from the establishment of industries such as the cement industry (which boosts the labor economy but continues to be polluting) to storage warehouses that bring greater insecurity because they are perceived as abandoned areas or do not allow the flow or connection of the community. A key point of community concern is the workers' union building (constructed entirely of asbestos), which some workers hope to sell. The soil contamination study with the presence of the contaminant corroborated all of these perceptions.
K4. Change from pedestrian to vehicular mode, affecting accessibility and safety. The increase of fast roads is a common and naturalized phenomenon of urban growth, causing isolation and connectivity between neighborhoods and

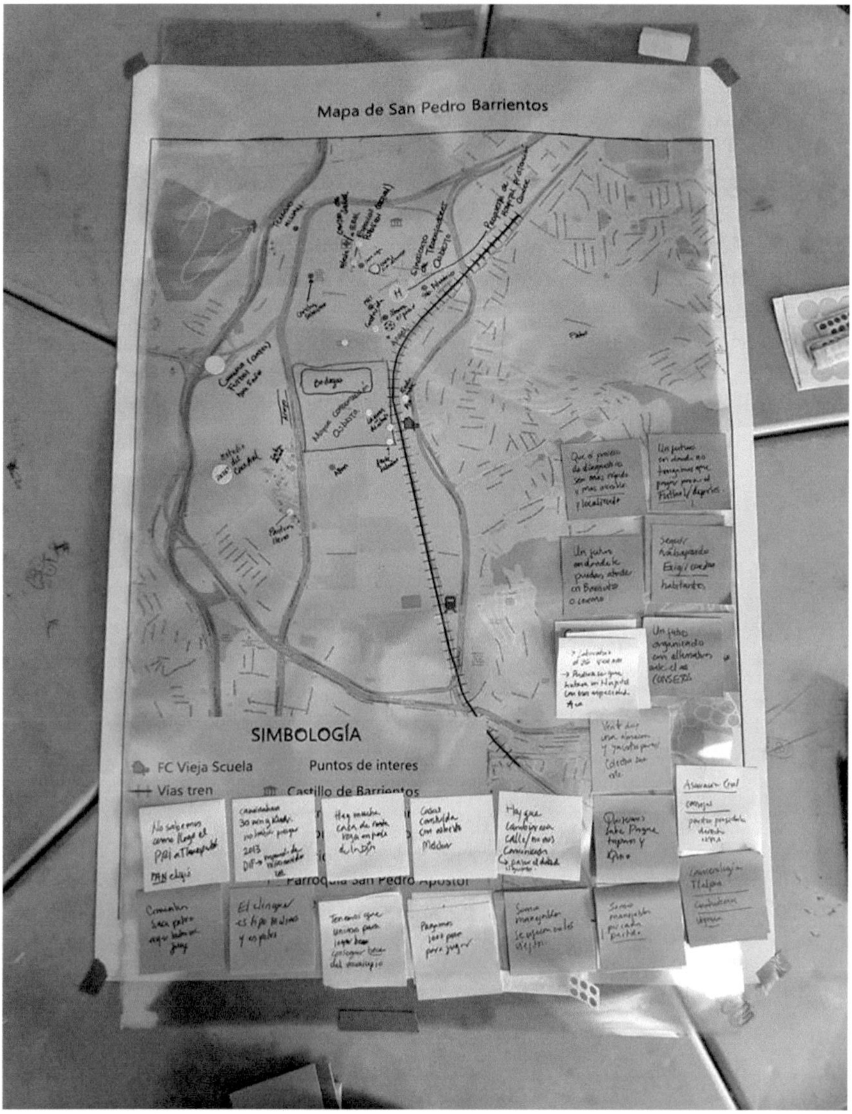

Fig. 8.2 One of the maps resulted from the activities of SPB's Workshop. Depicted as one of the main proposals for the future of SPR, we can see The General Cancer Hospital in the same place as the Workers Union building

populations; the large roads become unbreakable borders, practically preventing walkability and increasing the recurrence of road accidents. Workshop participants were concerned about the increase in the number of people being run over. Participants recurrently talked about their concerns about walking safely in streets either to do an errand to a grocery store or the market or to walk their children back from school (some participants noted that those trials have increased in distance since there were fewer schools for SPB children). They point out that neighbors tend to remove trees and build fences, diminishing accessibility and coexistence. Inhabitants tend to normalize the confinement state (reinforced by the pandemic) but also react to the lack of security and the increase of drug dealing in the area; this phenomenon also affects the low walkability in the streets and the problems of external migration and the increase of vehicular load.

K5. Preferential health access for the population. It is based on a story about the experience of a young participant in the workshop who told how he was diagnosed with prostate cancer and followed his treatment, going every 15 days to the cancer hospitals in the south of the city. Thanks to this, he was in remission. However, he had just received the news of lung cancer. He had decided to give up social security to accelerate his treatment and to look for alternative economic means impacting the family economy since this "new affectation" forced him to return to the beginning of a very complex and delayed process that represented the difference between life and death. They also commented on the problem of shortage and cost of medicines that forces them to access the black market.

8.7.4 Relationship Between Workshop Key Findings and Visions of the Future

Table 8.2 synthesizes the four scenarios previously generated with the findings of the prospective social mapping workshop, which reveal different values and shared beliefs among participants and researchers (Chevalier and Buckles 2019).

The scenarios generated by the workshop participants were taught from everyday life experiences and the incorporation of the spatial dimension, which brought a tangible component that encouraged participants to project an image for the future, allowing people to express themselves and communicate their ideas, fears, and hopes.

We could say that this workshop led facilitators and participants shared a collective intelligence process that uses the future differently. When a lack of knowledge of the future is assumed, a certain equity of ignorance forces reliance on the imagination to invent images of tomorrow. This methodology helped to gather evidence on how "people use the future" and, at the same time, citing Latour (1999), allowed us to learn from the inhabitants without imposing an a priori definition of the world and its capabilities.

Table 8.2 Scenarios linked to key findings from the social cartography workshop

Scenario	Findings to address/retake or reinforce	
S1. Recovery of green spaces and recreation	K2 K4	*Decrease in recreational venues.* This future is linked explicitly to the recovery of sports and recreational activities that are valuable and meaningful to the community (such as soccer) that strengthen social cohesion and reinforce the sense of identity *Change from pedestrian to vehicular mode, accessibility, and safety issues*
S2. (S.O.S) A call for help to urban authorities		
S3. Total ban on asbestos in Mexico		
S4. For a future free of environmental pollution	K3	*Localized contamination.* Where **remediation** and **detoxification of sites contaminated** by the mineral have been achieved, this aspect has caused greater concern and rejection by the scientific actors of the project since they understand the low feasibility and the long-term frame involved in such an initiative. Few case studies from industrialized and island countries demonstrate the potential of bioremediation or regeneration of massive amounts of soil contaminated by asbestos. Their results are not encouraging as they are influenced by cost and land availability (Wallis et al. 2020) Despite having few feasibility options in terms of initiatives, it reinforces Meadows' (2008) argument as it implies a rethinking of urban collective infrastructure that is still on the distant horizon

(continued)

Table 8.2 (continued)

Scenario	Findings to address/retake or reinforce	
NEW SCENARIO S5. Inhabitants with expertise in regeneration and removal of installed asbestos	K1 K3	*Political segmentation.* A workshop key finding could be defined as "**Community vs. political support,**" participants imagined a future that responds to the need to explore **alternative forms of organization** that do not depend on political affiliation to a particular party and the need to have alternative leaderships of socially committed citizens, who have recognition by the community and focus on everyday demands of welfare and health instead of individual benefit When exploring the exercise of imagination toward more distant temporal futures, common representations of the future emerge framed with the question: What would have to happen to the inhabitants of San Pedro Barrientos to reach this scenario? An interesting concept is related to the education and training of the inhabitants themselves regarding asbestos, first by establishing an asbestos museum and later by becoming **experts in remediation and care**; this new alternative opens up occupational and economic opportunities for the community (we can also extend this to the concept of citizen science)
NEW SCENARIO S6. The National Medical Cancer Center at SPB	K5	This scenario arises from the question posed in the mapping workshop from finding **H5**. *Preferential health access for the population.* In a few decades, the population of San Pedro Barrientos has changed from being an industrial corridor to a health complex specializing in cancerology; the first hospital was built in the location of the Asbestos Union building itself that the government bought from the workers and took care of removing the asbestos installed in a safe way for the population

8.8 Conclusions

We have addressed the SES and Transition Design (TD) approaches on describing, interpreting, and supporting desirable nonlinear social changes and complexity to facilitate a collaborative exploration of future scenarios in regard to a free-asbestos future. While doing so, we were able to identify diverse ways of practicing anticipation at the individual and collective levels.

To answer the question in regard to agency improvement and collective anticipation capacities in regard to the urban group of actors in the asbestos case, we were able to identify two kinds of anticipation: the first stage contemplates more the idea of anticipation as an activity that reinforces planning and programming toward the desired future. In fact, the discomfort reactions of the group of actors evidenced internalized anticipatory systems that are more normative and anchored to a way of creating knowledge. This way is difficult to modify at a conscious level, since the agency is framed in a series of decision-making and responsibilities that go according to a vision of the future that is preferable as a "goal to reach."

In that sense the first four scenarios generated for the research project helped us better understand the various social dynamics in the community around the problem and how to reconcile the present situation within different paradigms of anticipating the future. As Reed et al. stays, (2018): "Future scenarios generation processes go hand in hand with modes of engagement with stakeholders and span a continuum from information to consultation to co-production of transformative visions."

However, we can highlight that the participatory mapping workshop and the identification of hope as an orientation toward the future, gave an account of a different way of understanding anticipation. This second type of anticipation that Miller describes as: anticipation for emergence refers to making sense of aspects of the present, and the emergence of novelty. This type of anticipation relates to the ability to engage in improvisation and the acquisition of knowledge necessary to make sense of complexity including one that it's crucial specificity-unique and ephemeral (Miller 2018).

Based on the former, we could say that the prospective participatory mapping workshop, was a crucial step for the participants from SPB in regard of their literacy future capabilities, because even coming from already socially fragmented community they were able to embrace complexity rather than just lamenting it as some cursed and inescapable situation.

This tool in particular allowed the participants to project two novel desirable future scenarios for the research project. Which situates the potential of the spatial–temporal dimension of the creation of new images for the future and fostered some kind of anticipatory awareness of a potential possible future by allowing the participants to project their expectations.

Although the whole methodological experience was not able to generate concrete actions from the local actors in regard to a free-asbestos future, (that was part of the responsibilities of the research group). Following Miller's description of the Future Literacy Laboratories (FLL) model, which is seen as a continuing action learning process where people engage in a knowledge creation process designed to imagine emergent futures. We were able to identify three emergent capacities:

(1) Participants became explicitly conscious of how the future plays a central role in what they perceive and pay attention to, in the present, even if they do not immediately benefit from it.
(2) The capacity of imagining different futures lets participants be aware of their own capacity to invent and identify underlying assumptions. They became better at rooting their own history and specific socio-economic-cultural context.
(3) Participants started to reassess their perceptions of the present, depictions of the past and aspirations for the future.

Another assertive aspect that we would like to highlight, is the use of Bryant orientation of hope as complementary to anticipation in the asbestos case. "Hope emerges as a value that promises a new prosperity on the horizon"; this new horizon becomes feasible when the possibility of regeneration of the inhabited space becomes plausible from the direct message that reflects people's expectations.

The two new scenarios created by the community summarize a desirable or dreamed future and demonstrate the potential of a future blocked by industry and other actors like authorities discourses among years that reinforced a promise of economic development that has yet to be realized today.

Based on hope, the anticipatory assumptions reveal a potent past, present, and future narrative within local communities that can feed interventions at different system scales, functioning as acupuncture. These imaginaries are the first steps toward collective remediation actions by the community's inhabitants and could potentially transform into an urban regenerative public policy by influencing decision-makers and authorities.

We also agreed with Bryant when she argues that "the future often represents an unknown against which persons struggle to maintain stability clung to particular stories," that is what happened when the research and academic group felt confronted by the desired scenarios generated by the community.

We could conclude that anticipatory capacity is a flexible process that involves various actors at different scales. However, in Latin American cities, the asbestos situation is not an anticipatory remediation system, as it is characterized as a reactive, prohibitive, low agency, and zero regenerative system. Somehow, the generation of these new scenarios (S5) regenerative future and economic alternative and (S6) a cancerology center in SPB opened new possibilities to explore a regenerative urban aspect from top-to-bottom. The above reinforces to Bloch's concept of "dormant potentials," which in a way were triggered by the workshops, and benefit both the individual and the collective in the project.

Despite some challenges and barriers related to improving a more reliable structure that overcome restrictions and limitations of working with communities and integrating emotional aspects in foresight practice. We propose the need to explore tools that allow a diversity of actors within the three transformational spheres to imagine futures.

We believe our work adds up to the analysis of how societies and individuals conceive and work toward the future, and how these visions of the future could influence their actions and decisions in the present. We stress the importance of developing tools that facilitate collective imagination that could lead society to change or generate insights about different values and aspirations that come into play in societal systems at different scales.

Finally, we posit that the importance of developing anticipation capacity is perhaps the ultimate promise of future studies practice, which not only treasures and preserves expert findings and tools but also provides user-friendly generative tools that facilitate the expression of a larger population. Stuart Candy's suggestion of using experiential futures and, specifically, playful activities may help enhance this promise of democratizing tools (Candy 2018).

References

Becker E (2012) Social-ecological systems as epistemic objects. pp 37–59

Bloch E (1995) The principle of hope, vol 1 (N. Plaice, Ed.; 1. MIT paperback ed). MIT Press

Bloch E (2007) The hope principle/Bloch, Ernst. 1: The hope principle [1]: Written in the United States between 1938 and 1947. Revised in 1953 and 1959. Edited by Francisco Serra. Translation by Felipe González Vicén. Trotta

Bryant R, Knight DM (2019) The anthropology of the future. https://doi.org/10.1017/9781108378277

Candy S (2018) Gaming futures literacy. The thing from the future. In: Miller R (ed) Transforming the future. Anticipation in the 21st century, 1st edn. Routledge, pp 233–246. Gaming Futures Literacy|7|The Thing from the Future|Stuart Candy

Chambers R (2006) Participatory mapping and geographic information systems: whose map? Who is empowered and who's disempowered? Who gains and who loses? E J Info Syst Dev Countries 25(1):1–11. https://doi.org/10.1002/j.1681-4835.2006.tb00163.x

Chevalier JM, Buckles DJ (2019) Participatory action research: theory and methods for engaged inquiry, 2nd ed. Routledge. https://doi.org/10.4324/9781351033268

Deleuze G (2017) The anti-oedipus: capitalism and schizophrenia (1st ed, 1985, thirteenth printing). Paidós

Diez Tetamanti JM, Chanampa ME (2016) Perspectivas de la Cartografía Social, experiencias entre extensión, investigación e intervención social. Revista +E Version Digital 6:84–94

Dreborg KH (1996) Essence of backcasting. Futures 28(9): 813–828. https://doi.org/10.1016/S0016-3287(96)00044-4

Escobar A (2016) Autonomía y diseño: La realización de lo comunal (1st ed in Spanish). Editorial Universidad del Cauca

European Commission (2022) News: commission acts to better protect people from asbestos and ensure an asbestos-free future (2022). European Commission Site. Employment, Social Affairs & Inclusion. Retrieved 5 Nov 2023, from https://ec.europa.eu/social/main.jsp?langId=en&catId=89&furtherNews=yes&newsId=10418#navItem-relatedDocuments

Feola G (2015) Societal transformation in response to global environmental change: a review of emerging concepts. Ambio 44(5):376–390. https://doi.org/10.1007/s13280-014-0582-z

Folke C (2006) Resilience: the emergence of a perspective for social-ecological systems analyses. Glob Environ Chang 16(3):253–267. https://doi.org/10.1016/j.gloenvcha.2006.04.002

Folke C, Carpenter SR, Walker B, Scheffer M, Chapin T, Rockström J (2010) Resilience thinking: integrating resilience, adaptability, and transformability. Ecol Soc 15(4)

Gopar-Nieto R, Cabello-López A, Juárez-Pérez CA, Haro-García LC, Jiménez-Ramírez C, Aguilar-Madrid G (2016) Actualización sobre la epidemiología, fisiopatología, diagnóstico y tratamiento del mesotelioma maligno pleural*. Rev Méd del Inst Mex del Seguro Soc 54(6): 770–776

Gunderson LH, Holling CS (2012) Panarchy: Understanding transformations in human and natural systems. Island Press

González-Arellano S (2022) The future scenarios of cities: an analysis of their institutional construction. Int J Urban Reg Res 46(5):865–873. https://doi.org/10.1111/1468-2427.13103

Hadley B, Rennell T (2013) Asbestos the future risk (D Austin, ed). Iskaboo Publishing Ltd.

Hölscher K, Wittmayer JM, Loorbach D (2018) Transition versus transformation: what's the difference? Environ Innovation Societal Transitions 27:1–3. https://doi.org/10.1016/j.eist.2017.10.007

Irwin T, Kossoff G, Tonkinwise C (2015) Transition design provocation. Des Philos Papers 13(1):3–11

Irwin T, Tonkinwise C, Kossoff G (2020) Transition design: an educational framework for advancing the study and design of sustainable transitions. Cuadernos 105. https://doi.org/10.18682/cdc.vi105.4188

Latour B (1999) On recalling ant. Sociol Rev 47(1_suppl):15–25. https://doi.org/10.1111/j.1467-954X.1999.tb03480.x

Loorbach D, Frantzeskaki N, Avelino F (2017) Sustainability transitions research: transforming science and practice for societal change. Annu Rev Environ Resour 42(1):599–626. https://doi.org/10.1146/annurev-environ-102014-021340

Manzini E, Jégou F (2003) Sustainable everyday: Scenarios of urban life. Triennale di Milano, Milano. Ambiente

Meadows D (2008) Leverage points: places to intervene in a system 1999. The Sustainability Institute, Hartland, Vermont

Miller R (ed) (2018) Transforming the future: anticipation in the 21st century. Routledge

Morales Zaragoza NA, Barraza Gómez EC (2020) El Enfoque emergente del Diseño para la Transición: Acupunturas sociales, identidades y estilos de vida en relación con el problema del agua en la zona patrimonial de Xochimilco. DIS 5: 1–28. https://dis-journal.ibero.mx/index.php/DISJournal/article/view/71

Nussbaum MC (2013) Creating capabilities: the human development approach (First Harvard University Press paperback edition). The Belknap Press of Harvard University Press

O'Brien E (2022) The "Next" effect: When a better future worsens the present. Soc Psychol Personal Sci 13(2): 456–465. https://doi.org/10.1177/19485506211034972

O'Brien K, Sygna L (2013) Responding to climate change: the three spheres of transformation. Proc Conf Transform Changing Clim 16–23

Patterson J, Schulz K, Vervoort J, van der Hel S, Widerberg O, Adler C, Hurlbert M, Anderton K, Sethi M, Barau A (2017) Exploring the governance and politics of transformations towards sustainability. Environ Innovation Soc Transitions 24:1–16. https://doi.org/10.1016/j.eist.2016.09.001

Preiser R, Biggs R, De Vos A, Folke C (2018) Social-ecological systems as complex adaptive systems: organizing principles for advancing research methods and approaches. Ecol Soc 23(4). https://doi.org/10.5751/ES-10558-230446

Reed MS, Vella S, Challies E, De Vente J, Frewer L, Hohenwallner-Ries D, Huber T, Neumann RK, Oughton EA, Sidoli Del Ceno J, Van Delden H (2018) A theory of participation: what makes stakeholder and public engagement in environmental management work? Restor Ecol 26(S1). https://doi.org/10.1111/rec.12541

Steffen W, Persson A, Deutsch L, Zalasiewicz J, Williams M, Richardson K, Crumley C, Crutzen P, Folke C, Gordon L, Molina M, Ramanathan V, Rockström J, Scheffer M, Schellnhuber HJ, Svedin U (2011) The anthropocene: from global change to planetary stewardship. Ambio 40(7):739–761. https://doi.org/10.1007/s13280-011-0185-x

Valadez-Blanco O, Morales-Zaragoza NA, González-García DN (2020) Catalizando cambios sociales y bioculturales sobre el cáncer. Diseño de acciones transdisciplinarias contra la pandemia. Rev de Salud Pública 22(3): 1–10. https://doi.org/10.15446/rsap.v22n3.87147

Walker B, Holling CS, Carpenter SR, Kinzig AP (2004) Resilience, adaptability and transformability in social-ecological systems. Ecol Soc 9(2). https://doi.org/10.5751/ES-00650-090205

Wallis SL, Emmett EA, Hardy R, Casper BB, Blanchon DJ, Testa JR, Menges CW, Gonneau C, Jerolmack DJ, Seiphoori A, Steinhorn G, Berry T-A (2020) Challenging global waste management—bioremediation to detoxify asbestos. Front Environ Sci 8.https://doi.org/10.3389/fenvs.2020.00020

World Health Organization (2015) Chrysotile asbestos. World Health Organization. Retrieved from https://iris.who.int/handle/10665/178803

Zaragoza NAM, Ugarte BM, García CT (2021) Mapeos generativos, herramientas de participación interdisciplinaria para la transición socio-ecológica. Prototipado de iniciativas. Cuad Del Centro de Estud de Diseño y Comun 132. https://doi.org/10.18682/cdc.vi132.4989

Chapter 9
Anticipation in Three Restaurants of Mexico City: Imagining Edible Fungi in Urban Food Futures

Mónica Velázquez Téllez

Abstract This chapter analyzes the forms of anticipation of three mushroom-based food restaurants in Mexico City in the face of the horizon of urban food transitions in the context of the ecosocial crisis. The first part describes the food scenario proposed by international organizations and the importance of food systems for climate change. The second part describes the relationship between cities and food. The third part analyzes how the mushroom is being created as an emerging object from different fields, and its role in food and the Mexican food system. The fourth part deals with the case studies of the three restaurants (Tencüi, La Hongoteca, and Los Loosers). The fifth section discusses the anticipatory capacities of the restaurants and some little-known elements that the relationship between urban systems and wild mushrooms could make possible.

Keywords Food systems anticipation · Mushroom-based gastronomy · Urban food system futures

9.1 Introduction

This chapter will analyze the anticipation forms of three restaurants located in the Cuauhtemoc district of Mexico City whose gastronomic proposal employs mushrooms. The first is Tencüi, the second is La Hongoteca, and the third is Los Loosers. Anticipation occurs when the future is used in action (Poli 2020), which implies that, based on the reading or interpretation of a given context, the system that interprets this context can act or not act in certain behaviors to respond to a probable future, that which comes from the reading that the system itself makes of the context. Based on the fieldwork in the three restaurants mentioned above, we will analyze what forms of future around food are practiced in these restaurants by implementing a mushroom-based cuisine. Although no one has the last word on what will or will

M. Velázquez Téllez (✉)
Universidad Autonoma Metropolitana-Cuajimalpa, Mexico City, Mexico
e-mail: monica.velazquez@cua.uam.mx; monveltz@gmail.com

S. Gonzalez-Arellano and B. Gandlgruber (eds.), *Cities as Anticipatory Systems*, The Urban Book Series, https://doi.org/10.1007/978-3-031-78162-9_9

not happen in the future (Miller 2007), critical reflection on the future and, especially, conscious anticipatory thinking is vital to respond creatively to the challenges posed by the Anthropocene, a time in which the future is coded not only in terms of uncertainty—as it has always been—but also in terms of survival.

Given the dominant narratives about the multiple socio-environmental crises that this time implies, the role of urban food systems is crucial for imagining food futures that respond to a horizon in which multiple socio-environmental issues are looming. Likewise, the figure of the restaurant as a space for innovation and taste formation constitutes a site of transition that can articulate new food paradigms but also lead to the rise in price and gentrification of foods that were not previously part of the major market trends. The case of the three mushroom restaurants will allow us to observe the role of urban food systems in food transitions, as well as to reflect on the emergence of the mushroom as an object of scientific interest and a powerful metaphor that mobilizes imaginaries toward futures based on interspecies collaboration.

9.2 Scenarios of Food System in a Global Crisis Panorama

9.2.1 Scenarios of Food Crisis

International organizations and scientific groups such as EAT, FAO, and the Lancet have constructed a food scenario in which the multiple socio-environmental crises are strengthened. When considering the different elements that make up food systems and the influences they suffer and exert in a hyperconnected world, the anticipation linked to them is a fruitful topic for reflection. Food systems are the set of all factors and environmental elements, human inputs, processes, infrastructures, and institutions around food (Graziano da Silva et al. 2021). They are a crucial node in which elements of very different kinds intersect since the food field, species, geographies, climates, traditions, technological developments, symbols, governments, and economic powers, among many others are related. In a globalized planet, these relationships, with their asymmetries of power and cascading effects, are even stronger; as an example of this, the pandemic caused by SARS-CoV-2 that started at the end of 2019, probably in the province of Wuhan (Zapatero Gaviria and Barba Martin 2023), shows that there is no small problem when the factors of a local food system are disrupted (Nchanji and Lutomia 2021).

Recently, food has also become a lifestyle (Si 2020; Sexton et al. 2022; Vargas-Lopez et al. 2022) and even a political ecology attitude to embrace (Anguelovski 2015; Hebinck et al. 2018; Pasquier Merino et al. 2022) stance, especially as it relates to environmental justice (Kortetmäki and Oksanen 2021; Sexton et al. 2022) and the relationship between greenhouse gas emissions and animal protein consumption, particularly beef (Willett and Rockstrom 2018). These behaviors indicate that there is something in the food that is emerging with great power, as the decisions of some consumers whose economy allows them to prioritize consumption decisions, begin to conceive food functions and daily habits related to food as enclaves to promote transitions toward more sustainable, healthier or even, for some, more ethical food. The relationship between food choice and promoting some change is a novel behavior in food consumption. It is important to note that each of these food choices responds to an implicit idea of the future so that one decides to eat in one way or another, thinking, for example, that animal welfare will be achieved or that plant-based foods can help reduce greenhouse gas emissions, or that food should also be medicinal to prevent some ailments. Food futures are active in every food choice or omission, as seen below in the three mushroom restaurants.

We mean the superimposition of several negative factors around food systems by referring to the "food crisis" within the global environmental crisis. The first, related to hunger, is the continued malnutrition caused by deficient caloric intake. The second factor is the lack or low intake of vitamins, minerals, and other essential nutrients (hidden hunger). The third crisis factor has to do with the relationship between nutrition and the development of non-communicable diseases resulting from obesity, such as cancer, diabetes, or cardiovascular problems (Díaz 2021). To these crisis factors linking food and health, which are much more closely related to the weight of the food industry and the large corporations that control the food market, we should also add unexpected events or wild cards, such as the COVID-19 pandemic, whose impact on food systems has not yet been fully calculated (Adelodun et al. 2021).

Another crucial element involved in the crisis of food systems is the nexus between the production of space and the form of cultivation. Historically, food systems have produced diverse socio-ecosystems; today, extensive agriculture and monoculture contribute to reaffirming the physical and symbolic distance between the countryside and the city, spaces of production and consumption, producing agents and consuming agents, physical labor, and intellectual labor. Monoculture, allied with the global expansion of cities, produces a space that hinders food sovereignty and security in urban centers while conceiving cities, at least megacities, as spaces disconnected from the life-supporting elements that sustain global food. Food futures are, at least from a narrative, totally related to the socio-environmental crisis; therefore, food and the change in the human relationship with it is imagined as a space of incidence to shape, even the physical characteristics of the biosphere.

It is estimated that by 2050, the transformation to healthy diets will require substantial changes. Global consumption of fruits, vegetables, nuts, seeds, and legumes will have to double, while foods such as red meat and sugar must be reduced by more than 50% (Willett and Rockstrom 2018). The relationship between human and planetary health should be emphasized, as food systems are responsible for one-third of global greenhouse gas emissions (Crippa et al. 2015). For these reasons, food systems are crucial for optimizing interspecies life and safeguarding human health. In this scenario elaborated by international organizations assessing world food conditions, the discipline of anticipation becomes relevant in the transitions needed to articulate a radical transformation that allows the regeneration of the land and an effective integration of the socio-ecosystems that provide food or, at least, to visualize probable responses in the face of uncertainty. It is necessary to note that anticipation is not homogeneous since it depends on readings that belong to specific contexts. In this sense, it is important to analyze to which futures the practices that urban systems carry out in the present respond, for example, if they anticipate market trends, or an ecosocial crisis more linked to justice and the distribution of wealth, or a future in which those who produce food and those who consume it are conceived differently.

9.2.2 Cities and Food

Regarding the link between cities and food systems, it is worth mentioning that the urban population consumes 70% of the food grown in the world (Tefft et al. 2017); this implies that cities play a fundamental role in terms of the food production-distribution-consumption-waste relationship since it is expected that by 2050, 68% of the world's population is expected to live in cities, which makes it necessary to consider a scenario in which it would be favorable for the urban–rural continuum to evolve into cities centered on food production, as implied by the potential that Mexico City still has, as will be seen below.

In the urban context, food systems are particularly affected by six factors, which correspond to (1) demographic changes and migrations; (2) the paradox of rising incomes and widening inequality gap; (3) climate change and scarcity; (4) evolving consumption, nutrition, and health preferences; (5) rapid technological change and innovation; and (6) localized development and strong involvement of stakeholders and decision-makers (ibid.). In this sense, food systems in urban contexts represent a double challenge; on the one hand, they are not exempt from getting involved with the interconnection of apparently distant socio-ecosystems, as demonstrated by the 2020 pandemic or the conflict between Russia and Ukraine, because in cities, due to international mobility and tourism connections, local events are enhanced in a fractal way. On the other hand, in addition to the socio-environmental crisis that affects all food systems, in urban contexts, these effects are also combined with others of a social nature that deepen the inequality gap since "urban food insecurity in low-income countries is greater than in rural areas, reaching more than 70% in urban slums. Informal safety nets help boost access to food" (Ibid., p. 23).

In the present century, migrations of environmental refugees to urban areas produce permanent or transitory displacements in cities (Chung et al. 2022). Some 20% of the migrant population lives in the 20 largest cities in the world (Tefft et al. 2017). For urban food systems, voluntary or involuntary migration caused by climate change or other social crises generates demographic pressure that complicates the role of cities in terms of their ability to ensure food security; on the other hand, the migrant presence also influences the variety of urban gastronomies, as food is also a cultural heritage of peoples' identity (Reddy and van Dam 2020). Likewise, the number of megacities and cities (1–10 million residents) is expected to increase. The growth of urban forms as the hegemonic way of inhabiting the planet causes the abandonment of fields as cultivation areas. It makes it necessary to reorganize cities around food systems or, in other words, it is imperative to reimagine cities as spaces for living and food production.

Life on the planet is articulated around food systems, and food production shapes the planet. It is necessary to change extensive agriculture from monocultures to an ecological approach of family farming (Biel 2016). For these reasons, food systems are a key that could contain clues to reorder life in the face of the emergency derived from the global environmental crisis, since food systems encompass all natural resources of food not only for humans but for other species used in human consumption; thus, food systems also weave value chains at local and global scales while producing specific inequalities and designing the space of the habitable; the latter has important implications for the cities of the future in the key of resilience.

9.2.3 Food Systems in Mexico City

Even though Mexico City's productive food system has an agricultural potential of 49% (Fuentes 2023), the city, at least from the existing relationship between producers and government, does not seem to consider the current scenario of the global food crisis to activate its productive capacities. In government strategies, there are multiple impediments to promoting food productivity in the city, such as the absence of differentiated support that considers the characteristics of productivity per municipality, lack of an internal transportation network that links products to the core areas of consumption, pressure from food industries; generational abandonment of agricultural work, among others. That does not imply that, through other mechanisms, no measure is articulated that seeks to respond to the demands of civil society for food security and sovereignty in search of healthy and fair food that benefits local producers and consumers.

Even though urban agriculture in Mexico City is not widespread and that buying, rather than producing, seems the most attractive idea (Capron et al. 2017), it is true that, in the wake of the pandemic, a generational relay was observed of those unemployed grandchildren who returned to activate the productivity of the land that their parents had abandoned, this in response to the needs of local food supply (Fuentes 2023). As has a strong indigenous base in many aspects of food, traditional food systems based on pre-Hispanic methods coexist with urban ideas of food, for example, those executed from places such as urban gardens. These multifunctional spaces fulfill, at the same time, economic, cultural, and social demands with the capacity to promote education regarding the agro-food systems that survive in the city and contribute to urban environmental services that simultaneously provide food security and sovereignty (Alcántara and Larroa 2022).

Urban food systems can be classified into three types (traditional, modern, or emerging and informal), which generally exist in all large cities to varying degrees, as they overlap and share functions. The traditional system is the one that is articulated around the connections between large stores and markets in the city with the production of the countryside and its mediators. The global or emerging food system corresponds to the series of variants and actors around food, such as wholesale sales and elements that guarantee food safety, such as packaging, refrigeration chains, business services around food, and other technologies, such as food labeling, sales to restaurants, and the global interaction of food in cities. Informal food systems mainly supply the food demand of the city's impoverished populations. These types of systems are represented by transactions mostly in cash, and their forms of sale are generally itinerant, or else they are premises or businesses that are not registered with the formality of state institutions. The case that will be addressed here is the study of the modern urban food system, which involves value chains in which restaurants play an essential role, as seen below. Thus, it is possible to state that the three restaurants analyzed in this chapter are part of the modern urban food system in Mexico City.

About the three mushroom-based restaurants, it is crucial to consider the role that restaurants play in cities in shaping gastronomic taste by introducing and creating dishes that often differ from the most established gastronomic traditions or that use ingredients in other manners. Urban consumers of a medium–high socioeconomic stratum drive an emerging segment in restaurants that is characterized by demand for agroecological, non-GMO products with short production, distribution, and consumption chains (Bertran-Vilà et al. 2022; Tefft et al. 2017). Likewise, due to its purchasing power and cultural capital, this sector is in a position to promote the consumption of less industrialized food congruent with nutritional requirements prepared in urban restaurants that provide an environment or aesthetic experience around the conceptualization of food and the act of eating.

9.3 The Mushroom to the Future: Time to Revisit the Edible Past

9.3.1 Why Fungi Now?

The restaurants studied here concentrate the production of their dishes on the figure of the mushroom as a gastronomic protagonist; this idea is inserted in an international logic that emerges around the mushroom. Because mushrooms are today a food, pharmaceutical, and cosmetic trend, they are at the center of several discussions either from the scientific research of the international mycologist community (Van Court et al. 2022), from the rights of consumption of psychoactive substances or home growing and mycotourism clubs (Buntgen et al. 2017), or from the potential of the fungi kingdom to offer not only nutritious but also functional food. Although the relationship between edible and medicinal fungi has existed since ancient times among most of the world's populations (Pérez-Moreno et al. 2021),), still today, it is a kingdom that remains unknown since approximately 5% of an estimated 2.2–3.8 million global species are known to be familiar (Zhou and May 2023).

Research on the incipient development concerning the psychoactive potential of some mushrooms has been banned in the United States since the 1970s (Zuleta

2022). However, consuming mushrooms with these attributes was never extinct and remained low profile, as something marginal. The prohibitionist policy on the psychoactive attributes of some species had a turnaround because, in 2020, the John Hopkins Center for Research on Psychedelics and Consciousness was established. With this background, the reopening of research on the fungi kingdom was institutionalized, at least in the United States. At least since 2011, some American publications and newspapers referred to the "shroom boom" or "mushroom boom," highlighting the presence of mushrooms as the foods that were beginning to move "to the center of the plates" for being umami, that is, having a versatile flavor known in Japanese as "essence of delight" as it is simultaneously sweet, sour, bitter, bitter, and salty (Searles 2011).

With the confinements caused by the pandemic, the mushroom boom grew; Google search term charts show worldwide that, from April to August 2020, there was a peak in the increase in searches for the word "mushroom." The market for mushrooms (button: *Agaricus bisporus*; shiitake: *Lentinula edodes*; oyster mushroom: *Pleurotus ostreatus*) used in food, pharmaceuticals, and cosmetics was valued at $50.3 billion in 2021, and is expected to grow by 9.7% from 2022 to 2030 due to the increasing population demanding a plant-based diet or non-animal protein-rich food options around the world; in that sense, vegan and vegetarian diets are an influential factor for the development of the market around the mushroom. Mushrooms are also considered in the superfood category for their essential nutrients, such as selenium, vitamin D, glutathione, and ergothionein, which contribute to mitigating oxidative stress and preventing chronic diseases such as cancer, heart disease, and dementia (GVR 2023).

In 2020, Giuliana Furci, founder, and director of the Fungi Foundation, together with a team of Argentines, Chileans, and Brazilians, as well as the American mycologist Donald H. Pfister, proposed an initiative for the fungi kingdom, as well as flora and fauna, to be included in the biodiversity conservation goals from 2020 onward. Thus, the initiative of the three Fs (Flora, Fauna, and Funga) in international assemblies such as the International Union for Conservation of Nature (IUCN) and the United Nations Convention on Biological Diversity (CBD) exemplifies the rediscovery of the fungus and the future of conservation as an institutionalized interest in an international network of knowledge. The conservation of fungi not only implies the conservation of the fungi kingdom, but it is also a strategy focused on the conservation of an entire ecosystem since fungi fulfill specific alliance functions that sustain life, for example, "the vital importance of arbuscular mycorrhizal fungi colonizes 80% of the roots of all plants, a symbiosis that helps plants to establish themselves in the soil" (Bragança 2021, p. 403). Conservation based on the recognition of the fungal kingdom makes the fungus emerge as a prominent figure to influence the possibilities of conserving not only species but environments so that through the fungus and its present conservation, the conception of relationships between life in socio-ecosystems is expanded, thereby expanding the imagination of collaborations between species as well.

Around the fungi kingdom, new approaches are being concentrated to build it up again: the fungus as an object of resilience and forest health as a critical piece in the work of biodiversity conservation, as a potential adaptogen for the new psychiatric medicine that sees in the *psilocybes* an opportunity to treat mental conditions of modernity, such as anxiety, depression, obsessive–compulsive disorders or post-traumatic stress. The different actions motivated by the mushroom's potential seem to envision multiple changes in medical sciences, food, and conservation. However, as a metaphor for cooperation at different levels, the mushroom is also beginning to be constructed as a space for rereading the relationships between humans and non-humans. All these events seem to indicate that point to changes in the scientific, industrial, and cultural spheres around which the mushroom is imagined anew, at least in Western culture.

9.3.2 Mushrooms as a Food Alternative

Just as the fungi kingdom has acquired relevant functions to articulate conservation strategies and drive new research agendas, the food field has again looked toward these species. Edible fungi, the fruit of the mycelial fungal body, are beginning to be considered as an alternative to providing food security (Cuesta and Castro-Ríos 2017) in a world with a growing population because their production occupies less space for cultivation, uses forty times less water than beef protein and thirty times less than chicken protein (Kazer et al. 2022). Approximately 200 of the 2300 edible species are known (Vidal 2000). Likewise, options have begun to appear for home cultivation for self-consumption from food leftovers (Hellwig et al. 2023); in this case, mushrooms also represent an option that reduces the waste caused by leftovers since it is possible to reuse them to feed their growth. There are already companies, such as the U.S. company Quorn, that offer packaged foods made from mycoproteins of the *Fusarium venenatum* variety. Similarly, in 2019, the National Institute of Industrial Technology of Argentina and Enye Tech started cultivating the same variety under the local name "kernel" to focus on exploring new protein and food markets (Bär 2019).

The different interests in researching the mushroom, both for its potential as a food low in greenhouse gas emissions and for its high protein content, show that, although this species, in its edible varieties, has always existed in people's diets, it is now emerging on the horizon as a novelty that seems to be seen as a creative response to many current problems. The mushroom is considered an alternative not only for planetary health since its production reduces the carbon footprint and is an important option in the search for alternatives to supply quality proteins that do not come from animals, but also, the mushroom is part of the functional food, which considers food intake not only as a means of energy and supply of essential nutrients for the human body, but also as a nutraceutical source, since it provides elements to combat various diseases, whether exogenous, such as those caused by viruses and bacteria, or endogenous, such as cancer and diabetes, caused by a malfunctioning of the immune system and the combination of various factors, such as the intake of industrialized food, high in sugars and preservatives (Ishara et al. 2022; Shamim et al. 2023).

All these diverse interests and views on the fungal kingdom are enabling the re-emergence of the fungus as a new scientific object, in the sense that the food industry is interested in investigating it as an option for food transitions that propose diets focused on lower animal consumption in alignment with planetary and human health. Likewise, the mushroom shows that the concept of food is moving toward not only nutritious but also nutraceutical foods, which implies the emergence of a new food paradigm. For these reasons, the mushroom is interpreted as an emblem of resilience at different levels, which motivates images of futures centered on collaboration, in which the process of growth and cultivation is not only related to humans but to factors of cooperation between species, between mushrooms, forests, and humans, between what dies and what is born, as a metaphor for the degrading capacities of the mushroom. From this perspective, the form of food supply may have a sustainable, regenerative potential in ecological, social, and economic terms (Stöckelová et al. 2023).

9.3.3 Mushroom Gastronomic Tradition in Mexico

Mexico was the first country in Latin America to begin commercial production of edible mushrooms in 1933 near Texcoco, in the State of Mexico (Martínez-Carrera et al. 2000). The main variety, whose strain comes from abroad, is *agaricus*, also known as mushroom. Since 2016, researchers from the Colegio de Posgraduados in the state of Puebla initiated a project to strengthen the genetic varieties of this fungus by combining them with specimens of Mexican wild mushrooms to improve the emerging microbial agro-food chain of edible mushrooms in the country. Although mushroom production in small laboratories is currently obsolete at the industrial level because there is specificity in the production of seeds, substrates, and product processing, small laboratories are relevant at the restaurant production level, as evidenced by the case of the Tencüi restaurant, which is part of this study.

At the industrial or large-scale production level, the main Mexican states with consolidated production are Estado de México, Puebla, Veracruz, Oaxaca, Querétaro, Guanajuato, and Jalisco. The most cultivated varieties in Mexico are mushrooms (white and brown). However, the country also produces mushrooms (*Pleurotus*), shiitake (*Lentulina edodes*), deer mushroom (*Ramaria araiospora*), lion's mane (*Hericium erinaceus*), chicks (*Amanita caesarea*), maguey mushroom *(Pleurotus opuntiae*), oak mushroom (*Entoloma clypeatum*), and friar's head (*Boletus aereus*). Mexico ranks second in the diversity of wild edible mushrooms (Pérez-Moreno et al. 2021). Wild mushrooms have historically been used as part of the diet of indigenous and rural communities (Molina-Castillo et al. 2023), especially in the central and southern regions of the country. Regarding the use of edible mushrooms in the Mexican diet, it is common to find them in various preparations of up to 168 different dishes depending on the region (ibid.), such as dishes that include corn, chili, beans (linked to Mexican popular gastronomic culture) or in combination with elements that were introduced from the different moments of contact with foreigners, such as ingredients of animal origin, such as beef, pork, chicken, cheese, cream, butter, among others. In popular Mexican gastronomy, mushrooms can be considered the main dish, together with other vegetables, or a side dish stewed with *epazote* and other *quelites*.

Despite the Mexican gastronomic tradition around the mushroom, the innovations promoted by the restaurant sectors create the mushroom as a different object, which, although it uses traditional or indigenous discourses about it, is conceived more as a food associated with gourmet food and health. That also implies the emergence of new problems in a horizon of market and food industry interests, in which mushrooms also acquire a class distinction as an object of consumption, which may jeopardize the access that communities have to forest resources since, often, as happens in Japan with high-priced varieties such as matsutake (*Tricholoma matsutake*), they are collected for sale instead of being consumed internally.

9.4 Anticipation Forms in Three Mushroom-Based Restaurants

9.4.1 Tencüi: A Laboratory for the Kitchen

Tencüi, which means "to connect" in Nahuatl, is a restaurant in the Santa María la Ribera neighborhood in downtown Mexico City, in the Cuauhtémoc district. The area is considered an image of a glorious past of great architectural and historical value due to its museums and elements, such as the Moorish kiosk, designed by Ramón Ibarrola to represent Mexico at the Universal Exposition of 1884–1885 (Ramírez Crespo 2021). Toward the middle of the last decade, with the arrival of new inhabitants, Santa María La Ribera began to be a cultural spot (Vazquez-Medina et al. 2019) with which the consumption of goods and services also suffered readjustments, as an example of this, is the demand for prepared foods that formed a landscape of culinary cosmopolitanism (ibid.).

Tencüi, a gastronomic proposal inspired by mushrooms, is part of this landscape. Creating such a food concept responds to the combination of interior design trends proposed by Ricardo Casas, the cuisine of chefs Mario Espinosa and Rodrigo Morales, and the science of edible mushroom cultivation by Carlos Briones, an ethnomycologist. This restaurant employs, in situ, a mushroom cultivation chamber, mainly used to produce yellow mushroom (*pleurotus citrinopileatus*) and pink mushroom (*pleurotus djamor*), royal trumpet (*pleurotus eryngii*), resishi (*Ganoderma lucidum*), chestnut mushroom (*Pholiota adiposa*) and lion's mane (*Hericium erinaceus*), edible species that form part of the restaurant's dishes, either as an accompaniment to main courses or as part of beverages (Fig. 9.1). The mushroom laboratory or chamber began in April 2022, the year the restaurant opened.

Fig. 9.1 *Cerdo pelón* in mushroom sauce, inside the culture chamber, developing mycelium, restaurant exterior

Although in Mexico City there are restaurants whose products come from urban gardens or that have their cultivation methods for internal supply, Tencüi has incorporated a cultivation chamber to cover 30% of its requirements in fungal ingredients (Briones 2022), this implies a conception of the restaurant that goes beyond a space designed for the preparation and sale of food for the urban consumption of upper-middle class diners. However, adding the cultivation chamber to the restaurant infrastructure modifies the attributions of the restaurant as a space for food preparation. It turns into a place with hybrid attributes in which the functions of food production are shared, but also the spaces of the specialized science of ethnomycology linked to the cultivation of edible mushrooms. Carlos Briones, in charge of the laboratory until 2023, has trained the team of chefs in the kitchen so that they can also manage the laboratory; this implies new attributions regarding the capacities and spaces of the food circuits in restaurants and those involved in the circuit.

The mushroom chamber in the restaurant functions as a way to solve the supply, thus guaranteeing the origin and quality of its inputs while at the same time making it possible to respond to the problem that mushrooms are seasonal, that is, the operation of a restaurant based on wild mushrooms would depend on the rainy season, however, as an anticipatory practice, the small laboratory performs the artificial function of the natural ecosystem that allows the development of various mushrooms. The restaurant is making it possible to respond to its own demands thereby shaping itself as a small space that imagines restaurants as food producers and not only as traders or socializers. Certainly, integrating a laboratory into the restaurant is not a standard feature defining the "restaurant" space. Therefore, Tencüi proposes a new approach to the restaurant structure, in which food production is part of the food socialization chain through preparing various dishes. In addition, the cultivation chamber also has a kind of function as an experience for the diners who can visit it.

In this way, the Tencüi project can be thought of as a restaurant-laboratory that amalgamates economic interests and design and consumption trends through the creation of an innovative gastronomic option, in the sense that it uses various edible mushrooms for the design of main dishes, desserts, and even cocktails, contributing to the creation of a taste and a type of food option for consumption in restaurants linked to mushrooms. As a hybrid system, Tencüi employs self-cultivation of mushrooms thanks to the artificial conditions offered by the chamber equipped to humidify the environment, thus ensuring that the mushrooms will be available even outside the seasonal rainy seasons while at the same time making it possible for part of its inputs not to depend on external suppliers. Tencüi's anticipatory practices imply the consideration of a future in which its primary ingredient is unavailable at all times of the year, so the in-house laboratory anticipates cultivating various edible mushrooms. At the same time, creating a restaurant-laboratory implies an idea of the urban food future in which the food is produced in situ and functions as a space of mediation between specialized knowledge of mycology and gastronomy that becomes its different dishes.

9.4.2 *La Hongoteca: Gastronomy as Means to Divulgation*

Like Tencüi, La Hongoteca opened its doors in 2022, coinciding with the mushroom boom in other parts of the world, especially in the United States. However, the restaurant project became an effort to link the cuisine and the dissemination of the properties of mushrooms in other types of activities, so the restaurant, as such, is temporarily closed. The restaurant concept came from two partners who decided to start a mushroom-based food business in the Roma neighborhood, in the city's center and west of the Chapultepec Forest. The Roma neighborhood, similar to Santa María La Ribera, has had an increase in young population since 1990 (Mondragón Yépez 2015) and, in recent years, especially as a consequence of the pandemic and the possibilities that digital nomads from rich countries have to travel to cheaper countries and rent housing in cosmopolitan neighborhoods, the foreign population has also increased (Mariano-Juárez et al. 2023).

Ana Rosas, partner in charge of design, marketing, and administration of La Hongoteca, points out that she conceived the concept as a place that "stores or where there are several types of mushrooms" (Rosas 2023), hence the origin of the restaurant's name linked to the nouns with the suffix-*teca* in Spanish that serve to describe a place where something is stored or that gathers several species of the same thing: media library, library, discotheque, film library, among others. In this sense, La Hongoteca, as a restaurant, also aspired to build the idea of a place devoted to mushrooms, not only in their nutritional aspects but also in their cultural value, having gastronomy as a gateway. Rosas (2023) refers to the need to revalue mushrooms as socio-ecosystemic elements that, for some reason, were forgotten or are considered almost unknown and whose symbolic relationship for the bulk of the population lies in toxicity associated with death, psychoactive substances, or the caricatured image of pop culture from video games. These mentions show that the mushroom, especially in terms of its wide range as a culinary ingredient, remains unexploited, at least for urban populations whose daily life is independent of forest resources in an obvious way. For Rosas, the mushroom is not found in the popular imagination beyond a mushroom, a mushroom, or, in the case of central Mexico, a huitlacoche.

The emergence of La Hongoteca as a gastronomic project was possible thanks to the reading of several signs of the emerging context around mushrooms from 2018 to 2022, as Rosas, a mushroom aficionado and trained in marketing, knew how to interpret the culture of the fungi kingdom was going through a point of effervescence. The difficulties of introducing a vegetarian-vegan restaurant whose dishes were mushroom-based resulted in the creation of a menu in which the different varieties of mushrooms used had to go hand in hand with familiar flavors: Ensenada taco (gray mushroom-based); croquettes of the gods (huitlacoche croquettes with bechamel sauce); hamburger (fried chicken-style mushroom) (Fig. 9.2).To socialize the mushroom as the base of the restaurant's dishes, La Hongoteca used elements already consolidated in the popular taste of the public.

The Hongoteca assumes that gastronomy is a gateway to making mushrooms known, as the heading of its menu states: "Let's get to know the unknown, let's question and be amazed." Likewise, a legend on the menu announces: "Our store has curiosities, books, adaptogenic mushrooms with extract, powder and capsules, teas, stickers, art, jewelry, preserves, fresh and dried mushrooms." The restaurant also sought to generate a consumer experience focused on the dissemination of the nutritional, medicinal, and cultural attributes of mushrooms. With this intention, it is

Fig. 9.2 Buffalo wings made of mushrooms, Taco Ensenada of gray mushroom and blue corn, mushroom burger, cream of mushroom soup, restaurant exterior

evident the goal was to imagine the restaurant as a space of mediation in which food is a first step to creating knowledge, interest, and markets that offer experiences such as mycotourism or amateur workshops on the cultivation and properties of mushrooms.

Like the case of Tencüi, La Hongoteca was constituted as a hybrid space in which the dissemination of ethnomycological culture was part of the food sales space. Currently, the restaurant space, understood as a place for preparing and selling food on-site, does not have any explicit feature that characterizes it as an area where the public could go to get information on any subject. That is not a minor issue since, through the construction of hybrid spaces that merge the characteristics of an information or teaching center and the characteristics of a restaurant, La Hongoteca updates a specific idea of food futures, the one that conceives that eating is also knowing; that knowledge and scientific dissemination is part of the production, sale, and consumption of food in the restaurant. As an anticipatory practice, La Hongoteca updates a future idea of the restaurant concept in which education is mixed through gastronomy in the same space.

9.4.3 Los Loosers: Imagining the Mushroom as Resistance

Located today in the Roma neighborhood, the Los Loosers project began in 2011 as a ghost kitchen or dark kitchen that offered few options and whose owner, Mariana Blanco, cooked and delivered orders by bicycle. Unlike Tencüi or La Hongoteca, Los Loosers does not present itself as a mushroom-based, vegetarian, or vegan kitchen. However, its gastronomic bet is articulated through plants and mushrooms as a cuisine without animal-based elements. For Blanco, cuisine and stories play an essential role; the restaurant's menu is structured as a comic strip and manifesto in which the chef presents her life mission as a service to animals (Fig. 9.3). Hence, of her cuisine is based on mushrooms and plants, elements with which Blanco shares a life story, in the sense that, coming from a family in Oaxaca, mushrooms were an integral part of her diet long before the shroom boom.

Blanco (2023) considers that the work of a restaurant is collaborative; for her, to be self-sufficient in the supply of its raw materials in the manner of Tencüi would imply working against the mushroom itself, which collaborates in symbiosis to maintain the health of various elements that make up the entire ecosystem. For these reasons, Blanco is aware that the restaurant is indebted to the mushroom farmers (*hongueras*) of Puebla, Oaxaca, and Tlaxcala, who, with their community knowledge about the recognition, selection, and harvesting of mushrooms, form part of his restaurant's dishes. As an anticipatory practice, Blanco uses dehydration and fermentation techniques to preserve mushrooms out of season; for her at Los Loosers, the mushroom season is 365 days. From the owner's perspective, it is still difficult to make dishes without any element of animal origin to be entirely well received by her public because, even though most of her diners are omnivores, none of her dishes use any animal product. Blanco considers this difficulty to be due to a lack of awareness of animals as terrestrial inhabitants with the same rights as humans. In this sense, for the

Fig. 9.3 Menu. Entrance of the restaurant. Mushroom Ramen

chef, mushrooms do represent a way of eating "ahead of their time" because, in her opinion, the conditions are not yet in place for eating plants and mushrooms to be the standard basis of food. Blanco's comments allow us to observe how the scenario of a diet composed mainly of plants and mushrooms is part of the futures activated in the present practices of his restaurant by way of anticipation. Los Loosers participates in a food that is not yet generalized and thus puts into operation a specific idea about the future of food: scenarios in which the consumption of animals disappears.

As mentioned above, mushrooms are in vogue today. However, when Los Loosers began, the restaurant sought to appropriate those practices that were not commonplace and that belonged to trends that were little understood in the country, such as using the bicycle as the primary means of transportation or food based on plants and mushrooms. Los Loosers, from the choice of its name, uses a marginal discourse as a form of resistance. In that sense, Blanco's stories around the kitchen have a symbolic and primordial role in the act of cooking that she performs because, for Blanco, the dishes are created not only from the combination of ingredients but of stories that enclose memories about food that can remember territories, people, or moments. The chef and sole proprietor points out that her restaurant was gateway that allowed plant-based food to enter the city. Today, Los Loosers still does not proclaim itself as a vegetarian or vegan kitchen, but its owner considers the food space to be her way of remaining in the service of animals. From that perspective, her restaurant plays a role in generating spaces of transition toward food forms that do not use animals in the elaboration of their dishes as a way of embodying what for some is considered a more equitable political stance with lives similar to human forms.

9.5 Reflections on the Anticipatory Capacity of Restaurants

9.5.1 Anticipation Forms

Anticipation can be understood through two dimensions that involve, on the one hand, imagining the future and, on the other hand, implementing actions in the present that respond to what has been imagined of that future (See González-Arellano's Chap. 2 for a deeper discussion), i.e., anticipation is the practice of the future that is actualized in the present. In this sense, the three restaurants studied in this chapter have common and differentiated anticipation practices, generating different responses to emerging problems. On the one hand, from some horizons outlined by studies on food prospectives (Willett and Rockstrom 2018), the need to reduce the consumption of animal proteins is glimpsed. In that line, mushroom restaurants are strengthening a food trend by specializing in food produced mostly from mushrooms, an element that is not yet a common or generalized aspect of the food supply. The mushroom restaurants respond to an image of a food future that seeks to reduce the consumption of animal proteins; thus, the three restaurants implement the anticipation in the sense that their food, mainly made from mushrooms and plants, puts into action the idea of a food future more focused on dishes that do not use animals.

For the moment, this type of cuisine remains in areas of the city that concentrate a large influx of tourism and temporary or permanent residents of foreign origin, which favors the opening of these spaces and makes it possible for certain consumers to pay for these services. The use of wild mushroom varieties as main dishes begins to build an urban taste for foods rich in quality proteins that are not of animal origin; thus, in restaurants, aspirations of future food systems with less environmental impact materialize. In this sense, the restaurant, as a space for dish creation, improvisation, and mixing of elements, directs the taste of the upper-middle class consumer, whose carbon footprint is generally higher than that of the world's poor, toward food with a lower environmental impact that can prepare for major food transitions. This transitional approach can also be seen as an anticipatory practice that activates new forms of consumption focused on acquiring lower environmental impact tastes, thereby actively building for the future.

At the same time that restaurants are responding to food futures rich in vegetables and mushrooms, restaurants based on mushroom gastronomy have been able to interpret the economic trends that enable the mushroom boom by beginning to develop a specialized cuisine and, with it, the development of a specific gastronomic taste for mushrooms of multiple varieties, including wild mushrooms. In these restaurants, gastronomy is not only conceived as a space for food creativity but also as a gateway to socializing the virtues of the mushroom, not only as food but also as an essential part of the planet's health due to its regenerative capacities. On the horizon of species conservation, the cities' demand for wild edible mushrooms could generate positive pressure to strengthen the conservation of oak forests near Mexico City that supply part of the inputs of these restaurants. As explained above, including the fungi kingdom and the protection of flora and fauna implies that legal frameworks are built and enabled to strengthen the conservation of all the elements of the socio-ecosystem, which should include local knowledge communities around the management of mushrooms in forests.

The relationship that the restaurants establish with mushrooms has allowed them to become hybrid spaces in which diverse knowledge coincides, such as disseminating mushroom culture beyond its nutritional aspect or training restaurant personnel in handling and cultivating different mushrooms in the laboratory. The three restaurants extend the frontier of what is known by becoming educational spaces, self-cultivation, or personal projects that seek to embody a position in search of animal equity through the preparation of their dishes. These attributes are not a current trend that describes the typical functions of a restaurant; it means that Tencüi, La Hongoteca, and Los Loosers are inhabiting a transitory space that may or may not become a dominant trend in how food is conceived in cities. On a powerful level, the imagination of the mushroom as a being that cares for the forest and is a bridge between life and death by degrading the organic matter of the forest and nourishing the soils to enable life to continue to grow, feeding the social schemas of collaboration between cities and rural communities through how they eat.

9.5.2 Futures About Mushroom-Based Food

Although, as mentioned at the beginning of this chapter, no one has the last word on the future, the growing interest in wild mushrooms may lead to an over-exploitation of the forests, causing the communities that have historically passed on the knowledge of recognizing and collecting species to be displaced. However, the demand for these foods may also become a strategy that aims at ecosystemic conservation, with the mushroom as the center of all forest diversity. These actions open the door to multiple futures around mushrooms that, at this moment of juncture between specialized studies from the fields of ethnomycology, psychiatry, and food potentials around the fungi kingdom, simultaneously enable the path of forest devastation and the rise in the price of the mushroom as a luxury food or the articulation of value chains that include cities and the countryside in a fair way. In order to move toward these realities, it is essential to imagine what future we wish to achieve through the decisions of the mushroom as a new element in urban food systems.

The boom that mushrooms are experiencing in some food markets also implies unexplored areas for long-term nutrition. Although several studies have noted the nutritional and nutraceutical properties of edible mushrooms (Baruah et al. 2023; Gariboldi et al. 2023; Ghosh et al. 2021; Maury et al. 2011), there are also indications that due to their powers as organic matter degraders and their relationship with soils, edible macroscopic fungi can concentrate large amounts of heavy metals that may pose a threat to human health in the long term (De Lucio-Flores et al. 2021; Domínguez-Zúñiga et al. 2021). There are still unexplored areas about the relationship that mushrooms will have in food; for the moment, they emerge as a figure of interest that concentrates on the imaginary of a future food with sustainability and socio-environmental justice features.

9.6 Conclusion

In this chapter, the practices of three mushroom-based gastronomy restaurants were analyzed from the perspective of anticipation. It is possible to conclude that in all cases anticipatory practices articulate a food future in the present, in which the main ingredients are proteins obtained through the intake of mushrooms. Tencüi foresees and solves the problem of the seasonality of mushrooms with its integrated cultivation chamber; at the same time, this element activates a food future in which the food sold in restaurants comes from in situ production. La Hongoteca anticipates the figure of the restaurant as a hybrid space in charge of educating through the dishes; in the restaurant, notions of the food future are activated as an informative link between sellers and consumers, which is enabled through the figure of the restaurant, in this sense, La Hongoteca anticipates to update futures in which the restaurant is also a place to acquire knowledge. Employing the metaphor of the work of mycelium in ecosystems, The Loosers imagines a food future based on small-scale complex networks with local collaborations that empower wild mushroom gathering communities. It can be argued that, albeit in different ways, all three anticipatory strategies contribute to erecting food futures that transition to the exploration of a new taste in which the consumption of wild mushrooms and mushroom products is a critical factor in developing a new food future.

References

Adelodun B, Kareem KY, Kumar P, Kumar V, Choi KS, Yadav KK, Yadav A, El-Denglawey A, Cabral-Pinto M, Son CT, Krishnan S, Khan NA (2021) Understanding the impacts of the COVID-19 pandemic on sustainable agri-food system and agroecosystem decarbonization nexus: a review. J Clean Prod 318(February):128451. https://doi.org/10.1016/j.jclepro.2021.128451

Alcántara N, Larroa RM (2022) La multifuncionalidad de los huertos urbanos en la Ciudad de México. Espiral xxix(83):187–230

Anguelovski I (2015) Alternative food provision conflicts in cities: contesting food privilege, injustice, and whiteness in Jamaica Plain, Boston. Geoforum 58:184–194. https://doi.org/10.1016/j.geoforum.2014.10.014

Bär N (2019) Tecnología alimentaria: desarrollaron un hongo con alto contenido en proteínas. La Nación. https://www.lanacion.com.ar/sociedad/sin-titulo-nid2225441/

Baruah P, Patra A, Barge S, Khan MR, Mukherjee AK (2023) Therapeutic potential of bioactive compounds from edible mushrooms to attenuate SARS-CoV-2 infection and some complications of coronavirus disease (COVID-19). J Fungi 9(9):897. https://doi.org/10.3390/jof9090897

Bertran-Vilà M, Pasquier Merino AG, Villatoro Hernández JG (2022) Food producers in the Peri-Urban area of Mexico City. A study on the linkages between social capital and food sustainability. Sustainability (Switzerland) 14(23):1–20. https://doi.org/10.3390/su142315960

Biel R (2016) Sustainable food systems. The role of city. UCL, London

Blanco M (2023) Personal interview recorded on zoom
Bragança ACH (2021) Include all fungi in biodiversity goals climate lawsuits could protect Brazilian Amazon The hidden Olympic spectator. Sci Mag 373(6553):403–405
Briones C (2022) Personal interview recorded on zoom
Buntgen U, Latorre J, Egli S, Martinez-Peña F (2017) Socio-economic, scientific, and political benefits of mycotourism. Ecosphere 8(7). https://doi.org/10.1002/ecs2.1870
Capron G, González S, Wigle J, Diez AL, Monterrubio A (2017) The urban food system of Mexico City Mexico (Issue 7). Hungry Cities Partnership, Cape Town
Chung J, Buswala B, Keith M, Schwanen T (2022) Climate mobilities into cities: a systematic review of literature from 2011 to 2020. Urban Clim 45:101252. https://doi.org/10.1016/j.uclim.2022.101252
Crippa M, Solazzo E, Guizzardi D, Tubiello FN, Leip A (2015) Anthropogenic GHG emissions. Nat Food. https://doi.org/10.1038/s43016-021-00225-9
Cuesta MC, Castro-Ríos K (2017) Mushrooms as a strategy to reduce food insecurity in Colombia. Nutr Food Sci 47(6):817–828. https://doi.org/10.1108/NFS-03-2017-0039
De Lucio-Flores SA, Otazo-Sánchez EM, Romero-Bautista L, Gaytán-Oyarzún JC (2021) Hongos macroscópicos como bioacumuladores de metales pesados. Pädi Bol Científico Cienc Básicas Ingenierías Del ICBI 8(16):60–65. https://doi.org/10.29057/icbi.v8i16.5823
Díaz E (2021) El rol de la innovación institucional, tecnológica y financiera en la transformación de los sistemas alimentarios en América Latina y el Caribe, en el contexto de la agenda 2030 y de la pandemia de covid-19. In: Graziano da Silva et al (eds) Sistemas alimentarios en América Latina y el Caribe. Desafíos en un escenario pospandemia, FAO, CIDES, Panamá, pp 71–94. https://doi.org/10.4060/cb5441es
Domínguez-Zúñiga LI, Puente-Valenzuela CO, Estrada-Arellano JR, Aguirre-Acosta E, Aguillón-Gutiérrez DR (2021) Concentración de metales pesados en hongos de la zona metropolitana de la Comarca Lagunera, México. Sci Fungorum 52(November 2020):e1389. https://doi.org/10.33885/sf.2021.52.1389
Fuentes M (2023) Lecture on Agroecosistemas: relevancia socioambiental y alimentaria. Planeación diferenciada. La transición del sistema alimentario en la CDMX. https://www.youtube.com/watch?v=Ww2vWiPoIJw&list=PLqGZhNa0qbM1sgzp-k-MmzgG6dhmyEMmW&index=3
Gariboldi MB, Marras E, Ferrario N, Vivona V, Prini P, Vignati F, Perletti G (2023) Anti-cancer potential of edible/medicinal mushrooms in breast cancer. Int J Mol Sci 24(12). https://doi.org/10.3390/ijms241210120
Ghosh S, Nandi S, Banerjee A, Sarkar S, Chakraborty N, Acharya K (2021) Prospecting medicinal properties of Lion's mane mushroom. J Food Biochem 45(8):1–13. https://doi.org/10.1111/jfbc.13833
Graziano da Silva J, Jales M, Rapallo R (2021) Sistemas alimentarios en América Latina y el Caribe—Desafíos en un escenario pospandemia. Panamá, FAO y CIDES.https://doi.org/10.4060/cb5441es
GVR (2023) Market analysis report. Mushroom market size. 2022–2030. ReportLinker, Lyon
Hebinck A, Vervoort JM, Hebinck P, Rutting L, Galli F (2018) Imagining transformative futures: participatory foresight for food systems. Special feature on designing transformative spaces for sustainability in social-ecological systems. Ecol Soc 23(2):19
Hellwig C, Bolton K, Häggblom-Kronlöf G, Rousta K (2023) Aspects affecting food choice in daily life as well as drivers and barriers to engagement with Fungi-based food—a qualitative perspective. Sustainability (Switzerland) 15(1). https://doi.org/10.3390/su15010317
Ishara J, Buzera A, Mushagalusa GN, Hammam ARA, Munga J, Karanja P, Kinyuru J (2022) Nutraceutical potential of mushroom bioactive metabolites and their food functionality. J Food Biochem 46(1):1–31. https://doi.org/10.1111/jfbc.14025
Kazer J, Orfanos G, Gallop C (2022) Quorn footprint comparison report. Carbon Trust, United Kindom, pp 1–33
Kortetmäki T, Oksanen M (2021) Is there a convincing case for climate veganism? Agric Hum Values 38(3):729–740. https://doi.org/10.1007/s10460-020-10182-x

Mariano-Juárez L, Medina FX, Conde-Caballero D (2023) Popular cuisines at foodie tables. Social class, distinction, and food transformation and incorporation in contemporary gastronomies. Int J Gastron Food Sci 33:100791. https://doi.org/10.1016/j.ijgfs.2023.100791
Martínez-Carrera D, Larqué-Saavedra A, Aliphat M, Aguilar A, Bonilla M, Martínez W (2000) La Biotecnología De Hongos Comestibles en la seguridad y soberanía alimentaria de México. Puebla, II Foro Nacional sobre Seguridad y Soberanía Alimentaria, AcademiaMexicana deCiencias-CONACYT *Y Seguridad Alimentaria.* Pp 1–12
Maury GL, Humberto Joaquín Morris Quevedo D, Albear JM, Chibás LC, Rosa Catalina Bermúdez Savón D (2011) Plantas y hongos comestibles en la modulación del sistema inmune. Rev Cubana Inv Biomedicas 30(4):511–527
Miller R (2007) Futures literacy: a hybrid strategic scenario method. Futures 39(4):341–362. https://doi.org/10.1016/j.futures.2006.12.001
Molina-Castillo S, Espinoza-Ortega A, Thomé-Ortiz H, Moctezuma-Pérez S (2023) Gastronomic diversity of wild edible mushrooms in the Mexican cuisine. Int J Gastronomy Food Sci 31(September 2022). https://doi.org/10.1016/j.ijgfs.2022.100652
Mondragón Yépez M (2015) Proceso de gentrificación de la Colonia Roma.pdf. Universidad Nacional Autónoma de México
Nchanji EB, Lutomia CK (2021) COVID-19 challenges to sustainable food production and consumption: future lessons for food systems in eastern and southern Africa from a gender lens. Sustain Prod Consumption 27:2208–2220. https://doi.org/10.1016/j.spc.2021.05.016
Pasquier Merino AG, Torres Salcido G, Monachon DS, Villatoro Hernández JG (2022) Alternative food networks, social capital, and public policy in Mexico City. Sustainability (Switzerland) 14(23). https://doi.org/10.3390/su142316278
Pérez-Moreno J, Guerin-Laguette A, Rinaldi AC, Yu F, Verbeken A, Hernández-Santiago F, Martínez-Reyes M (2021) Edible mycorrhizal fungi of the world: what is their role in forest sustainability, food security, biocultural conservation and climate change? Plants People Planet 3(5):471–490. https://doi.org/10.1002/ppp3.10199
Poli R (2020) Handbook of anticipation. https://doi.org/10.1007/978-3-319-31737-3
Ramírez Crespo FS (2021) Santa María la Ribera : gentrificación y producción de fronteras sociales. El Colegio de México
Reddy G, van Dam RM (2020) Food, culture, and identity in multicultural societies: Insights from Singapore. Appetite 149(September 2019):104633. https://doi.org/10.1016/j.appet.2020.104633
Rosas A (2023) Personal interview recorded on zoom
Searles BR (2011) A boom in ' Shrooms. Psychology Today, New York, pp 45–46
Sexton AE, Garnett T, Lorimer J (2022) Vegan food geographies and the rise of Big Veganism. Prog Hum Geogr 46(2):605–628. https://doi.org/10.1177/03091325211051021
Shamim MZ, Mishra AK, Kausar T, Mahanta S, Sarma B, Kumar V, Mishra PK, Panda J, Baek KH, Mohanta YK (2023) Exploring edible mushrooms for diabetes: unveiling their role in prevention and treatment. Molecules 28(6). https://doi.org/10.3390/molecules28062837
Si LW (2020) Trending foods and beverages.Food Soc 305–321.https://doi.org/10.1016/b978-0-12-811808-5.00016-7
Stöckelová T, Senft L, Kolářová K (2023) Sympoietic growth: living and producing with fungi in times of ecological distress. Agric Hum Values 40(1):359–371. https://doi.org/10.1007/s10460-022-10366-7
Tefft J, Jonasova M, Adjao R, Morgan A (2017) Food systems for an urbanizing world.https://doi.org/10.1596/32502
Van Court RC, Wiseman MS, Meyer KW, Ballhorn DJ, Amses KR, Slot JC, Dentinger BTM, Garibay-Orijel R, Uehling JK (2022) Diversity, biology, and history of psilocybin-containing fungi: suggestions for research and technological development. Fungal Biol 126(4):308–319. https://doi.org/10.1016/j.funbio.2022.01.003
Vargas-Lopez A, Cicatiello C, Principato L, Secondi L (2022) Consumer expenditure, elasticity and value of food waste: a quadratic almost ideal demand system for evaluating changes in

Mexico during COVID-19. Socio-Econ Planning Sci 82(PA):101065. https://doi.org/10.1016/j.seps.2021.101065

Vazquez-Medina JA, Hernandez Cordero A, Lagos SI, Rangel Trujullo FR (2019) Gentrificacion y alimentacion en Santa Maria La Ribera. Estud Sociales. https://www.scielo.org.mx/pdf/esracdr/v30n55/2395-9169-esracdr-30-55-e20838.pdf

Vidal DR (2000) Ciencia y tecnología de alimentos. Arbor 166(653):127–139. https://doi.org/10.3989/arbor.2000.i653.1004

Willett W, Rockstrom J (2018) Alimentos Planeta Salud. La Comisión EAT-Lancet 32

Zapatero Gaviria A, Barba Martin R (2023) What do we know about the origin of COVID-19 three years later? Rev Clin Esp 223(4):240–243. https://doi.org/10.1016/j.rce.2023.02.002

Zhou LW, May TW (2023) Fungal taxonomy: current status and research agendas for the interdisciplinary and globalisation era. Mycology 14(1):52–59. https://doi.org/10.1080/21501203.2022.2103194

Zuleta P (2022) Interview about: uso terapéutico de los hongos alucinógenos: "todavía tenemos mucho por recorrer". France 24. https://www.france24.com/es/programas/salud/20220826-uso-terap%C3%A9utico-de-los-hongos-alucin%C3%B3genos-todav%C3%ADa-tenemos-mucho-por-recorrer

Chapter 10
Anticipatory Systems and Security in Mexican Cities. A Case-Based Review

José Luis González Madrigal

Abstract This chapter aims to present a reflection on some examples of urban security anticipation. Some attempts at anticipation based on urban actors are reviewed, as well as an analysis of their influence on the anticipatory capacity of the city in general. This will allow us to address the central question of the analysis: can a city by itself constitute a security anticipation system? In terms of security, it could be considered that one of the main problems in anticipating the future in cities is the existence of actors claiming both security and insecurity conditions. The discourses, actions, and mechanisms that allow individuals to anticipate a situation of insecurity are confronted with opposing intentions, making anticipation a difficult process. The chapter recovers concepts of anticipation, such as future literacy, identifying their relevance within an anticipatory system. From there, two cases of Mexican cities are taken up, which allow us to review these ideas in terms of security and anticipation. This discussion allows an analysis of the different types of urban actors involved in security, as well as an approach to their anticipation systems. Finally, it concludes with some considerations about security anticipation at the urban level.

Keywords Anticipatory systems · Urban security · Cities · Urban studies · Anticipation · Futures

10.1 Introduction

To speak of cities today is to think not only in terms of the space they occupy, the inhabitants who occupy them or the socio-economic dynamics they present, but also in terms of the potential that cities represent for individuals. On the one hand, security is an inalienable promise of the city, as well as one of the most challenging urban conditions. Between the search for security and the existence of a market of insecurity, the anticipation of security in cities is in constant dispute.

J. L. González Madrigal (✉)
Universidad Autonoma Metropolitana, Mexico City, Mexico
e-mail: jose.gonzalez@cua.uam.mx

S. Gonzalez-Arellano and B. Gandlgruber (eds.), *Cities as Anticipatory Systems*, The Urban Book Series, https://doi.org/10.1007/978-3-031-78162-9_10

This chapter aims to approach a reflection on the urban capacity to anticipate in terms of security. To achieve the above objective, the first section develops the relationship between cities, their aspiration for security and their need for anticipatory capacities. To do this, it draws on concepts such as future literacy, as well as its relevance within an anticipatory system. In general, some elements that allow the individual to anticipate are identified, in addition to linking this with elements related to security, in such a way that it is possible to approximate what is the basic structure for an anticipatory system in terms of security.

In a second section, we present the cases of two Mexican cities where an exercise of anticipation is intended, based on the experience of proposals and interventions for public security. The first corresponds to a deliberate exercise on the part of security agencies to break with the current situation of insecurity for the inhabitants of Juárez, Chihuahua: the *Todos Somos Juárez* program. The other case corresponds to a coordinated exercise between state agencies and citizens, in which the development of more desirable futures, different from what the prospective marked: the project Red *Conéctate por tu Seguridad*, of the Azcapotzalco district in Mexico City. From this, a third section shows an analysis of the anticipatory capacity for different actors, relevant to a security scheme, will be developed. The first of these is that of security promoters, among whom the governmental structure and its supporters are identified. The second group corresponds to the beneficiaries of security services, including citizens who have no authority to provide or reduce security. Finally, detractors of security, who benefit from breaking the law and violating others, as well as from the deterioration of the security situation, such as those who are not in favor of the governmental structure, the promoters of insecurity. The main delimitation criterion for these groups is the intention to change the security situation in the future.

The document concludes with an analysis of the state of literacy and the future in terms of security that present the cases analyzed. It is based on the above that the document aims to reflect on how the interaction of these different perspectives and aspirations in terms of security contributes to the possibility of integrating an urban anticipatory system. The question that will guide the discussion in the following: how do cities form an anticipatory system in terms of security?

10.2 About Cities, Security and Anticipatory Systems

When we talk about cities, we are not just talking about a community of people with common goals. Nor is it just the network of buildings and infrastructure that allows industries and services to operate. The dynamics and influence of cities go beyond these delimited concepts and are also interpreted as an instrument of human use; the urban has been consolidated as an environment that contributes to transcendental transformations that affect the environment and individuals (Brenner 2013).

In addition, a rhetoric of promise, of a shared desirable future, is usually introduced into the discussion when talking about cities. Among the most common expectations attributed to an urban society are ideas of development, quality of life, economic growth, among others. In general, it can be seen that for each of these promises, the underlying idea has been modified in order to add desirable elements, always with a view to what a city can be and what a city will be. This implies an important complexity about what we can expect from urban life.

In this sense, cities are always in a state of comparison between what they are and what they can be for their citizens. Their decision-makers and inhabitants have configured an expected future within their routine functional logic. This future serves as a guide for the design and operation of cities. In this way, the exercises of anticipation have specialized in the collection and analysis of statistics and metrics, in the creation of thematic observatories, as well as in the exercise of strategic planning in all areas that concern cities: environmental, economic, labor, among others.

10.2.1 An Uncertain Future: Urban Insecurity

One of the promises whose future is more difficult to predict relates to an essential requirement for living in cities: security. As a desirable ideal, security has been one of the motivations for the concentration of a large part of the population in large urban settlements. However, contemporary societies have seen that security remains a promise that seems out of reach, at least in the short term. Moreover, social expressions of insecurity seem to be increasing significantly, not only in frequency, but also in the nature and vulnerability they represent for the inhabitants.

The phenomena of (in)security have changed and become more complex as the urban dimension has grown. This is due to the increasingly specific demands that societies are placing on their social constructs. At the beginning of the discussion, the concept of security referred only to national security, which sought to protect territory from threats between states (PNUD 1994). A second approach is public security, which is the instrumentalization of government policies of territorial control based on the preservation of order in society (Montero 2013). In an alternative sense, internal security rejects the premise that it is a country, nation or territory that needs to be defended or from whom it needs to be defended; it is society that deserves protection and its dimensions of vulnerability are diverse (Montero 2013).

According to Krauze (1998), security approaches began to broaden to include new elements of threat and to be grouped into three main analytical segments. First, those that seek to protect a nation within state boundaries. A second approach remained within the spectrum of state action, but focused on the strengthening of inter-institutional and inter-state arrangements. Finally, those that broaden the protection agenda, taking the protagonism away from the state by including more individual issues, but extending the boundaries of attention to regional and global spaces.

The complexity of the notion of security takes into account not only differences in the criteria for government action, but also the objective it pursues. In this sense, human security emerges as an alternative to the notion of risk reduction and proposes as an additional objective that of peaceful coexistence in a multicultural world (Montero 2013). The duality of human security thus corresponds to the search for peace and development (PNUD 1994). According to Carreón and García (2014), the first approach corresponds to the solid modernity approach, which involves competition for political, territorial, legal, rational, bureaucratic, productive and authoritarian legitimacy, while the second type of approach represents the society of liquid modernity, which has transformed the legitimizing discourse through skepticism, which has brought risk, uncertainty, chance, and insecurity to the social individual.

The diversity of approaches from which the notion of security has been adapted points to a transformation of what is considered threatening. The individual's perception of his or her social and physical environment is what he or she identifies as an enemy or threat. This is particularly relevant in increasingly diverse societies, since the greater the complexity and size, the greater the uncertainty about the different environments and other inhabitants.

An example that synthesizes the above can be seen in the General Security Strategy in Mexico. In it, the Gobierno de México (2018) proposes various actions to achieve security, including the reorganization of police structures, operational guidelines, and the redesign of the functions of federal, state, and municipal agencies. However, the document also recognizes actions aimed at strengthening the ethical perspective of citizens, regaining the trust of citizens, formulating a culture of peace, among other actions that include issues not directly related to crime and violence.

In this sense, it is possible to recognize that the threat has approached the individual from various fields of action. When an individual or group does not behave like the rest of the population (due to lack of incentives to accept the institutional social arrangement), they will seek and provoke the disruption of the system (Foucault 2006), becoming a foreigner (risky) in their own territory. In this way, violence has become increasingly close to the immediate environment of the individual, forcing him to understand his counterpart no longer as an outsider, but as part of his environment; the disruption of peace is made by the neighbor, the community itself (Bauman and Donskis 2015).

It is also important to consider that the complexity of these risks is not absent from the intentionality of some individuals. The idea of criminal and violent acts as an economic instrument has favored the creation and maintenance of criminal groups that, organized or not, find profit in maintaining a climate of insecurity. This is particularly relevant at a time when criminal groups are professionalizing the work of insecurity in order to increase profits and compete for the power that violence brings.

On the basis of the above, it can be said that urban insecurity has become a complex construction in analytical terms, as well as in terms of anticipation. If the threat has diversified in terms of its causes, in addition to the fact that its proximity to the individual has increased, people living in cities find it more difficult to anticipate in a real way the risk to which they are exposed. In addition, the constant intervention of individuals or groups that seek to perpetuate crime and violence in the city creates a crossroads of security strategies.

10.2.2 Security and Anticipatory Systems

It is a fact that security is experienced differently in each city, but in general they show increasing rates of insecurity. This condition is particularly relevant at the Latin American level, since these are the cities that present the highest risk situation (Muggah et al. 2016). However, despite the complexity that cities present to understand and anticipate risk and threat conditions, the expectation generated by the city itself forces to rethink and reformulate new ways of approaching security. Security organizations are forced to broaden their scope in terms of policy areas and expand their strategic capacity to also anticipate developments and risks (Habegger 2022).

This section recognizes the existence of multiple efforts by city agencies to technify their work to provide security in cities. The mythification of technology and the so-called smart cities has generated an interesting debate on the use of technology for security actions. Among many alternatives, information technologies, cybersecurity and the Internet of Things have intervened in security processes (Vergel-Clavijo and Guerrero-Bayona 2023; Lechner 2016).

However, even though the proposals pretend to prevent, control, and deter the criminal acts, some of the results are not as conclusive as expected. Some of the initiatives only represent a way to technologize the security department, without representing an effective change in the security situation or an effective use of this technology (Alvarado 2018). In other cases, it's possible to observe that the initiative itself has brought unwanted risks (Williams et al. 2016).

In terms of the future, those technology initiatives are focused on rather reactive aspects, almost all of which are designed to combat violence, focusing on the effects of insecurity rather than providing security conditions for citizens. Even for those alternatives that could be interpreted as disruptive proposals, there is no indication of what kind of radical change or paradigmatic shift is being sought (Csernatoni and Oliveira 2023). In this sense, another approach that has taken place when talking about security and the use of futures has to do with thinking not in terms of a prospective logic that prevents insecurity, but in terms of resilience and adaptation to imminent risk phenomena (Habegger 2022).

Talking about the prospective of security in cities is not a new topic, since for several decades there has been evidence of cities that have developed initiatives to improve security conditions based on two key considerations: time projection and an orientation of the strategy to address risk (Muggah et al. 2016). But, as discussed in Chap. 1 of this book by González, talking about cities and futures requires both understanding them from the complexity of their system, as well as thinking about the mechanisms through which cities implement visions that have not yet materialized and transform them into processes of action. For this, it is possible to rely on the concepts of anticipation systems and future literacy.

With regard to anticipatory systems, these are tools (technological, social, cognitive, etc.) that allow actors to think about the future and act in the present, based on what the system allows them to observe. For Aquilué and Ruiz (2021), the anticipation of a system corresponds to the capacity to anticipate changes in the environment; to reduce uncertainty, understood as the complexity of the environment.

In a system such as the city, this logic corresponds to deliberate exercises of anticipation by the urban collective, subsystems that are formed to read, interpret, and give meaning to environmental, social, and economic dynamics in order to anticipate (a) what is likely to occur or (b) what may occur based on certain behaviors. It is from this reading of what the city will be in the future that efforts are proposed within the city to access or, as the case may be, move away from that possibility.

As Poli (2019) recognizes, anticipation occurs when the future is used for action. For the author, the above implies a fundamental notion of anticipation, but also the capabilities to take advantage of future information, starting from the intervention of a present situation. In this sense, cities use anticipation exercises with different levels of complexity, ranging from foresight exercises to trend analysis. Cities that present successful security interventions are those that are able to establish strategies thought in short and long-term horizons (Muggah et al. 2016).

According to Boer et al. (2018), it is possible to articulate capabilities, making sense of future literacy. This concept can be disaggregated from an analytical perspective into three considerations: the articulation of anticipatory systems; the use and exploitation of futures (contingent, optimizing, and exploratory); the construction of a sense of use for the information resulting from anticipatory systems.

González-Arellano (2022) considers that the construction of futures is a regular, cumulative, and collective process, since they are the result of an interactive process of narratives. This construction corresponds to the identification of a possible imaginary for those who think about it. If this is true, then the capacity of cities to imagine futures depends directly not only on the outcome of the interaction between individuals, but also on the capacity of its members, including their biases or intentions, in the process of producing future imaginaries.

More generally, this chapter argues that the anticipatory capacity of cities in relation to security cannot be interpreted in a comprehensive way for a city as a whole. It is noted that some cities propose anticipation as a decision-making tool, but it is also observed that there is more than one competing anticipation system, which together inhibit the proper implementation of each one. In this sense, Aguirre (2016) points out that the anticipatory capacity of state agents in security matters is insufficient as long as they are unable to anticipate the operation of criminal agencies.

In this sense, an urban security anticipation system is understood as one that allows, in a clear and simple way, (1) the future literacy (of its actors) on security, (2) to translate this ability to read (foresee) possible scenarios into actions in the present for the transformation of some of the causes that trigger the phenomena of insecurity, as well as (3) to give a sense of use to the information of the future. In this sense, it is not enough to take measures to prevent crime; it is necessary to implement actions that promote a climate of security.

From the perspective of anticipation proposed in this section, it could be argued that one of the main problems in developing anticipatory systems in relation to urban security is the existence of actors who seek security, but also those who prefer insecurity. In this sense, the discourses, actions and mechanisms that allow individuals to anticipate an insecure situation are confronted with opposing intentions and actions, thus hindering the process.

10.3 Urban Security in México. Examples of Futures-Based Action

In addition to the complexity of understanding what insecurity is, the identification of causes and motivations are diverse in the way they manifest themselves and in the social sphere to which they belong. Some of the factors that foster a climate of insecurity are poverty, precarious infrastructure, domestic violence, youth gangs, impunity in the administration of justice, low social cohesion, presence of weapons and drugs, corruption in government agencies, poor education system, child abuse, among others (Arriagada 2001; Gobierno de México 2021; Herrera-Lasso 2013). In this sense, the way in which insecurity manifests itself in the city varies as much as the factors that cause it are present in the city.

In particular, the scope and intentions of urban security schemes vary from city to city. From a forward-looking perspective, crime prevention as an approach is one of the simplest mechanisms used by local governments. It corresponds to a strategy that anticipates an outcome and integrates actions in the present to either promote or prevent it. Since the 1990s, the formulation of strategies that accompany the fight against crime with prevention programs has had favorable results in Latin American cities, compared to those that seek only to combat crime (Arriagada and Godoy 1999).

This chapter seeks to deepen the analysis of the use of futures for urban security, in order to contribute to the reflection on anticipation. In this sense, the next section examines some proposals that, in addition to prevention approaches, recognize the review of future scenarios, and transform them into present decisions that seek to modify the observed trend in a much more active way. Each of the examples will briefly develop what the program consisted of and then review it from the perspective of anticipation.

In order to provide evidence for the central discussion of the chapter, one case was selected in which the exercise corresponds primarily to the governmental sphere, and the second case corresponds to a civil society effort. The purpose of bringing these perspectives together is to provide information about different positions and the construction of futures by different urban actors.

10.3.1 Enough of This Present: Governmental Program "Todos Somos Juárez"

As a government response to the significant increase in violence and crime, the *Todos Somos Juárez* (TSJ) program was implemented in Ciudad Juárez (Juárez City), in the state of Chihuahua, México. Although the program was short-lived (the efforts were concentrated mainly in 2010 and 2011), there was a significant decrease in the number of manifestations of insecurity, as well as an increase in citizen intervention in these matters. It even went so far as to try to replicate it in other cities, such as Acapulco (Rosas 2012), and to use similar mechanisms in other Mexican cities (Arratia 2017).

Although the program reacted to the critical conditions of insecurity experienced in the city, it structured a redesign of security policies in a transversal way, looking for the present causes of a future problem (increase or maintenance of the level of violence). This made it possible to outline preventive measures to deal with the problem, but also to implement emergency measures to deal with the problems experienced. In general, the structure of the effort was intended to encompass the three levels of government, to work intersectorally, to operate intensively, and to actively involve citizens in the diagnosis of the problem and in the operation of the program (Muggah et al. 2016).

A fundamental part of the program consisted in the creation of working groups to identify alternative solutions to the problem of insecurity through intensive citizen participation (op. cit.). Based on the intention of restoring order and the social fabric, resources were used for the implementation of 160 actions in six areas: employment, social development, education, economy, health, and public security (Paniagua and Camargo-González 2017).

In terms of anticipation, the operation of the program highlights the integration of working tables with the purpose of (a) diagnosing the problem (present situation) and (b) establishing routes for action (visions of the future). In addition, the intensive intervention of resources, objectives, and instruments on the city suggests the intention to bring to a convulsive present some tranquility of what the program could offer in the future.

According to Arratia (2017), the results obtained by *Todos Somos Juárez* have been beneficial for the city in two ways. The first is related to the improvement of conditions that allow social development: investments in education, health, economic incentives such as social assistance and credit, among other actions. In terms of security, there has been a decrease in six types of crime: homicide, car theft with and without violence, kidnapping, extortion, and commercial robbery.

Some of the deliberations on the operation of *Todos Somos Juárez* are coinciding with an effective application in the short term, but failed in its execution in the medium and long term. It is even said that once the program was implemented, the longer-term horizon was forgotten (Muggah et al. 2016). Some analysts attribute this trend simply to the political and governmental changes that occurred due to the presidential transition that took place in 2012. However, there is evidence of security effects that were not intended by the policy, one of which was an increase in cases of violent acts by local security agencies (Arratia 2017).

According to Sumano (2018), violence and crime are still manifested in Juarez after the results of the *Todos Somos Juarez* operation, but what has changed is the way they are expressed in the city. For example, the reduction of extortion and rent-seeking crimes in the city is acknowledged; however, the number of homicides, while decreasing in the short term, increased again after the operation. In addition, the city has seen an adjustment in criminal strategy, as there has been a change not only in the cartel present in the area, but also in the drugs consumed by a significant portion of the population, as well as in the type of crime committed in the city.

10.3.2 Proposing Community-Based Futures: Citizen Empowerment Projects and "Red Conéctate por Tu Seguridad"

Another example of a response supported by the review of the futures can be seen in the citizen empowerment projects of the civil association *Mexicanos Unidos Contra la Delincuencia* (MUCD), and more specifically in the *Red conéctate por tu seguridad* (RCS) project, implemented in different municipalities in Estado de México state. The duration of these projects is about 3 years, with most of the actions identified between 2019 and 2021. Projects of similar intent have been carried out in other municipalities of the same state.

Based on the initiative of MUCD, a series of projects have been implemented in the years around 2019 to strengthen social cohesion through the promotion of citizen competences, improve community coexistence, and promote the recovery of public space (MUCD 2019a, b; Alcaldía Azcapotzalco and MUCD 2021). Although the objective of these programs is directly expressed as improving citizen security, since they do not operate on the basis of government action, the focus was not on fighting crime, but on transforming the conditions in which citizens operate.

The general methodology from which the program operated was the training of "facilitators" who would enter into a dynamic of training more trainers and creating networks of citizens to (a) increase social cohesion by locating and strengthening leadership in the community, as well as (b) activate citizens in the implementation of collective actions and citizen empowerment. The idea is the operation of actions from the citizenry, as well as the strengthening of a citizen criterion of respect and care for the environment and the conditions of legality (MUCD 2019a).

The results of the programs do not focus on the actual reduction of crime, but on the transformation of social dynamics. Relevant for the present reflection is the replication of the working groups, not by MUCD, but by the citizens who participated in the primary sessions of the project. That is, the adoption of alternative methodologies to understand and confront the problems present in the community, as well as the proposal and execution of projects by citizens.

In terms of anticipation, the program did not operate with the search for or identification of a specific future, but rather favored the formation of citizens' capacities to visualize and differentiate between possible futures based on their interventions in the community. According to the results reports, the program seeks to train citizens who can lead their community to an environment of legality, not by sharing specific topics and information, following a rigid scheme of citizen training, but by identifying, strengthening, and fostering citizen capacities.

The long-term goals of the project are to achieve peaceful coexistence, the prevention of violence and crime, and the orderly and peaceful use of public spaces (op. cit.). This objective is set not only without a clear roadmap, but also without specific attributions, as could be observed in projects implemented by the government. The bet corresponds to the formation of networks that are adaptive and proactive, based on projects for the community itself.

The identification of a future to be achieved and the recognition of a limited capacity for action in the medium and long term led to the integration of the project into a series of subprojects which, although not articulated, correspond to proposals for present actions for this desirable future. They respond to a kind of "echo" from which present actions modify the state of the question and, in the following deliberation, the network will be able to integrate new action alternatives for this new present, based on the new future identified.

In the specific case of the municipality of Azcapotzalco, the project was carried out in coordination with the mayor's office, which has made it possible to link the citizen's exercise with government action. The goal of this coordination is not only to build community networks, but also to take advantage of other mechanisms for monitoring insecurity by the security agencies in order to improve the understanding of the criminal phenomenon (Vistazo al futuro 2021). For this part of the project, no open evidence of results is identified due to the modifications or adjustments made to the task of combating insecurity.

10.4 Security Anticipation Capacity

As mentioned above, the anticipatory capacity of a city is usually interpreted as the prevention and foresight mechanisms and tools developed by security agencies. However, it has also been argued that this usually does not effectively refer to the anticipatory capacity of the city itself, since it is possible to identify actors who intervene in the security situation, making the anticipation process difficult.

In order to analyze these capacities on the basis of the anticipation literature, it is necessary to recall that anticipatory capacity can be summarized as that literacy about the future of security that allows foreseeing possible scenarios, thus enabling present actions and decisions. This type of capacity has a longitudinal scope, so that it is possible to speak of short-, medium-, and long-term anticipatory capacity. In addition, it should be remembered that not all uses of futures are equally relevant in terms of anticipation, since there are uses of contingent futures that have to do with a short- and medium-term reaction. There is another type of use of the future that has to do with the optimization of processes that are currently taking place in cities. Finally, there are exploratory uses of the future, which aim to significantly modify the trend of urban functioning and thus achieve future processes that are different from the current trend.

On the basis of the above, it would be possible to categorize the capacity of anticipation according to the temporal scope and the type of use of the future. In this line of thought, the peculiarities of the contingent use of the future force urban actors to make proposals with a short-term scope, conditions that limit their capacity to a low level. A medium capacity, on the other hand, has to do with the intention of optimizing the efforts made in the present in the future, which is why it tends to focus on the short to medium term. Finally, a high capacity for anticipation has to do with the use of the future, which seeks an exploratory future different from that proposed by the present trend, so it always pretends the long term.

The scope of anticipation offered by the *Todos Somos Juárez* program is medium to high. As mentioned above, the program seems to correspond to a process of breaking out of an undesirable present and the articulation of joint efforts (between government and citizens) to orient toward a better future; this implies that the focus was taken away from contingent anticipation and a style of exploratory anticipation was adopted, at least at the time. However, this intentionality was rather short-lived, and it was implemented in a rigid manner, so that once it was implemented, it stopped anticipating, causing a setback in the progress that had been made. In addition, the visualization process, although it tried to include as many aspects as possible, seems to have ignored the reaction of the criminal groups, so that when they changed their behavior, the project ceased to be effective and failed to achieve good results.

In terms of the anticipatory capacity of the Red Conéctate por tu Seguridad, this is identified as medium–low. Although the project does not intend to make a radical break with the present, there is a clear intention to modify the situation based on what citizens seek to improve in the future. It seems to recognize that the intended ideal is rather long-term, so no clear actions are established to achieve it, but it is established as a lighthouse that will allow to read the situation again in the future and to break this new present with another advocacy project. In other words, it corresponds to a long-term vision, but with contingent actions, in recognition of a limited advocacy capacity. In this sense, it favors an approach of replicable anticipation that encourages repeating the process as long as initiatives exist in the city. Unfortunately, there is no evidence of a third replication of the initiative after the projects that followed the implementation of the network (by MUCD).

Thus, in order to answer the guiding question of whether it is possible to speak of an urban anticipatory security system, this section examines the future literacy of different types of urban actors in the cases described above, as well as the application of this capacity to determine actions to be developed in the present. To this end, an analysis will be made based on three types of actors that can be found in a city: promoters of security, beneficiaries (or victims) of the urban condition, as well as detractors, those who benefit from a climate of insecurity. Each of these actors may have different capabilities to anticipate security in the same city, and these competencies are complicated by each other.

10.4.1 Security Promoters. Anticipation Capacity Medium–High

In order to interpret the anticipation capacity of the security promoters, it is necessary to recognize that the information on this task corresponds to strategic information of the States, so that even from reports and reports such as those presented above, they correspond only to a partial vision of the intelligence exercise carried out by the security promoters. However, the indicators of insecurity are recognized as an approximation of the efficiency of such processes.

In this sense, according to the information observed, it is possible to say that the capacity for anticipation, in terms of constructing future security scenarios, is also strong in terms of contingent future literacy and optimization. The security forces and governmental structures in charge of security build their scenarios from first-hand information, in addition to classified information, so they are the actors with the greatest amount of formal information on the phenomenon. Their anticipatory systems are strong in terms of medium- and long-term foresight.

Among its peculiarities, it is possible to mention that its prospective capacities formalize strategic information for the construction of possible scenarios and possible policies, anticipating situations of risk and vulnerability to crime and violence. By networking with other actors, it is able to extend its anticipatory capacity toward an exploratory logic. In addition, the way in which it determines unsafe and safe spaces depends mostly on the actions and strategies of criminals, so it is naturally forced to keep a large part of its anticipatory processes in reaction.

Its ability to use futures remains, at best, in the medium term. Its obligation to maintain the climate of peace, in addition to establishing itself as the agency that seeks to monopolize violence, keeps it in a permanent function of attention to expressions of insecurity. This keeps security planning permanently in the short term. This does not deny the possibility of establishing long-term scenarios by state actors, but given their short duration (based on political processes), it requires additional incentives to maintain this vision; a situation that in most cases does not happen.

On the other hand, the promoters of security are the actors who primarily collect information on acts of insecurity, as well as on measures and means to combat them. However, in a formal way, they can only relate to the beneficiaries of the programs, since the detractors act informally, in addition to establishing themselves as agents of competence in security matters. This leads to a partial identification of information.

This is particularly relevant in terms of anticipation. By referring to a construction process that goes beyond empirical information, operating from segmented information (intentionally) only complicates the anticipation process, since it is constantly confronted with anticipatory processes that seek to counteract the actions that seek specific scenarios.

10.4.2 Beneficiaries. Low Anticipatory Capacity

Security is not so much a result as a state that is experienced. In this sense, there are people who benefit from its expression in cities: the inhabitants. For them, their ability to anticipate security is linked to their ability to observe whether this condition is met in the environment in which they live. In this sense, the signals that individuals receive to interpret their future security situation are weak and diffuse.

The influence on the individual perception of urban security is constructed from incomplete information, as well as from unverifiable anecdotal information, so that the construction of the scenario of insecurity cannot be clearly formed. In addition, he is aware that moments of vulnerability (risk and threat) depend on the will of others rather than on foreseeable conditions, which prevents him from clearly determining the (un)safe space in the living environment. Thus, their capability to identify future scenarios is weak.

Their capacity to use and take advantage of their futures remains in the short term, because in the absence of specific incentives for this purpose, there are no incentives in the environment to operate them. On the one hand, the inability to formalize a clear scenario based on the signals he receives prevents him from proposing viable alternatives or modification mechanisms. In addition, the individual is limited in his ability to influence security; not only is he unable to use force when insecurity occurs, but he also does not have the resources or power to comprehensively address the structural problems that cause it.

For the beneficiary, it is impossible to give a sense of use to the information of the future, since one version is constantly pitted against the other. Understanding that security depends on a collective construction, the individual constantly observes how his perception of security is constantly confronted with information and scenarios proposed by other citizens, but also by promoters and detractors of security.

10.4.3 Detractors of Security—Anticipatory Capacity Medium–High

Although when we talk about the city and its capacities, we usually refer to the formal structures that make it up, the existence of individuals and groups that operate in a formal manner is also a reality. In terms of security, Ferrato (2014) recognizes the strategic capacity (both intelligence and structural) of criminals to influence conditions within a state. Within each city, this works in the same way. According to Arratia (2017), in spaces where insecurity phenomena operate, those who promote them and those who fight them find themselves in a confrontation of actions, strategies, and alternatives to win the loyalty of the inhabitants.

According to the National Security Strategy, the authorities in charge of security in México have enough problems to stop and dismantle criminal strategies, while the results seem to be an increase in insecurity. In addition, the document repeatedly acknowledges the infiltration and even complicity of public officials and government agencies. "As long as any authority struggles with political conflicts, crime advances in an organized manner" (Gobierno de México 2018, p. 10).

The above is not insignificant. Criminals operating in cities seem to act from processes very similar to those of formal structures, that is, they generate and make sense of strategic information. In addition to establishing actions and operations that reactively counter the actions of security promoters. In this sense, the prospective information with which they operate allows security detractors to exercise a strong capacity of anticipation.

With respect to the observed evidence, it is not possible to identify capabilities for future scenario building. However, the rapid response, as well as the crime and violence indicators obtained after the projects studied, point to a contingent and optimizing futures literacy. In addition, this type of actor has the capacity of cohesion over other actors. By linking with other actors, they are able to expand their anticipatory capacity toward an exploratory logic.

In the specific case of Ciudad Juarez, several analysts have considered the possibility that the efforts, resources, and actions developed by the Todos Somos Juarez program were not the reason for the change in the insecurity scheme, but that this could be explained by the activities of criminal groups (Arratia 2017; Sumano 2018). For its part, the evidence shows that in Azcapotzalco, despite the projects developed, the expressions of insecurity remained relatively stable.

The proactive nature of these actors means that their ability to use futures remains in the medium term, since security promoters, whatever their strategy, are forced to reduce it. This keeps security planning permanently in the short term, but also in the search for alternatives to expand its influence on the (in)security situation.

10.5 Anticipating the Future of Urban Security

As mentioned above, the reference to the future of security is usually interpreted as those actions of technification and prospective that the security agencies carry out in order to optimize their capacity for action. It is recognized that there are proposals in which artificial intelligence, the use of satellites and video surveillance systems represent the future of security, because in this way police agencies could improve their performance.

These approaches correspond to the traditional logic of understanding the anticipation of security in cities. However, the chapter also recognized that this approach does not consider the city as a complex system in terms of its actors, the objectives they pursue and the powers they have to achieve them. While there are government agencies that are strengthening their technological foresight capabilities, there are citizens who remain expectant of what is being done in terms of security. Moreover, there are actors who deliberately try to block the anticipatory capacities of security promoters by professionalizing a system of anticipating insecurity.

From the review of the evidence discussed in this chapter, the existence of anticipation by cities in terms of security is undeniable, but it is not as effective as the traditional approach shows. From the interaction of these three actors observed in the cases presented, it is possible to reflect on the anticipatory capacity of cities, which is more limited than we would like. As can be seen, the anticipatory capacity of the three different actors seeks to achieve different (in)security objectives, including limiting, hindering, or preventing the ability of the counterpart to anticipate its own actions.

In this sense, it is possible to confirm that security as a phenomenon is not only subject to very different structural social conditions, but also to the will of the actors involved in acts of insecurity. The formal responsibilities of state actors, as well as the interests of their opponents, allow these actors to exercise discretion in their response to (in)security. It is convenient for them not to provide information about their actions (e.g. high-impact security strategies or the strategies and actions of criminals). This favors a literacy about the future that varies between contingent and for optimization in general.

In general, the possibility of anticipating and understanding the future in terms of security by the urban system has been incapacitated by several factors, among which stand out: (a) an exponential growth of diffuse and sometimes contradictory signals, (b) a distancing of the motivations and capacities for intervention in security by the actors involved, and (c) a constant confrontation between promoters and detractors of security by the situation of crime and violence in the present and very short term.

The interplay of diffuse signals, as well as the inability to formalize a structure for constructing and using knowledge about the future in terms of security, has made it difficult not only for the authorities but also for the city's inhabitants to become literate in these terms. This makes it difficult to modify or even anticipate the security dynamics in which cities live; thus, the proposals and alternatives for intervention remain in the short term, in addition to favoring attention to the effects over the causes that generate them.

Although the cases studied present information that allows us to confirm the above, they also provide interesting evidence for reflection in terms of anticipation. One of the main peculiarities of *Todos Somos Juárez* was its implementation out of necessity. Faced with the recognition of an undesirable present for the city, it seeks to break with it by anticipating both desired and undesired possibilities.

The integration of different urban issues refers to the prediction of more than one outcome of the intervention. This is possible by examining the interaction of possible futures in order to select the most desirable one. The involvement of citizens and governmental actors at different levels also refers to a deliberation on the feasibility of the futures, which refers to an assessment of alternatives and an interpretation of the real scope that the futures would have from actions in the present. This allowed the practice of a more explorative reading of the future, although it was limited in its capacity to make use of this resource due to political restrictions.

On the other hand, the *Red Conéctate por tu Seguridad* project, although born out of the need to improve security conditions, does not represent such a marked break with the present, but rather a need to adapt the living conditions of the city. In this sense, the project was characterized by visualizing futures (short and medium term) and, based on a desirable future (long term), articulating actions to act in the present. Since we were not able to clearly visualize the results in the medium and long term, the tactic used was to sequentially rethink the logic of the project and adapt to the new present once it was achieved.

An important feature of the project has to do with its ability to establish the culture of legality as a guiding, but not central, element in the construction of intervention initiatives. That is to say, although the association promoting the project intended to train the participants in legality, it made it possible for the successive interventions and projects (now implemented by the beneficiaries of the workshops) to be adapted by each one of them. This opened up the possibility of seeing and proposing futures based on the observations of each of the participants.

Although the Municipality of Azcapotzalco was one of the promoters of the project in the city, there is little evidence of the use and application of the information generated in terms of futures. In this sense, although an alliance between the anticipatory capacities of both the promoters and the beneficiaries of security is identified, it is not possible to verify that they have worked together and have had results.

Moreover, speaking of an urban system of anticipation forces us to think not only about state interventions, but also about the interaction that these interventions represent in the city as a whole. In this sense, the evidence allows us to observe a very similar capacity for anticipation between the security detractors and the security promoters. Although security benefits were obtained during the implementation of the program, the quick reaction and adaptation of the criminal groups allowed a setback in this situation. Their adaptation of strategy is evidence of their ability to explore new futures based on contingency, which overcame the initial strategy of the security promoters.

10.6 Final Thoughts

The available information on the complex relationship between crime increase and decrease in both cities only reflects the high adaptive capacity of both security promoters and detractors. In both cases, after the change in the (in)security situation, a change in the opponents' strategy was observed, leading to a return to the initial situation. Beneficiaries, on the other hand, remain with a low anticipatory capacity, as they generally continue to search for alternatives in the short term and begin to operate in a contingent manner.

According to the evidence presented, this is related not only to their anticipation capacities, but also to their formal and informal functional structures. The former operate on the basis of rigid mandates and during periods that interrupt medium- and long-term processes. On the other hand, the second type prefers to operate on a discretionary basis and therefore has an interest in maintaining a dispersed and unclear system of signals about the future.

It can be said that the cities studied have been able to structure and give meaning to the anticipatory systems in a functional way for their operation, but keep them in a medium capacity. However, the peculiarities found give important information about the possibility of expanding and operating urban anticipatory systems in terms of security. The positive results in both cases allow us to reflect on the virtues and contributions of anticipation to urban security.

Recognizing the complexity of actors, objectives, and mechanisms that exist within cities to address security strategies is relevant to improving urban anticipation. An urban anticipation system cannot consider only one of its elements, but must include them or, in the worst case, consider the existence of the rest. The issue of urban security is particularly challenging, since the actors not only have different objectives, but there are also marked oppositions, which implies the will to obstruct and deny the possibilities of the other to achieve the desired future.

In this sense, it is undeniable that the technological tools that have been introduced for security agencies are bringing us closer to this goal. Information management mechanisms and new institutional structures allow us to imagine scenarios in which anticipation can be strengthened throughout the city. Security agencies must continue to strengthen their prospective and adaptive capacities in order to visualize scenarios, so that today it is possible to develop actions that are neither reactive nor traditional prevention, but rather transform urban conditions.

An indispensable consideration for a dialogue on the urban future of security has to do with observing the models and tools of prospective with a lens that recognizes the distortion generated by the existence of a strategy that pretends to anticipate in turn, but with the opposite ends in mind. That is, to recognize the game of anticipation that both security agencies and those who seek insecurity seek. In this aspect, it seems essential to involve the inhabitant actor in order to provide additional information on these future scenarios.

References

Aguirre J (2016) La tecnología de información y comunicación en prevención del delito. URVIO 18:90–103

Alcaldía Azcapotzalco & MUCD (2021) Reporte final de actividades de intervención y resultados 2019–2021. Retrieved 15 Nov 2023, from https://www.mucd.org.mx/wp-content/uploads/2021/10/Reporte-RCxTS-Azcapotzalco-2019-2021.pdf

Alvarado N (2018) Tecnología contra el crimen: Entusiasmo con cautela y criterio. Sin miedos. Seguridad Ciudadana. https://blogs.iadb.org/seguridad-ciudadana/es/tecnologia-contra-el-crimen-entusiasmo-con-criterio/

Aquilué I, Ruiz J (2021) Ciudad, complejidad y cambio: fundamentos para el análisis de la incertidumbre en sistemas urbanos. Revista INVI 36(101)

Arratia E (2017) Todos Somos Juárez. Competititon in state-making y la guerra contra el narcotráfico (2006–2012). Rev Esp Cienc Polít 43:83–111. https://doi.org/10.21308/recp.43.04
Arriagada I (2001). Seguridad ciudadana y violencia en América Latina. In: LASA 2001 XXIII international congress session, Washington
Arriagada I, Godoy L (1999) Seguridad Ciudadana y violencia en América Latina: diagnóstico y políticas en los años noventa. CEPAL. Serie de Políticas Sociales 32, LC/L. 1179-P
Bauman Z, Donskis L (2015) Ceguera moral. La pérdida de sensibilidad en la modernidad líquida. Paidós
Boer A, Wiekens C, Damhof L (2018) How futures literate are you? Exploratory research on how to operationalize and measure futures literacy. In: 6th international conference on Future-Oriented Technology Analysis (FTA)—future in the making Brussels
Brenner N (2013) Tesis sobre la urbanización planetaria. Nueva Soc 243
Carreón J, García C (2014) Teorías de la seguridad pública y percepción del delito. Margen 71
Csernatoni R, Oliveira B (2023) Disruptive technologies for security and defence: temporality, performativity and imagination. Geopolitics. https://doi.org/10.1080/14650045.2023.2224235
Ferrato A (2014) Crimen organizado e inteligencia estratégica: desafíos para el siglo XXI. URVIO 15:55–70. https://doi.org/10.17141/urvio.15.2014.1587
Foucault M (2006) Seguridad, territorio, población: Curso en el Collège de France: 1977:1978. Fondo de Cultura Económica. Buenos Aires, Argentina
Gobierno de México (2018) Estrategia Nacional de Seguridad Pública. Retrieved 15 Nov 2023, https://www.gob.mx/cms/uploads/attachment/file/434517/Estrategia_Seguridad-ilovepdf-compressed-ilovepdf-compressed-ilovepdf-compressed__1_.pdf
Gobierno de México (2021) Factores de Inseguridad. Observatorio de reclutamiento. Retrieved 15 Nov 2023, https://www.gob.mx/sspc/observatorioreclutamiento/documentos/factores-de-inseguridad
González-Arellano S (2022) The future scenarios of cities: an analysis of their institutional construction abstract. Int J Urban Reg Res 46(5):865–873. https://doi.org/10.1111/1468-2427.13103
Habegger B (2022) Securing the future: the use of strategic foresight in the security sector. Geneva Centre for security policy. Strateg Secur Anal 23. https://doi.org/10.21256/zhaw-28391
Herrera-Lasso L (2013) Factores que propician la violencia y la inseguridad: apuntes para una estrategia integral de seguridad pública en México. CEI-El Colegio de México, México. Retrieved 15 Nov 2023, https://cei.colmex.mx/archivos/213/Luis_Herrera-Lasso__Factores_que_propician_la_violencia_y_la_inseguridad.pdf
Krauze K (1998) Theorizing security, state formation and the 'Third World' in the post-Cold War world. Rev Int Stud 24(01):125–136. https://doi.org/10.1017/S0260210598001259
Lechner M (2016) Tecnologías aplicadas a la seguridad ciudadana: desafíos para la justicia transicional ante nuevos mecanismos de control social. Divulgatio 1(1). https://doi.org/10.48160/25913530di01.9
Montero J (2013) El concepto de seguridad en el nuevo paradigma de la normatividad mexicana. Reg Soc XXV(58). https://doi.org/10.22198/rys.2013.58.a128
MUCD (2019a) Cultura de la legalidad Centro Histórico. Polígonos Santa María de la Redonda y la Antigua Merced. Retrieved 15 Nov 2023, https://www.mucd.org.mx/wp-content/uploads/2017/10/ESTADO-DE-ME%CC%81XICO-2015-TUI.pdf
MUCD (2019b) Azcapotzalco. Plataforma de Participación ciudadana 2019–2021. Retrieved 15 Nov 2023, https://www.mucd.org.mx/azcapotzalco/
Muggah R, Szabó de Carvalho I, Alvarado N, Marmolejo L, Wang R (2016) Haciendo de las ciudades lugares más seguros: Innovaciones sobre seguridad ciudadana en América Latina. Artículo estratégico 20 de junio 2016. Ed. Instituto Igarapé, Banco Interamericano de Desarrollo y Consejo sobre Fragilidad, Conflicto y Violencia de la Agencia Global del Foro Económico Mundial
Paniagua A, Camargo-González I (2017) La implementación de las políticas públicas para la prevención de la violencia en el municipio de Ciudad Juárez, Chihuahua: el involucramiento de la

sociedad civil de los liderazgos comunitarios. Acta Universitaria 27(4):92–108. https://doi.org/10.15174/au.2017.1457

PNUD (1994) Informe sobre desarrollo humano 1994. Fondo de Cultura Económica, México

Poli R (2019) Introducing anticipation. In: Handbook of anticipation theoretical an applied aspects of the use of future in decision making. Springer. https://doi.org/10.1007/978-3-319-31737-3_1-1

Rosas T (2012) Replican Todos Somos Juárez en Acapulco. El Economista. 31 de mayo de 2012

Sumano J (2018) La violencia en Ciudad Juárez: a seis años de la estrategia Todos Somos Juárez. Nexos—Paz Seguridad, Julio 2018. Retrieved 15 Nov 2023, https://seguridad.nexos.com.mx/la-violencia-en-ciudad-juarez-a-seis-anos-de-la-estrategia-todos-somos-juarez/

Vergel-Clavijo G, Guerrero-Bayona A (2023) Ciudad inteligente: mejoramiento de la seguridad a través del uso de nuevas tecnologías. Rev Ingenio 20(1). https://doi.org/10.22463/2011642X.3510

Vistazo al futuro (2021) Se activa "Red Conéctate por tu seguridad" Azcapotzalco. 5 de marzo de 2021 Retrieved 15 Nov 2023, https://vistazoalfuturo.com/se-activa-red-conectate-por-tu-seguridad-en-azcapotzalco/2785/

Williams M, Axon L, Nurse J, Creese S (2016) Future scenarios and challenges for security and privacy. In: IEEE 2nd international forum on research and technologies for society and industry leveraging a better tomorrow (RTSI), Bologna, Italy, pp 1–6

Chapter 11
Anticipating Zero Plastics Futures in Urban Contexts in Mexico

Bruno Gandlgruber and Marco Jaso

Abstract The production and consumption of plastics is closely related to the prosperity, but also the calamities of the modern socioeconomic system in terms of the generation of excessive waste. And cities have been considered the primary origin of plastic contamination of all land and marine ecosystems. In response, different stakeholders at a global and local level have used diverse images and narratives to create scenarios of the future to illustrate the long-term consequences of plastic pollution, but also strategies of mitigation and adaptation. In this chapter we explore some of the main contributions, following the conceptual distinction between foresight, forecast, and anticipation in contemporary futures studies. We are particularly interested in narratives related to long-term efforts to radically reduce plastic pollution in urban contexts, also called zero plastics futures, and their influence on international regulatory efforts to cope with this challenge. In the last section we develop some elements for a fundamental approach to conceive the urban plastic economy as an anticipatory system to achieve a deeper understanding on the characteristics of the role of urban anticipatory communities. We use a brief overview of important advances in narratives of regulation for the mitigation of plastics on a national and city level in Mexico, with particular emphasis on the relation to zero plastics.

Keywords Anticipatory systems · Anticipatory communities · Future scenarios · Plastics economy · Zero plastics policies

B. Gandlgruber (✉) · M. Jaso
Department of Institutional Studies, Universidad Autónoma Metropolitana-Cuajimalpa, Mexico City, Mexico
e-mail: bgandl@cua.uam.mx

M. Jaso
e-mail: mjaso@cua.uam.mx

S. Gonzalez-Arellano and B. Gandlgruber (eds.), *Cities as Anticipatory Systems*, The Urban Book Series, https://doi.org/10.1007/978-3-031-78162-9_11

11.1 Introduction

Plastics have many important properties for a wide variety of applications. Still, the socially and economically established practices of their use, mainly because of the intensive processing of fossil fuels and the inadequate forms of plastic waste disposal, have generated a wide variety of health and environmental risks, which has led to a growing consensus on the importance to reduce its production and consumption and to transform the sociotechnical system that it sustains toward a configuration that mitigates these risks. The transformation of this system toward a significant reduction of the negative impacts is a complex problem on a global scale that involves many different actors in different social spheres.

This chapter analyzes different aspects of this problem. It develops an initial analytical proposal based on the urban plastic economy as an anticipatory system that creates images, narratives, and models of a future with a strictly reduced and modified form of plastic use in urban contexts. We critically analyze the use of the future in documented proposals of this kind of transformation process. Capacities of anticipation refer not only to strategic planning and foresight by core decision-makers like multinational companies and environmental agencies in different levels of government but also to the required societal potential of imagining and applying changes on the social institutions or rule systems as guided by future narratives. As the primary origin of plastic contamination of (land and sea) ecosystems, cities play a central role in these transformation processes.

The chapter includes four main sections. We begin in the first section with a general explanation of the expansion of the plastics economy as a complex social system of production and consumption in recent years and the growing risks to health and the environment because of the often uncontrolled diffusion of plastic debris into the environment. In the second section, we analyze the particular uses of images, narratives, and models of the future related to the plastics economy, following the distinction of the three levels of future studies: foresight, forecast, and anticipation. At the end of the second section, we develop a fundamental approach to conceive the urban plastic economy as an anticipatory system. In the third section, we give a brief overview of important advances in narratives of regulation for the mitigation of plastics on a global and national level in Mexico, with particular emphasis on the proposal of zero waste or zero plastics as a scenario where plastics are completely irradicated at some moment in the future. The fourth section looks at this process of building rules as the institutionalization of futures in regulating plastic use at the level of Mexico City. The information we gathered allows us to characterize different aspects of the urban plastics economy in Mexico as an anticipatory system, like elements of its structure and specific anticipatory capacities of visualization and transformation.

11.2 Plastics, Plastics Pollution, and the Plastics Economy

Plastics as non-natural polymeric systems[1] like polyethylene, polyester, or polypropylene are considered one of the major human inventions in modern times, and they are ubiquitous in contemporary production and consumption systems for purposes like storage, transportation, packaging, or insulation. Their flexibility, adaptability, and resistance have made it possible to substitute wood, metal, or glass at lower production costs (Andrady and Neal 2009). These societal benefits of plastics explain their high popularity, and new properties and applications in areas like energy conservation, robotics, or transplantation medicine are continuously added.

However, the exponential growth of plastic applications in modern times[2] has severe downsides. The unconstrained disposal of plastics has created growing levels of environmental stress and has become a global threat to all living and non-living systems (Iroegbu et al. 2021, p. 19344). We are talking about different forms of **plastic pollution** (macro-, micro-, and nano-plastic debris) as "the intrusion or invasion by plastic materials (i.e., polymeric systems), either through direct introduction or degradation processes of environments (to which they are not native) to negatively or undesirably impact such environments" (ibid.). Plastic pollution is an anthropogenic form of pollution through the introduction of plastic materials in different spheres of the environment that are not "natural" to them. They cause (reversible or irreversible) modifications, disturbances, and disruptions of ecosystems and balances between ecosystems (ibid.).

The hazardous nature of plastic pollution is a consequence of the scale of production, its biological impact at low levels of concentration, the diffusion and degradation of plastic debris in air, water, and solid textures, its long-term endurance, and the diverse consequences for human health and the stability of the invaded ecosystems. Due to the complex nature of these processes and the growing visibility of diverse repercussions, the number of scientific publications related to "plastic pollution" has grown significantly in recent years to reach almost 460 thousand in the last decade (ibid.).

---

[1] There are also natural polymers like cellulose, carbohydrates, proteins or rubber, and human DNA or skin (Iroegbu et al. 2021, p. 19345).

[2] Garbage, in particular household garbage, is quite a recent phenomenon. Until the nineteenth century, leftovers and broken objects had usually specific uses or were repaired: "Leftover food was boiled to make soup or fed to animals. Things that lost their use for adults became playthings for children, broken objects were repaired, and durable goods and items, like furniture, clocks, and watches, were handed down from generation to generation. In agriculture and animal husbandry, manure was used as organic fertilizer, straw was reused in construction, rags found a new purpose in papermaking, bones were valued in producing soap, and ashes were used to control pests and fertilize soils. Even dog excrement was a sought-after commodity, as it served a purpose in tanning leather. For thousands of years, economies were circular economies: they produced very little waste (…). This began to change with urbanization and rising populations and, more importantly, with the advance of assembly-line manufacturing that made cheap, mass-produced items available worldwide. Consumerism created a throw-away society, particularly in the United States, that produced an excess of easy-to-use, disposable items" (Mauch 2016, p. 5).

We can distinguish three categories of plastics in the global "plastics ecosystem": plastics in use, managed plastic waste, and mismanaged plastic waste (Lebreton and Andrady 2019). For example, 11% of the global mismanaged plastic waste generated in 2016 (19–23 million metric tons) found its way into different aquatic environments, with growingly disastrous consequences (Jambeck 2015). Continuous growth of these amounts to approximately 53 million metric tons per year in 2030 is projected, even in scenarios of effective regulation. Without the application of restrictive policies, it could even reach 80 million metric tons by 2030 (Borrelle et al. 2020, p. 1516), with an intense concentration in specific regions in Asia and Africa with the highest levels of demographic and urban growth (Lebreton and Andrady 2019).[3] One of the main problems is the insufficient solid waste infrastructure, even in industrialized countries with high levels of waste generation,[4] and the massive exportation of plastic waste to countries with low-quality standards and recycling capacities, which often just dump or burn plastic garbage indiscriminately (Law et al. 2020). All these factors contribute to the fast growth of plastic intrusion into the environment.

The main **sources of plastic pollution** are different types of plastic waste and debris, mainly contained in textiles and fibers, cosmetics, paints and coatings, and many other basic consumer products (plastic bags, toothbrushes, shoes, electric cables, writing pens, mattresses, wipes, computer casings, etc.). One of the several particularly invasive forms of microplastic debris is tire abrasion ("wear and tear") from all kinds of vehicles. On average, the global per capita emission of microplastics through tire abrasion is estimated to have reached more than 800 g yearly (Kole et al. 2017). An estimated 5–10% reaches the oceans from the microplastic debris generated through tire abrasion, and 3–7% is diffused in the air (ibid.). Tire abrasion is also a form of contamination that has been difficult to address, partly due to the lack of comprehensive studies and partly due to the intense lobbying of the powerful tire and car industries against regulation (ibid.). This is also illustrated through a study by the Swedish government from 2020 that showed that more than 50% of Sweden's microplastic pollution is a consequence of tire abrasion. However, a broad counter-initiative by these industries has blocked stricter measures proposed by the European Union (European Commission 2021, Iroegbu et al. 2021, p. 19348).

The specific consequences of plastic pollution are often still not known in detail because of a lack of long-term impact studies, despite the growing research interest, but what we already know so far indicates a wide range of calamities from inhalation of microplastics, intrusions from apparel and home textiles, as well as microplastics contained in drinking water. There are mainly two types of problems: **health risks and environmental risks**. In general, 20% of global deaths are estimated to be related to pollution-caused health complications (Prüss-Ustün et al. 2016). In particular, a rather long list of health problems, like thyroid dysfunction, obesity, diabetes, and reproductive impairment, have been related to different forms of plastic pollution

[3] More details on future scenarios in the following sections.

[4] The US as the country with the highest amount of plastic garbage generated 42 million metric tons in 2016 (Law et al. 2020).

(Iroegbu et al. 2021, p. 19346). Some authors have recently pointed to the negative impact of nanoplastics on microorganisms in the digestive system, which are, among other things, responsible for establishing chemical balances in the endocrine, immune, and nervous systems (Teles et al. 2020). Additionally, there are indications that micro- and nanoplastics can cause specific forms of genotoxicity (ibid.). The main environmental risks related to plastic pollution refer to the acceleration of climate change through its fossil origin and plastic contamination of aquatic environments, the acidification of seawater through dichlorination of PVC, and other (sometimes hard to anticipate) effects of plastic debris in ecosystems (Iroegbu et al. 2021, p. 19347).

To provide and distribute the vigorously growing quantity and variety of plastics, an extensive and diversified **plastics economy** has emerged as a complex sociotechnical system that organizes interconnected value chains of innovation, production, consumption, and recycling of plastics, including the formation of institutional frameworks of diverse social rules (habits, laws, regulation, and others).

Various **policies** and tools to reduce plastic pollution have been developed and partially applied in recent years. The common objective of all these efforts has been to control and mitigate the generation and diffusion of plastic debris into the environment. Some policies focus mainly on the production side, promoting reducing the use of polymers and their substitution with green(er) materials like bioplastics and biodegradability (UNEP 2018; WRI Mexico 2020; OECD 2022a). A related approach aims to strengthen extended producer responsibility with voluntary contributions and different forms of circular economies (Lindhqvist 2000; Ellen McArthur Foundation 2021; GAIA 2022). The third group of mitigation measures emphasizes consumer knowledge and responsibility through promoting a green lifestyle, enhanced education efforts, and a wider diffusion of scientific studies on the matter (Iroegbu et al. 2021, pp. 19350–19351). Most of these efforts to control, mitigate, and regulate plastic pollution are related to urban contexts of production and consumption. A comprehensive example of an integrative policy is the EU program of phasing out the main single-use plastic products,[5] with specific targets for separation, recycling, and bans[6] (EU 2021).

Several obstacles can be associated with the application of existing policies and tools. First, the narrow focus of measures on more visible topics like plastic bags and other single-use plastic items has obscured the public view on other equally or even more critical problems. As mentioned, the second problem has been the trivialization through apparently diverging scientific positions regarding the causes of environmental affectations and the associated levels of risk. Finally, the asymmetry

[5] "Target the top single-use plastic products found on our beaches: bottles (and caps); cups (and lids); cutlery, plates, straws, and stirrers; food containers; lightweight plastic bags; packets and wrappers; cigarette butts; tampons, applicators, sanitary pads, and wet wipes; and cotton buds, balloons and balloon sticks. Also, aim for oxo-degradable plastic products and fishing gear containing plastic" (EU 2021).

[6] "Specific targets include a 77% separate collection target for plastic bottles by 2025—increasing to 90% by 2029, incorporating 25% of recycled plastic in PET beverage bottles from 2025, and 30% in all plastic beverage bottles from 2030" (EU 2021).

of power of the different stakeholders involved, with particular interests of multinational companies and their strong influence on the research agenda and also on the political decision-making process, has weakened the responses (Iroegbu et al. 2021, pp. 19347–19348; Dumbili and Henderson 2020).

11.3 Conceiving Futures of the Plastics Economy

Due to the widespread and growing consequences of plastic pollution and the associated (mostly hidden or external) costs of today's approximately USD 1 trillion related to the loss of ocean productivity alone and USD 2 trillion of the total cost (OECD 2023, p. 24), the topic has been included in the catalog of UNESCO's social development goals (SDG) for 2030, generally expressed in SDG 14: "conserve and sustainably use the oceans, seas and marine resources for sustainable development" and the specific objective SDG 14.1, that demands "the prevention of plastic pollution through sustainable development" (UN 2015, p. 23). The formulation of these specific objectives that describe pathways of change toward lower pollution levels is closely related to the construction of future scenarios of adapted production and consumption systems based on different assumptions and methods of futures thinking involved.

Plastic pollution is a long-term global issue[7] that can only be tackled with a transformation at all levels of the plastics economy. The impact of plastic pollution is already widely perceived and documented today. However, the central health and environmental risks will unfold due to the projected growth of plastics production and the prevalence of low capacities to reduce uncontrolled environmental diffusion (mismanaged plastic waste). The development of the plastics ecosystem in general, the evaluation of the associated risks, and the capacity to create and apply measures for a sustained transition process are closely related to open future developments and to the diverse **representations of possible futures**, including the disposition to enhance the collective capacity to push the system toward more positive, sustainable futures. By creating and using different images of its future, the plastics economy or the diverse coalitions and communities, particularly in urban contexts, can be considered anticipatory systems. When we analyze the characteristics of an anticipatory system, following the general conceptual framework that guides this book (see Chap. 1 by Gonzalez), we are interested in these images and in the social processes of creating and diffusing them, as well as the consequences for individual and collective decision-making and the related construction of narratives about alternative futures and possibilities to build preferable futures. In this process, formal institutions and regulations play a vital role in conveying those narratives to stakeholders by setting goals and points of arrival.

That is why we are recurring to different approaches of contemporary futures studies that recognize the substantial prevalence of uncertainty in all efforts to deal

[7] Even if we would stop the use of plastics immediately, the consequences of past pollution persist.

with the future and that combine a whole variety of methods as applied in the three basic levels of analysis: forecast, foresight and anticipation (Poli 2019). There is no precise temporal or conceptual separation or temporal succession of these three approaches in the canon of futures studies since there are different overlaps and combinations of diverse methods.

Forecast, as the first and basic level of futures studies, is characterized by efforts to predict the specific behavior of a phenomenon or a social system in the future, based mainly on quantitative analysis from historical data related to selected variables. The extrapolation of behaviors from the past to the future expresses (at least implicitly) strong assumptions about the continuity of the corresponding social systems with a somewhat limited capacity to consider systemic uncertainty, the possibility of ruptures in the future, and the mental and social capacities of pushing alternative agendas. Well-known examples of the forecast type of future studies include the oil peak studies about the prediction of the moment of maximum fossil consumption or business-as-usual climate scenarios that predict a specific rise of the global and regional temperatures as a consequence of the linear growth of greenhouse gas emissions. Forecast models represent the early stages of futures studies in the 1960s and 1970s and are still extremely popular in many scientific studies and policy proposals. They are, of course, important inputs for contemporary futures studies and decision-making because they allow us to visualize the consequences of linear dynamics in stable systems. Nevertheless, additional tools are needed to take into account possible ruptures and changes in the mental models of the agents that compose the social systems, and these are crucial elements to understand and guide all kinds of transition and transformation processes.

Projections on the level of foresight exercises are widespread in many studies on the plastics economy. The statistical data that estimates future developments based on static economic models and linear anticipatory assumptions generally predict strong growth of plastics production for the next decades. Following OECD data (OECD 2022b), global production from 2020 will approximately double until 2040 at 800 million tons per year and reach 1.2 billion tons in 2060. Foresight calculations for mismanaged plastic waste show similar tendencies, reaching 200 million tons in 2060 (Lebreton and Andrady 2019). This linear overall tendency in plastic debris results from a moderate reduction in the Global North combined with a relative growth in the Global South, with a high concentration in African and Asian countries, following the estimated population growth rates and urbanization in these regions. One lesson from these projections is to promote large-scale investment in waste management in these countries (Kaza et al. 2018). These measures could have a determining impact in vulnerable countries where the populations are strongly affected by health and environmental risks (ibid.) and also allow for reducing CO_2-equivalent emissions. Projections are an important form of creating images about the future because they allow us to understand what might happen if nothing changes. For that same reason, the associated anticipatory value is limited because there is very little space to conceive alternations to linear extrapolation. Scenarios outside these limited trajectories, which can be important to understand risks and opportunities, are often blind spots in the futures map.

Strategic foresight as the second level of future studies has incorporated several qualitative methods, like exploratory scenario analysis or weak signals, among many other innovative methods (Schwartz 2023), that allow for a much broader perspective in long-term analysis, with many implications for a more robust understanding of possible pathways in the future development of all kinds of social systems. More awareness of less probable existential risks can strengthen individual and social resilience; imagining diverse nonlinear opportunities helps conceive preferable futures and liberate significant potentials for their realization.

Foresight exercises are also quite common in conceiving future images of the plastics economy, generating alternative future scenarios with mixed methods. For example, comparing possible scenarios for plastic emissions into aquatic systems allows us to anticipate the consequences of the transition process of the plastics economy and the guiding rule systems that must be developed (Borrelle et al. 2020, see Fig. 11.1). These approaches imagine different pathways of the plastics economy regarding production levels, types of plastics used, and recycling capacities. The foresight exercises, like the ones we introduced in the chapter before, are closely related to the "business-as-usual" (continuity) type of scenarios. Borrelle et al. (2020) additionally use intervals of confidence that reflect the critical effects of dispersions that are part of a more profound scenario analysis, where uncertainty is taken for granted and specific anticipatory assumptions are included. The baseline scenario adapted from the World Bank leads to growth dynamics of aquatic plastic pollution, with approximately 50 and 90 million tons in 2030. A second scenario (called "ambitious") estimates a reduction of plastic debris that reaches aquatic systems to 20–50 million tons in 2030 if all countries comply with a series of global and national commitments, like those included in the G7 Plastics Charter. This would mean reaching national targets of between 30 and 90% of managed waste, depending on the classification in the Plastics Charter. In terms of futures studies, the "ambitious" scenario implies that a systematic and effective "formal" regulation of plastic use on a global level would have to be put in place, but also that images and narratives of the underlying future images are being shared and accepted. This point is even more compelling with the third scenario, the so-called "target" scenario, which is based on a *backcasting* logic of formulating pathways and specific strategies that allow reaching a pre-defined target of 8 million tons of pollution per year as an amount that is necessary for the long-term recovery of marine ecosystems. The future images and anticipatory assumption of this type of scenario not only follow a different logic of construction but also question the direction of the long-term development of the plastics economy and the economic model in general, introducing goals of a general reduction of 85% of plastics production and high levels of recycling, that would require a reorganization of many related economic subsystems toward effective sustainability and alternative forms of growth.

When we analyze the plastics economy with strategic foresight tools, we can recognize alternative visions of transformation and outline particular strategies and tools (regulation, bans) or specific roles for different actors in futures scenarios delineating alternative scenarios (Greenpeace 2019a, b). Other proposals have created narratives of the future that connect technological innovation related to smart cities

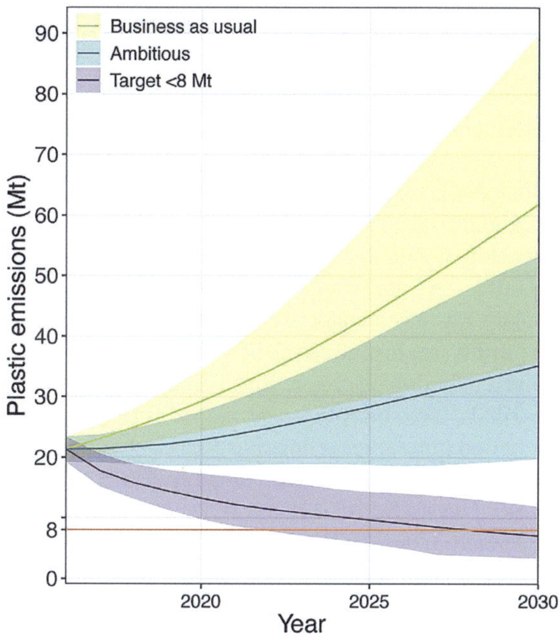

Fig. 11.1 Strategic foresight and future scenarios for the plastics economy. *Source* Borrelle (2020), p. 1516. Reprinted with permission from AAAS

with new forms of urban waste management (Circular Economy 4.0) (Mavropoulos and Nilsen 2020).

From an alternative epistemic perspective, the **theory of anticipation** (Poli 2019), as the centerpiece of the third level of futures studies, systematically analyzes different uses of the future in anticipatory systems that are part of any existing social system and explores the conditions for generating alternative futures based on the collective intelligence of empowered and more "future literate" actors (Miller 2018). The most considerable differences between forecast, foresight, and anticipation lie in the human conceptualization of uncertainty and complexity, the importance of an open and emergent construction of desirable futures, and the participatory and interactive governance of visualization and transformation processes. Anticipation theory revises mental models, imaginaries, and narratives of the future with novel analysis and practice tools. The first step in the participatory analysis of anticipatory systems is characterizing their images and the main processes of their creation, including mental models with all their biases and prejudices. Frequently, participating actors recognize the prevalence of low anticipatory capacities, linear anticipatory assumptions, and narrow pathways for evolving transition processes. In the second step, existing knowledge repositories and institutional rule systems are modified and reframed to explore trajectories that lead to scenarios with more positive perceptions. One of the main differences with strategic foresight is that there is no preestablished "best" alternative.

11.4 Zero Waste and Zero Plastics Futures Initiatives

The development of images of the future related to the reduction of plastic pollution has led to dynamics of imagining futures without plastic use, a significant reduction in plastic use, or without plastic pollution. The differences in the scenarios and the velocity of change depend on other elements of economic and social change related to the processes of socioecological transition (Fischer-Kowalski and Haberl 2007). This section will illustrate some of these narratives and their differences, particularly regarding zero waste and zero plastics movements.

Zero waste is a concept coined in 1970 by Paul Palmer, the founding director of the Zero Waste Institute in California, highlighting the economic value of waste in the context of high-tech production in Silicon Valley (Mauch 2016, p. 6). Later on, during the 1980s, the concept was developed further by a group of European experts with the notion of total recycling. Further, Canberra in Australia was the first city that, in 1995, began to legislate based on these principles and set itself the goal of generating no more waste by 2010 (Zero Waste Revolution 2024). Also, in San Francisco, in the United States, legislation based on the same principles was introduced in 2002 and was crucial for reducing waste by more than half in a decade (EPA 2024). The zero waste movement created a global platform called the Zero Waste International Alliance (ZWIA) to share experiences in different cities worldwide and establish common standards to guide the development of zero waste projects worldwide (Zero Waste International Alliance (ZWIA) 2019). It evokes a different scenario and has begun integrating a broad international movement to redirect public policy initiatives toward a comprehensive and preventative approach inherent in the circular economy paradigm (ZWIA 2017; Hannon and Zaman 2018; Brandao et al. 2020; Baskar et al. 2022).

In recent years, zero waste has become a catchword for waste reduction in companies, cities, and countries worldwide. However, the specific applications refer to rather different problems and objectives that sometimes differ from a systematic strategy of avoiding plastic debris. In all these cases, zero waste refers to an image of a future free of (unnecessary) waste and particularly free of plastic waste. Zero waste is the central part of a future narrative to build social systems capable of organizing production and consumption without the negative impact of plastic pollution. Many different actors have contributed to creating and diffusing this narrative and developing and implementing different strategies for plastic waste reduction through the transformation of the plastics economy on different levels, particularly in urban contexts, with the creation of communities like grassroots foundations, zero waste communities or citizens' movements in different parts of the world (Kaza et al. 2018; UNEP 2018, 2020).

The basic narrative of the zero waste concept promoted by a diverse group of urban communities and movements is related to changing production and consumption practices that allow the reduction of waste with different forms of feedback into the economic cycle, preventing it from being released into the environment. Zero waste means in the first place designing and managing products and processes to systematically avoid and eliminate the volume and toxicity of waste and materials, conserve and recover all resources, and not burn or bury them, and is, in that sense, closely related to the idea of a circular economy. A systematic "zero waste" approach implies not just building technical solutions for recycling plastic waste but also redesigning products and production processes, allowing the reuse of all materials along the whole value chain. In that sense, it is related to images and narratives of the future economy associated with transforming large parts of the production and consumption system. This means creating a different mental model for producers and consumers or a "zero waste philosophy" (Mauch 2016, pp. 6–7). "Zero-waste is a goal that is ethical, economical, efficient and visionary, to guide people in changing their lifestyles and practices to emulate sustainable natural cycles, where all discarded materials are designed to become resources for others to use" (Zero Waste International Alliance 2024). These initiatives are closely connected to other processes of institution building at different levels.

At a global level, the United Nations Environment Organization (UNEP) and the World Economic Forum have used future risk assessment methods to mitigate the global problem of the distribution of inadequately managed urban plastic waste through bodies of water and air currents. For example, a 2017 UN resolution signed by nearly 200 nations stands out, emphasizing the importance of tackling the future growth of ocean plastic pollution (UNEP 2022). When China suspended the massive imports of plastic garbage from major industrialized countries in 2018, efforts to promote circular plastic management accelerated. The Ocean Plastic charter mission, signed by the G7, stands out. The European Commission published the "European Strategy for Plastics in the Circular Economy," and the OECD held the "Global Forum on Plastics in the Circular Economy." One of the most relevant milestones within the framework of the 5th UNEP Assembly was the resolution "End Plastic Pollution: Towards an internationally legally binding instrument." The Intergovernmental Negotiating Committee is expected to arrive at a legally binding treaty by the end of 2024, with a comprehensive circular economy approach, encouraging the participation of all stakeholders and promoting cooperation at all levels. In the meantime, stakeholders are engaged in periodic discussions and negotiations on the final draft intended to be voted on in the 6th UNEP Assembly in early 2025 (UNEP 2022).

Public pressure has pushed federal and local authorities related to environmental, industrial, and waste management policy to pay more attention to plastic pollution as an environmental policy priority. Policymakers have experienced the progressive emergence in their agendas of the measures to stop plastic contamination (UrbanShift 2023a, b). At the same time, relevant sectors of the population, with greater capacity for agency, showed their concern in social networks and news media regarding the gigantic islands of plastic parked in the ocean gyres, the images of marine fauna suffocated by plastic, the pieces of plastic found in the stomachs of fish and birds, as well as the videos of severe affectations of marine fauna (2015) (Barkela et al. 2021; Teh et al. 2022; Pop et al. 2023).

These initiatives have become more present, but they have had a rather limited impact, particularly in countries in the Global South. Among different challenges, we note that international policies and social norms are pending to be introduced and applied more systematically. Formal institution building is advancing but is not enough. We argue here that the existing anticipatory systems are weak, with scarce elements of visualization and transformative capacities. We want to show these problems by analyzing some elements of the normative pillars of institutionalization in Mexico and their weak embeddedness in social contexts. Anticipatory communities do not exist or do have low capacities of anticipation (González-Arellano 2025).

11.5 The Urban Plastic Economy in Mexico City as an Anticipatory System

The capacity to anticipate the consequences for the future and build more robust long-term strategies is crucial, as well as the main arena where plastic is used, as plastic pollution originates in the urban arena. We want to analyze the potential of these efforts, introducing the capacity to imagine and build less invasive futures in cities as anticipatory systems. In this section, we analyze the images of the future and the anticipatory assumptions related to the efforts to control and regulate the mitigation of plastic pollution in Mexico. To analyze the types of anticipatory approaches that have been coexisting, it is important to understand the tendencies in the national political context on these topics as reflected in federal and local regulations passed by Congress. Both arenas are important to understand the uses of future images in the buildup, application, and evaluation of these efforts.

During the last three decades, the interest of environmental groups and parliamentary groups in addressing the problem of plastic pollution brought 16 initiatives to the Chamber of Deputies and 22 to the Chamber of Senators, mainly focused on regulating bags and adding other single-use plastics in 2018 (WRI 2020). Since 2003, the main institutional channel for this regulation has been the General Law for the Prevention and Comprehensive Management of Waste (LGPGIR). The last modification of this law occurred on December 1, 2018, and gave rise to important guidelines related to plastic control. Article 7, for example, establishes that it is the Federation's responsibility to issue official **standards** for plastics. Article 28 states that both generators and producers, as well as importers, exporters, and distributors of plastic containers, including expanded polystyrene, as well as importers and distributors of used tires, are obliged to formulate and execute **plastic management plans** under the principles of valorization and **shared responsibility**. Our review of state legislation on environmental and waste management allowed us to identify that in the first three years of the new federal administration, between 2018 and 2020, 18 entities modified their environmental or waste management legislation to mention the management of plastics and bioplastic explicitly.

The current initiatives in Mexico are related to the political cycle that started with the victory of Andrés Manuel López Obrador and his MORENA movement in the presidential elections in 2018 and also in the state elections in México City.

The majority in the local parliament has allowed MORENA to push the approval of bills with the support of other political parties in their coalition (INE 2018, Oxford Analytica 2018). The announcement of Mexico City's new Solid Waste Law in 2019 represented a significant step in implementing a regulatory framework for reducing plastic pollution. It prohibited the selling and distributing of plastic bags since December 2020 and included other single-use plastics from January 2021 onward. The Mexican Ministry of Environment (SEMARNAT) produced in 2019 a policy document entitled "A National Vision towards Sustainable Management: Zero Waste" (SEMARNAT 2019) that embraced **zero waste** futures. It has been intended as a general guideline to transition toward a circular economy. The document establishes general **goals toward 2030**, related to specific images that the waste management policy has been aimed to reverse the traditional priorities that concentrated efforts in waste management, to transition to policies that prioritize prevention, minimization, and reuse, followed by alternatives such as secondary recycling, economic recovery, and final disposal.[8]

To put this vision into practice, actions related to monitoring, communication, education, co-responsibility of stakeholders, government coordination, and the promotion of citizen participation were proposed. These could, in principle, be operators of change in the anticipatory urban system, but for the moment, only their importance and statements are recognized in a very general way. However, incipient and subtle changes in plastic manufacturing and education campaigns have begun to be introduced due to national agreement among plastics producers and the Mexican Senate. They have placed their vision of the future exclusively on targeting the increase in PET collection and recycling content for 2025 and 2030 (Senado de la República 2019).

Two crucial regulatory documents were updated at the federal level as part of the strategy to implement this approach. On the one hand, the General Law for the Prevention and Comprehensive Management of Waste, and on the other, The National Program for the Prevention and Comprehensive Management of Special Management Waste. Both insist on the importance of generating **information systems** on waste's health and environmental effects, including plastics. Along these lines, one of the most relevant strategies and actions is formulating a **national inventory of sources of plastic pollution**. This inventory was presented in 2023 as part of the National Action Plan on Marine Waste and Plastic Pollution (SEMARNAT 2023). Additionally, the concept of shared responsibility, in which producers and users are involved, has been

[8] This policy document proposes expanding the coverage of collection services and infrastructure development, increasing the capacity for waste recovery, strengthening the separation of waste at source, promoting recycling, and establishing an effective inspection and surveillance body. The document proposes a general road map, which conceives six actions, among them: a comprehensive diagnosis (of waste, its management, infrastructure, and the regulatory framework); closure of final disposal destinations that do not comply with regulations; getting international expert advice to promote private investment for the development of technologies that allow the use and recovery of waste; local development of appropriate management models that benefit from best practices; transformation of open-air dumps into material banks; as well as avoiding food waste and taking advantage of its organic and energy potential (SEMARNAT 2019).

incorporated, intending to transcend a waste management approach and conceive it as a more systemic problem that involves producers.

Implementing the legal provisions for managing plastic and its waste was part of a broader program with a **circular approach**. The Government of Mexico City, through its Environment Secretariat (SEDEMA), presented the "Zero Waste Action Plan towards a Circular Economy" in May 2019. It is a set of actions aimed at reducing waste generation and no longer solely focused on waste management at the end of the product lifespan. The Plan has set goals for the short (2024) and medium term (2030). The Plan consists of actions aimed at separating and collecting waste, training and information campaigns, infrastructure development, and the dignity and recognition of the work of the cleaning system employees. In a transversal way, the issue of plastic crosses these lines of action in a very relevant way. Their promotional slogans mention: "A city with zero waste in 2030."

In response to the release of the governmental Zero Waste program, Greenpeace, along with three other relevant environmental organizations,[9] positioned itself by stating that "the 'National Vision of Sustainable Management: Zero Waste' does not solve the root problem," as it highlights waste management instead of waste reduction and overproduction. There is also concern that the official position does not ban the burning of waste, including plastics, for energy generation, enabling the release of toxic substances and greenhouse gases (Greenpeace 2019a, b). In this context, the burning of plastic and other waste is contemplated in the prospective scenarios of the cement industry to replace traditional petroleum coke. It is intended to increase these alternative fuels to 20% in 2025, 25% in 2030, and 35% in 2050 (Rappo 2020). A couple of years later, Marisa Jacott, representative of Common Frontiers, insisted during an interview that "There is no clear policy on the part of the government to implement a true zero waste policy" (Greenpeace 2021).

Mexico City has also been the center for environmental civil society organizations concerned about the excessive production of plastics and their poor management. In 2019, several of them integrated a broad network called "Mexico Zero Plastic Alliance." This network brings together recognized organizations such as Greenpeace but is open to other organizations and citizens. It currently includes 241 members and 125 organizations nationwide (https://alianzamexicosinplastico.org/). These organizations have been very skeptical regarding the scope of the legislation discussed above (Greenpeace 2024). They have criticized, for example, that the responsibility for the reduction of debris is still placed mainly on the side of the consumers and that it is very soft concerning limiting production or placing greater responsibility on the side of plastic producing and distributing companies (AMSP 2021).

Analyzing the structure of the urban anticipatory system, we can distinguish three anticipatory communities: plastics industrialists, a group of legislators and public officials, and environmental organizations. The established plastics industry, aware of public discontent and the incoming policy initiatives to regulate plastic production

[9] Center for Analysis and Action on Toxics and their Alternatives (CAATA), Common Borders, Community Development Research Laboratory (LIDC).

and consumption, has promoted campaigns to encourage recycling as a solution to the problem in the future. Likewise, it has sought to discredit the use of substitute materials as a feasible solution at the present. Policymakers and government officials have needed to legitimize their government programs in line with the current international vision of zero waste environmental policy and, at the same time, try to reconcile the competing interests of plastics industrialists and the pressure of local environmental organizations.

Concerning the regulatory scope, we identify that this is very varied and covers a broad spectrum of future transformations of the urban plastic economies ranging from (a) calls to encourage the use of alternative products, including biodegradable ones, (b) promote recycling, (c) sanction poor management of waste, and even (d) prohibit the marketing of bags and other single-use plastics. This change at the regulatory level did not immediately generate significant changes in production and consumption patterns due to the lack of complementary legal elements in lower levels (e.g., adequate norms and certifications) and to the slow changes in the availability of raw materials and alternative products, its technical standards, adequate surveillance and sanction mechanisms, among others.

We can observe that there has been a lack of visualization of alternative futures and also a low transformative capacity of guiding transition inside the communities of actors. It is also important to note that the health measures related to the COVID-19 pandemic relaxed the regulations. One of the central issues is the divergence in the future images between the pressures of business actors and environmental groups. Incumbent plastic producers have tried to place responsibility on public administration and consumers, suggesting the improvement of municipal collection systems and establishing recycling as the primary measure. At the same time, environmental groups and some academics insisted that responsibility must be shared with producers, promoting the substitution of materials, prohibiting unnecessary or single-use plastics, discouraging the burning of plastics to generate energy, and only opting for recycling when there is no other alternative (GAIA 2024).

11.6 Conclusions

Plastics have become an integral part of modern production and consumption systems with many important properties and applications that make replacing or reducing their usage difficult. This is another example of the rigidities in modern economies' technical, economic, cultural, institutional, and social configurations. Nevertheless, at the same time, the accelerated use of resources involved, particularly single-use plastics, has led to enormous resource waste and is causing significant risks, as plastic pollution is creating severe risks to human health and the environment. Moreover, these risks are only enhancing, as forecasts based on stable conditions show that the plastics economy will still grow substantially in the next decade.

The diverse representations of the plastics ecosystems and the diverse intervention proposals show a limited and strongly biased use of future images, narratives, and scenarios with frequently very narrow anticipatory assumptions. Individuals and society generally have a limited capacity to visualize futures and build solutions. However, the successful transition of systems of production and consumption of plastics toward less invasive or even sustainable models depends, among other things, on the capacity to create, compare, and apply long-term scenarios with diverse intervention options on different levels. Current futures studies offer more inclusive forms of revision, formulation, and application of alternative narratives. Anticipation theory and futures literacy promote individual, organizational, and social capacities of anticipation and pathways to multi-level governance.

Strong anticipation lies in the human conceptualization of uncertainty and complexity, the importance of an open and emergent construction of desirable futures, and the participatory and interactive governance of visualization and transformation processes. Anticipation theory revises mental models, imaginaries, and narratives of the future with novel analysis and practice tools. The first step in the participatory analysis of anticipatory systems is characterizing their images and the main processes of their creation, including mental models with all their biases and prejudices. Frequently, participating actors recognize the prevalence of low anticipatory capacities, linear anticipatory assumptions, and narrow pathways.

In our analysis of the urban plastic economy in Mexico City as an anticipatory system, we observed low capacities for visualization of futures for the plastic economy and very weak capacities for transformation. Many initiatives include images of short-term futures (related to the political cycle) that are fragmented and empty. Existing anticipatory communities cultivate futures narratives connected to basic concepts of plastic waste reduction but lack consistent anticipatory assumptions and governance mechanisms. For these reasons, the urban plastic economy in Mexico City has had a deficient transformative capacity.

We have also identified that the Mexican policy cycle allowed the reframing of customary waste programs narratives and that in such endeavor, urban anticipatory communities have played an essential role by introducing the international wording—even heterogeneous—about different expectations on a new plastics economy, but still struggle to translate alternative visions on new operative models to mature anticipatory urban capabilities.

References

AMSP (2021) Senadores simulan ley 'antiplásticos' para favorecer a la industria. https://alianzamexicosinplastico.org/wp-content/uploads/2021/10/Comunicado-de-prensa-AMSP-oct-2021.pdf

Andrady AL, Neal MA (2009) Applications and societal benefits of plastics. Philos Trans R Soc B Biol Sci 364:1977–1984

Barkela B, Werling K, Filler SE, Heidbreder LM (2021) Fasting plastic—the role of media reports in a 'window of opportunity' to reduce plastic consumption. Sustainability 13:11807. https://doi.org/10.3390/su132111807

Baskar et al (eds) (2022) Handbook of solid waste management. Sustainability through circular economy. Springer, Singapore

Borrelle SB, Ringma J, Law KL, Monnahan CC, Lebreton L, McGivern A, Murphy E, Jambeck J, Leonard GH, Hilleary MA, Eriksen M, Possingham HP, De Frond H, Gerber LR, Polidoro B, Tahir A, Bernard M, Mallos N, Barnes M, Rochman CM (2020) Predicted growth in plastic waste exceeds efforts to mitigate plastic pollution. Science 369(6510):1515–1518. https://doi.org/10.1126/science.aba3656

Brandao M, Lazarevic D, Göran F (eds) (2020) Handbook of the circular economy. Edward Elgar, Cheltenham

Dumbili E, Henderson L (2020) The challenge of plastic pollution in Nigeria. In: Plastic waste and recycling. Elsevier Academic Press, pp 569–583

Ellen McArthur Foundation (2021) Responsabilidad Extendida del Productor. Una parte necesaria de la solución para los residuos y la contaminación de los empaques y envases. https://plastics.ellenmacarthurfoundation.org/epr#Position-paper

EPA (United States Environmental Protection Agency) (2024) Zero-waste case study: San Francisco. https://www.epa.gov/transforming-waste-tool/zero-waste-case-study-san-francisco

European Commission: Directorate-General for Environment (2021) Turning the tide on single-use plastics. Publications Office of the European Union. https://data.europa.eu/doi/10.2779/800074

Fischer-Kowalski M, Haberl H (2007) Conceptualizing, observing and comparing socioecological transitions. In: Fischer-Kowalsk M, Haberl H (eds) Socioecological transitions and global change: trajectories of social metabolism and land use. Edward Elgar Publishing, pp 1–30. https://www.e-elgar.com/shop/usd/socioecological-transitions-and-global-change-9781847203403.html

GAIA (2022) La responsabilidad extendida del productor (REP). https://www.no-burn.org/wp-content/uploads/2022/06/UNEA-publication-packet_EPR_ES.pdf

GAIA (2024) Tratado Mundial sobre Plásticos: Menos Plásticos, Más Vida. https://www.no-burn.org/es/unea-plastics-treaty/. Access Feb 23

González-Arellano S (2025) Cities as anticipatory systems. In: González-Arellano S, Gandlgruber B (eds) Cities as anticipatory systems. Springer. https://link.springer.com/book/9783031781612

Greenpeace (2019a) Throwing away the future: how companies still have it wrong on plastic pollution 'solutions.' https://www.greenpeace.org/usa/research/how-companies-still-have-it-wrong-on-plastic-pollution-solutions/

Greenpeace (2019b) La visión nacional para gestionar los residuos no resuleve el problema de raíz. https://www.greenpeace.org/mexico/noticia/1726/la-vision-nacional-para-gestionar-los-residuos-no-resuelve-el-problema-de-raiz/. Access 5 Feb 2024

Greenpeace (January 16, 2021) Cuando la basura nos alcanzó y nos rebasó. https://www.youtube.com/watch?v=1aqC_xinDgo

Greenpeace (2024) ¡Libérate del Plástico! ¡Exijamos leyes que prohiban los plásticos de un solo uso!. https://www.greenpeace.org/mexico/participa/liberate-del-plastico/

Hannon J, Zaman AU (2018) Exploring the phenomenon of zero waste and future cities. Urban Sci 2(3):90

INE (2018) Cómputos de entidad federativa y circunscripción 2018. Elecciones Federales. https://computos2018-entidad.ine.mx/#/senadurias/nacional/1/2/1/2

Iroegbu AOC, Ray SS, Mbarane V, Bordado JC, Sardinha JP (2021) Plastic pollution: a perspective on matters arising: challenges and opportunities. ACS Omega 6(30):19343–19355. https://doi.org/10.1021/acsomega.1c02760

Jambeck JR, Geyer R, Wilcox C, Siegler TR, Perryman M, Andrady A, Narayan R, Law KL (2015) Marine pollution. Plastic waste inputs from land into the ocean. Science 347(6223):768–771. https://doi.org/10.1126/science.1260352

Kaza S, Yao L, Bhada-Tata P, Van Woerden F (2018) What a waste 2.0: a global snapshot of solid waste management to 2050. World Bank Publications. Urban Development Series. Washington, D.C. https://doi.org/10.1596/978-1-4648-1329-0

Kole PJ, Löhr AJ, Van Belleghem F, Ragas A (2017) Wear and tear of tyres: a stealthy source of microplastics in the environment. Int J Environ Res Public Health 14(10):1265. MDPI AG

Law KL, Starr N, Siegler TR, Jambeck JR, Mallos NJ, Leonard GH (2020) The United States' contribution of plastic waste to land and ocean. Sci Adv 6(44):eabd0288. https://doi.org/10.1126/sciadv.abd0288
Lebreton L, Andrady A (2019) Future scenarios of global plastic waste generation and disposal. Palgrave Commun 5(1):1–11
Lindhqvist T (2000) Extended producer responsibility in cleaner production: policy principle to promote environmental improvements of product systems. PhD Thesis, Lund University. https://portal.research.lu.se/en/publications/extended-producer-responsibility-in-cleaner-production-policy-pri
Mauch C (ed) (2016) A future without waste?: Zero-waste in theory and practice. Rachel Carson Center for Environment and Society
Mavropoulos A, Nilsen AW (2020) Industry 4.0 and circular economy: towards a wasteless future or a wasteful planet? Wiley
Miller R (2018) Transforming the future: anticipation in the 21st century. Taylor & Francis, p 300
OECD (2022a) Global plastics outlook: economic drivers, environmental impacts and policy options. https://www.oecd-ilibrary.org/environment/global-plastics-outlook_de747aef-en
OECD (2022b) Global plastics outlook: policy scenarios to 2060. OECD Publishing, Paris.https://doi.org/10.1787/aa1edf33-en
OECD (2023) Towards eliminating plastic pollution by 2040. A policy scenario analysis. Interim findings. https://www.oecd.org/environment/plastics/Interim-Findings-Towards-Eliminating-Plastic-Pollution-by-2040-Policy-Scenario-Analysis.pdf
Oxford Analytica (2018) Morena triumph changes the face of Mexican politics. Expert Briefings.https://doi.org/10.1108/OXAN-DB235869
Poli R (2019) Handbook of anticipation. Theoretical and applied aspects of the use in decision making. Springer
Pop V, Ozunu A, Petrescu DC, Stan AD, Petrescu-Mag RM (2023) The influence of media narratives on microplastics risk perception. PeerJ 11:e16338. https://doi.org/10.7717/peerj.16338
Prüss-Ustün A, Wolf J, Corvalán C, Bos R, Neira M (2016) Preventing disease through healthy environments: a global assessment of the burden of disease from environmental risks. World Health Organization, Geneva, Switzerland
Rappo S (February 28, 2020) Quemar residuos municipales e industrials en Cemex, parte de los mecanismos de mitigación al cambio climático. Jorn Oriente. https://www.lajornadadeoriente.com.mx/puebla/quemar-residuos-municipales-e-industriales-en-cemex-parte-de-los-mecanismos-de-mitigacion-al-cambio-climatico/
Schwartz JO (2023) Strategic foresight. An introductory guide to practice. Routledge. https://www.routledge.com/Strategic-Foresight-An-Introductory-Guide-to-Practice/Schwarz/p/book/9781032299235
SEMARNAT (2019) Visión nacional hacia una gestión sustentable: Cero residuos. https://www.gob.mx/cms/uploads/attachment/file/435917/Vision_Nacional_Cero_Residuos_6_FEB_2019.pdf
SEMARNAT (2023) Inventario Nacional de Fuentes de Contaminación Plástica
Senado de la República (2019) Acuerdo Nacional para la Nueva Economía del Plástico en México. https://cespedes.org.mx/wp-content/uploads/2021/02/Acuerdo-Nacional-de-la-Nueva-Economia-del-Plastico-en-Mexico-version-final-Senado.pdf
Teh PL, Piao S, Almansour M, Ong HF, Ahad A (2022) Analysis of popular social media topics regarding plastic pollution. Sustainability 14:1709. https://doi.org/10.3390/su14031709
Teles M, Balasch JC, Oliveira M, Sardans J, Peñuelas J (2020) Insights into nanoplastics effects on human health. Sci Bull 65(23):1966–1969. https://doi.org/10.1016/j.scib.2020.08.003
UN (2015) Transforming our world: the 2030 agenda for sustainable development. Resolution Adopted by the General Assembly on 25 Sep 2015, 42809, pp 1–13
UNEP (2018) Límites legales de los productos de plástico desechables y los microplásticos: examen a escala mundial de las leyes y los reglamentos nacionales. https://wedocs.unep.org/handle/20.500.1/1822/27113

UNEP (2020) Tackling plastic pollution: legislative guide for the regulation of single-use plastic products. https://wedocs.unep.org/handle/20.500.11822/34570?show=full
UNEP (2022) Draft resolution on an internationally legally binding instrument on plastic pollution. https://wedocs.unep.org/bitstream/handle/20.500.11822/37395/UNEA5.2%20Global_Agreement_Explanatory%20note%20and%20Resolution%2027%20October.pdf?sequence=1&isAllowed=y
UrbanShift (2023a) Foro Internacional de Paris para Acabar con la Contaminación por Plásticos en las Ciudades. https://es.shiftcities.org/event/paris-international-forum-end-plastic-pollution-cities
UrbanShift (2023b) Cómo las Ciudades Lideran la Lucha contra la Contaminación por Plásticos. https://es.shiftcities.org/post/how-cities-are-leading-fight-against-plastic-pollution
WRI (2020) Mecanismos de política para disminuir los residuos plásticos de un solo uso: Revisión de las opciones disponibles y su aplicabilidad en México. https://wrimexico.org/publication/mecanismos-de-pol%C3%ADtica-para-disminuir-los-residuos-pl%C3%A1sticos-de-un-solo-uso#:~:text=%C3%89stas%20describen%20el%20estado%20actual,4)%20responsabilidad%20extendida%20del%20productor
Zero Waste International Alliance (ZWIA) (February 19, 2019) Working towards a world without waste. https://zwia.org/
Zero Waste International Alliance (February 23, 2024) Zero waste definition. https://zwia.org/zero-waste-definition/
Zero Waste Revolution (February 23, 2024) Our mission: zero waste. https://zerowasterevolution.net/

Part V
Anticipatory Elements for Urban Climate Policies

Chapter 12
Recife and Anticipation of Extreme Weather Events: Paths to Advance in Climate Justice

Pedro Roberto Jacobi, Marcos Tavares de Arruda Filho, and Rafael Calderón-Contreras

Abstract Anticipatory governance represents the need for changes in a long-term vision based on multiple strategies that prevent future crises, especially those arising from the global environmental emergency. In Brazil, around 80% of people already live in urban centers with multiple dimensions of vulnerability, where the issue of anticipation and inclusion in local government agendas is fundamental for identifying and implementing initiatives for reducing future impacts. Recife, located in the state of Pernambuco, is the 16th city most threatened by climate change in the world due to its low altitude and the large number of watercourses across its territory. This work aims at discussing the anticipatory nature of Recife's urban policies to prevent further extreme climate disasters that already affect the city. The presence of a framework of public policies that form an anticipatory system for extreme weather events means that coastal cities like Recife may open a wide array of possibilities for reducing the risks and impacts associated with climate change-related crisis. Anticipatory systems rely on the partnership between municipal institutions and other cooperation networks for inserting debates such as climate justice, which are sometimes limited to just the academic sphere within the decision-making arena.

Keywords Anticipatory systems · Socio-ecological transition · Cities · Recife · Climate justice

P. R. Jacobi · M. T. de Arruda Filho
Instituto de Energia E Ambiente—Universidade de São Paulo, Recife, Brasil
e-mail: marcostavares@usp.br

R. Calderón-Contreras (✉)
Department of Social Sciences, Universidad Autónoma Metropolitana-Cuajimalpa, Mexico City, Mexico
e-mail: rcalderon@cua.uam.mx

S. Gonzalez-Arellano and B. Gandlgruber (eds.), *Cities as Anticipatory Systems*, The Urban Book Series, https://doi.org/10.1007/978-3-031-78162-9_12

12.1 Introduction

Debates on local governments and climate change, especially in the urban context, have become fundamental in several areas, such as public management. Anticipatory governance represents the need for changes in a long-term vision based on multiple strategies that prevent future crises, such as the one arising from the global climate issue. In Brazil, around 80% of people already live in urban centers. Cities increasingly demand more attention to climate change due to their high vulnerability to the effects of climate change worldwide.

In the context of Northeast Brazil, where populations and territories are vulnerable in multiple dimensions, including climate change, the issue of anticipation and inclusion in local government agendas is fundamental to identifying and implementing initiatives to reduce their impacts. Most state capitals in the region have climate adaptation plans, mainly due to the exposure of these cities to rising sea levels and their location close to the coast.

Recife, located in the state of Pernambuco, is the 16th most threatened city by climate change in the world due to its low altitude and the large number of watercourses that cross its territory. This article aims to show how, in the case of the city of Recife, there is some level of internalization of the theme of climate change in municipal urban policy and the challenges that exist in deepening actions that present the scope of public policies that in collaboration with cooperation networks such as ICLEI, are configured as a climate governance arrangement, whose primary role is to contribute to promoting local actions to combat the climate crisis.

This work aims to discuss the anticipatory system within urban policies focused on climate issues in Recife to prevent extreme climate disasters that already affect the city. The text is divided into five sections: this introduction addresses the theme analyzed and raises the main topics to be discussed. The second section aims to frame the importance of anticipatory approaches to dealing with the climatic transition in cities. The Socio-ecological Transition in Cities discusses how cities prepare for a new ecological paradigm. The following section addresses the Anticipatory Climate Governance in Brazil, which provides a history and a discussion on how multilateral governance has been working on the climate issue in the country. The fourth section deals with the case of the Brazilian Northeast: Signs of Change in the Climate Transition, discussing how the states in this region address climate change within their public policies. The chapter then discusses the issue of Climate Justice in Recife: Transition to Anticipation, which brings the evolution and construction of a history of public policies focused on the climate and the final considerations in which there is closure and direction so that more cities can work on anticipatory systems for global climate change.

12.2 The Social-ecological Transition in Cities

As mentioned elsewhere in this book, cities are social-ecological systems entitled to permanent shifts and complex dynamics of change. Some of those changes are based on a disruption of some of the components of these systems. Global environmental change has become one of such transgressive factors that originate profound transitions in the structure of cities as systems. Global environmental change has associated risks that pose severe vulnerabilities to city populations and their biophysical grounds. Furthermore, those risks also negatively impact the social, political, and economic traits that constitute cities' social organization and governance of cities worldwide. Vulnerability has become one of the normalized states in which cities develop and grow. However, a growth process that often does not include a formalized process of anticipation translated into a general reduction of risks associated with global environmental change.

Anticipation in the context of cities' social-ecological transition is a necessary goal for governments and social institutions alike. As this chapter illustrates, the city of Recife has implemented a series of mechanisms to deal with scenarios related to the future conditions expected due to climate change. As analyzed by González, in this volume, imagining the future is one of two dimensions of anticipation that cities as anticipatory systems must implement to deal with the future consequences of external risks. Given the vast array of threats that climate change implies for Recife, considering the future is one of the critical aspects that public policies and plans for facing climate change must consider.

The second dimension of anticipation in cities as social-ecological systems is the possibility of incorporating the future into the system's transformation. In light of the ever-increasing challenges associated with climate change, cities, like any social-ecological system, react to different stimuli to change specific traits at determined scales. These changes are associated with transitions, which are chained to other transitions in different system qualities and may lead to social-ecological transformations that may lead to a better state of resilience in the system (Sievers-Glotzbach and Tschersich 2019). In the case of cities, this chain of transitions may imply the construction of social-ecological resilience to climate change and its impacts.

Building social-ecological resilience to withstand the effects of climate change in cities requires a process of institutionalization of anticipation (González-Arellano 2022). This process of institutionalization may often depart from the urban governance derived from a wide array of policies designed. The case of Brazil is illustrative when it comes to this effort. From national policies to adopting and adapting such policies in urban contexts, the potential for anticipating the conditions that cities will face due to climate change is paramount for creating resilience. As this book specifies, the regulatory pillar for which it is possible to enforce future scenarios in cities must be derived from multi-scalar governance, but also on the local actor's ability to foresee the impacts of climate change. This chapter uses the case of Recife to illustrate how the multi-scale governance in Brazil derives into an anticipatory system with the potential for building resilience in the face of climate change.

The transition of Recife, as a social-ecological system and a more resilient city, implies a wide array of anticipatory processes. Such processes characterize the social-ecological transition that cities must endure to promote more resilient futures. The case of Recife is illustrative of a factor often denied in the analysis of social-ecological transitions in cities, and it is the need for justice as the core for the construction of imaginaries of the city's future. Justice becomes not only the final goal of a process of anticipatory governance but the process by which a city, understood as a social-ecological system, may rely on its regulatory and normative governance structure and biophysical structure. Combining these traits (the ecological and the social) provides the means for creating anticipatory systems that may derive a better future for cities: one that appeals to a safe and just social-ecological system.

12.3 Anticipatory Climate Governance in Brazil

Actions aimed at environmental issues, according to the Federal Constitution, the National Environmental Policy—PNMA (1981), and the National Adaptation Plan—PNA (2016), require integration of policies considering the different levels of government, sectors and a diverse set of territorial characteristics that go beyond the borders of states and municipalities. In large urban agglomerations, the environmental governance of urban space must be innovative to face social pressures in the context of urbanization and the vulnerabilities of populations associated with occupation in areas at risk of landslides or subject to flooding. These are risks associated with poverty and inequality in urban logic, exposing the lack of urban planning and, notably, the use and occupation of land. However, when addressing the issue of climate change, it becomes necessary to address the risks; as Beck (2010) states, disasters are announced and cannot be seen as fatalities and can, in most cases, be predicted and avoided.

The main aspects (variables) that directly or indirectly influence the climate adaptation capacity of cities are systematized by studies such as that by Teixeira et al. (2020). The authors emphasize access and use of information; capacity to allocate economic, institutional, and technological resources; the intersection between adaptation, mitigation, and sustainable development; synergies between generic and specific capabilities; share capital; trust relationships; presence of public policy entrepreneurs; and participatory governance processes. Risk reduction and adaptation management is a transversal theme, and this demands that city formulators and planners understand how climate adaptation needs to be incorporated or integrated into urban planning practices (ICLEI, CEPS, 2013).

At the federal level, the central normative frameworks are the National Climate Change Policy of 2009, the National Civil Protection and Defense Policy in 2012, and the National Adaptation Plan in 2026. The PMMC defines in its Article 5 the objective of contributing to stabilizing greenhouse gases in the atmosphere "within a sufficient time to allow ecosystems to adapt naturally to climate change." It determines its implementation in sectoral axes, and adaptation focuses on health and land use axes, emphasizing the elimination or minimization of risks and adaptation measures related to climate change. Its guidelines include coordinated action between the Union, States, and Municipalities, with the Union coordinating and structuring policy instruments and supporting other entities in mapping risk areas and prevention and response actions, as well as the obligation to transfer resources to them in case of disaster.

The National Plan for Adaptation to Climate Change (PNA), established in 2016, presents the guideline that adaptation to climate change includes the improvement of urban planning models, integrating the planning of land use and occupation with environmental preservation to prevent the risks of natural disasters (CAF/WAYCARBON/ICLEI 2019). It is observed that the occurrence of extreme events, such as water scarcity, exposes the authorities' lack of preparation to deal with these issues and the lack of a culture of prevention in society. Preventive management of climate risks is still more in theory than practical actions (Jacobi and Sulaiman 2016). It is worth noting that many municipalities still do not have an adequate civil defense structure for preventive risk management. In the face of uncertainty and multiple risks, this requires that the preventive management of climate risks and the promotion of actions guided by a sustainability agenda are highlighted, thus reducing vulnerability with innovative management and urban planning. In this sense, adaptation measures can be adopted to reduce socio-environmental costs and losses in vulnerable social groups.

In 2016, in an effort led by the federal government that included the collaboration of civil society, the private sector, and state governments, the National Plan for Adaptation to Climate Change (PNA) was established. Its main objective is to promote the management and reduction of climate risk in the country in the face of the adverse effects of climate change, to take advantage of emerging opportunities, avoid losses and damages, and build instruments that allow the adaptation of natural, human, production, and infrastructure.

In national communications (2016), an index of vulnerability to landslide and flood disasters was created. Subsequently, a second indicator was created on vulnerability to droughts, information that supported the PNA (Camarinha 2019). These indicators make the relationship between exposure and sensitivity (or susceptibility), resulting in potential impacts. To this end, a map of potential impacts was prepared in the current situation, and estimates for 2040 to estimate future impacts.

Brazil's Fourth National Communication to the United Nations Framework Convention on Climate Change (UNFCCC), sent to the UNFCC in December 2020, assesses vulnerability, impacts, and adaptation measures in an integrated manner (Nexus + approach), considering water security, energy, food and socio-environmental (Coutinho et al. 2019). In the Third Communication in 2016, an index of vulnerability to landslide and flood disasters was created, and subsequently, a second indicator was created on vulnerability to droughts, information that subsidized the PNA (Camarinha 2019).

The IPCC proposes new emission scenarios (IPCC 2022) that allow new projections of global climate change to be defined, with global Earth system models becoming the main tools for studies. Based on this innovation, several Brazilian government institutions that develop research in science and technology in convergence with global models from England and Japan have developed regionalized models for the Brazilian territory with a national model developed by the National Institute for Space Research-INPE (Chou et al. 2014). The indicators mentioned above make it possible to verify potential impacts based on the mapping of possible actions that municipal governments could implement to mitigate future impacts.

Among the problems observed, some of the most pressing ones are faced by municipal and regional governments with measures based on how infrastructure is being used for reducing risks. Urban infrastructure is often directed to reducing risks caused by heat islands reduced air quality and accelerated stormwater runoff. Therefore, municipal administrations, being responsible for implementing adaptation to the effects of climate change, must promote changes in the governance logic, stimulating intersectoral actions; promoting housing policies for the most vulnerable sectors and strengthening preventive and transversal actions incorporating adaptive responses to climate change in master plans and land use and occupation legislation.

Romero-Lankao and Gnatz (2019) highlight the importance of incorporating equality in the definition of urban adaptation actions, as inequality in exposure to risks shapes populations' mitigation and adaptation capacity. In this sense, there are no possible goals for plans or policies related to climate justice without local co-creation work. On the contrary, what is suggested is an inversion in how plans and policies are constructed with the inclusion of local actors, their training to influence these processes autonomously, and the structural themes described above, such as structural racism, among others.

Research carried out in the context of the *Latino Adapta project* with decision-makers related to adaptation to climate change systematized a set of aspects that require concrete actions and advances regarding (i) investments directed to research, maintenance of instruments for monitoring climate variables, technological advances, facilitation of the production process and dissemination of knowledge, among others; (ii) efforts to stimulate dialogue, raise awareness of the topic and translate data into applicable information; and (iii) promoting partnerships and synergies between actors and activity coordinators, to better use of information, as well as other resources, at all stages of the process of developing and implementing public policies. Finally, it is concluded that adaptation to climate change needs to be closely linked to local development. Identifying impacts and vulnerabilities and evaluating the various adaptation options must be part of the public manager's routine. Thus, the knowledge produced can clarify any measure selected based on the scenarios presented to generate benefits (with or without climate change) and avoid poor investments and actions that cause regret (Jacobi et al. 2019).

Osório (2016) points out that concerning the PNA, numerous challenges are to be faced and reinforces that the adaptation process demands enormous commitment from national leaders. Concerning municipal plans and strategies, Margulis (2017) states that the main challenges are in the planning and executing adaptation and mitigation strategies and that they are not yet prepared for the occurrence of disasters resulting from events related to climate change.

Multiple forms of environmental degradation occur, predominantly where lower-income populations, black communities, and indigenous groups live. There is a need to address issues of gender, social inequalities, and human rights together with social inclusion strategies and integrative responses that promote dialogue and reciprocity. This capillarity is also essential when implementing adaptation and loss and damage policies, as NGOs often already operate at the local level and in communities where public authorities are often not present. This creates a rapprochement with the population (Lopez and Abreu 2014). However, it is also important to highlight that an unequal distribution of resources makes it challenging to promote easy adaptation (Travassos et al. 2021).

To this end, the development of vulnerability indices must include the engagement of directly affected populations, facilitating access to information that enables their more effective participation; as stated by Teixeira et al. (2020), "municipal adaptation plans and strategies need to define parameters for decision-making at different adaptive scales: social vulnerability, urban development standards, and infrastructure, urban planning, and delimit which regions, sectors, areas and populations of the municipality should have their adaptive capacity increased to reduce future impacts related to projected climatic and meteorological events, and indicate which sectoral public policies should increase this adaptive capacity."

12.4 Brazilian Northeast: Signs of Change in the Climate Transition

Brazil presents an exclusion of the theme focused on the climate emergency from the decision-making process, making it difficult to debate the inclusion of the human element within the crisis. However, social concern about the issue is latent and urgent. This political lack of interest in the agenda reflects the scarcity of specific plans for adapting to climate change in cities and the growth of socio-climatically vulnerable communities. Local adaptation policies are fundamental in reducing socio-environmental vulnerability in each part of the country since the impacts of climate change are different for each location (Barbi and Rei 2021). However, as Coutinho et al. (2021) analyzed, Brazilian municipal adaptation plans and strategies need to define parameters for decision-making at different adaptive scales.

In the country's northeast region, Teixeira et al. (2020), when analyzing policies aimed at climate issues in the region's capitals, detected that only the cities of Salvador and Teresina incorporated concerns about climate issues into their municipal master plans. Fortaleza and Recife, capitals of the states of Ceará and Pernambuco, respectively, have strong integration of climate change into their government agendas, with the capital of Pernambuco playing a leading role in the climate emergency in the northeast and having an action policy considered successful for climate prevention and adaptation. (Oliveira et al. 2014; Debortoli et al. 2017). In this deficit of an anticipatory system of public policies, the greatest weakness of the climate transition in the northeast region of Brazil lies. The lack of policies focused on the social agenda within the climate issue exposes all previous social vulnerability aggravated by climate change when faced with an extreme climate event.

In another direction, cities on the coast of Brazil, mainly due to their exposure to rising sea levels and the fact that they are already feeling this effect along the entire coast, are more robust in the construction of public policies focused on the climate. Moraes (2020) considers Recife, one of the cities in this situation, as the city most sensitive to climate events and the triggering of disasters, which, consequently, presents this character of anticipation of upcoming events.

12.5 Climate Justice in Recife: Transition to Anticipation

Although extreme weather events affect the entire population, they do not affect everyone similarly. Different social groups are more exposed to its effects and consequences and have a lower capacity for resilience and restructuring immediately after the catastrophe. The inclusion of social justice in the climate change agenda was called climate justice. According to Sultana (2021), climate justice is fundamentally about paying attention to how climate change affects people differently, unequally, and disproportionately, as well as redressing the resulting injustices in fair and equitable ways. Climate justice means different things to different groups and individuals, and the potential implications of climate change for justice are varied and complex. It is not simply a quick reference for acting ethically but rather a source of reasoning for what is considered legitimate, which relates to more normative issues of equity and justice (Huntjens and Zhang 2016).

Recife is currently the Brazilian capital most exposed to extreme weather events; and due to its low altitude and also the large number of rivers and waterways, the 16th most threatened city by rising sea levels in the world (IPCC 2022) (See Fig. 12.1). The city comprises more than 100 rivers and streams that make up the hydrographic basins of the Capibaribe, Beberibe, and Tejipió rivers. This number of bodies of water and the composition of islands and bridges interconnecting them make the city known as "Brazilian Venice."

Recife's geography is compounded by poor urban planning, directly affecting low-income populations. The lack of adequate public policies that prevent residents from the consequences of rain means that the city is exposed to flooding when rainfall increases. Hill regions and flooded areas that comprise the city's relief suffer most from extreme weather events. The biggest climate catastrophe recorded in recent years occurred in May 2022, when it rained 679.1 mm, double the amount predicted for the same period (APAC 2022), and led to the death of 120 people from poor neighborhoods and hilly areas of the city.

Another climate event that worries the city is the advance of the sea over the coast. According to the Brazilian Network for Research on Global Climate Change (Rede Clima 2022), the sea has advanced an average of 0.54 cm from the coast

Fig. 12.1 Hydrography of Recife, Pernambuco, Brazil. *Scale* 1000 m *Source* Recife Data, *Google Maps, 2022*

of Recife per year since 1940. The city's coastline is already 45.7% of its length as having high vulnerability to advance and, according to the Climate Risk and Vulnerability Analysis, 81.8% of urban constructions that are less than 30 m from the coastline and on land below 5 m will be directly affected by changes in the water level (Recife 2023). PLAC-Recife (2020) states that places floods, landslides, heat waves, communicable diseases, meteorological drought, and rising sea levels are the leading climate threats to the city.

This problem is aggravated by the urban occupation of the city's coastline, which means that the sand strip is reduced yearly (Figs. 12.2 and 12.3). At the end of the 70s and beginning of the 80s, major anthropogenic changes occurred on Boa Viagem beach, the city's most famous tourist spot and one of the main urban beaches in Brazil. The construction of a seaside avenue (lanes, sidewalks, walls, kiosks, bathrooms, and leisure infrastructure) had a strong influence on the waterproofing of the land and immobilization of the dunes, which could have led to a severe displacement of the fragile sedimentary balance towards the erosion side of the beach. Oliveira et al. (2014) warn against limiting the future use of areas on Boa Viagem beach, as, due to the socioeconomic pressure occurring in the state of Pernambuco, the construction of large-scale works increases erosion at the site and produces a deficit in sand sedimentation, with a risk of reducing the coastline.

The fact that the region is in a privileged location in Recife means that the communities that inhabit Boa Viagem beach, together with their Special Zones of Social Interest (ZEIS), are targets of intense expectations of changes brought about by real estate and commercial appropriation (De Souza et al. 2020).

Historically, the municipal management of the city of Recife has been concerned about climate change and extreme events in its public environmental policies in

Fig. 12.2 Aerial photograph of the Recife waterfront in 1930. *Source* Joaquim Nabuco Foundation-FUNDAJ, 2022

Fig. 12.3 Aerial photograph of the Boa Viagem waterfront in 2021. *Source* G1 Pernambuco, 2022

recent years. In 2009, the Municipal Environment Fund-FMMA (Law No. 16,047) was created, which guarantees resources for the operation of the Secretariat for the Environment and Sustainability (Smas) and projects in the environmental and climate area. Then, in 2013, the Recife Sustainability and Climate Change

Committee—CONCLIMA and the Sustainability and Climate Change Group—GECLIMA (Decree No. 27,343/2013) were created, whose mission is to establish policies and actions to combat the climate emergency locally. CONCLIMA comprises six municipal departments and is coordinated by the Secretariat for Sustainable Development and Environment. In 2014, the Recife Sustainability and Climate Change Policy was created (Law No. 18,011 of 2014). The following year, the first inventory of greenhouse gas emissions in Recife was carried out.

In 2019, the Municipality of Recife's Analysis of Climate Risks and Vulnerabilities and Adaptation Strategy was launched in partnership with the City of Recife and ICLEI—Local Governments for Sustainability. This document is considered a critical turning point in the anticipatory management of extreme weather events in the city, as the risk is internalized for the first time, and the human and social component begins to be addressed within the climate issue.

The most recent of the public policies is the Local Climate Action Plan of the city of Recife (PLAC-Recife), which was created again by the partnership between Recife City Hall and ICLEI and internalizes and prepares the city for future extreme climate-related events; especially extreme rain and the consequences of rising sea levels. Being the result of a participatory process and bringing actions aimed at marginalized and vulnerable population groups. This document has climate justice as one of its basic principles, being one of the first Brazilian public policies to include this concept within its text. One of its characteristics is to address an anticipatory perspective on the current problems arising from climate change in the city through the axes of Energy, Sanitation, Mobility, and Resilience.

The presence of a framework of public policies that form an anticipatory system for extreme weather events means that coastal cities like Recife can assess vulnerable regions and parts of the population that may suffer the consequences in advance. That is important in decision-making and constructing urban infrastructure works for climate adaptation.

A specific anticipatory system for the city of Recife must permeate a predictive model about the main problems related to precipitation, especially during its period of greatest concentration, which has the greatest impact on the city. The social element must be at the heart of this debate since, with the inequality of resources that is present in this territory, the promotion of climate justice within this sphere will only happen by looking at the poorer sections of the population and their losses and damages through the growing extreme weather events in the city.

The existing plans within Recife's current public climate policy still have gaps regarding the social issue of climate change. The construction of anticipatory climate governance is necessary to strengthen climate justice in society and the state's practical actions towards those most affected. A strengthened anticipatory system for the city could take the form, for example, of a specific plan for hill areas and communities living close to the coast and riverbanks.

12.6 Final Considerations

Recife, like other cities with high climate vulnerability in the world, ended up emerging as an example in the formulation of an anticipatory system for extreme climatic events through the formulation of an extended framework of public policies focused on the issue and the promotion of an intense debate around them, with a strong insertion of the agenda within society.

The partnership between municipal management and cooperation networks such as ICLEI is essential in inserting debates such as climate justice, which are sometimes limited to just the academic sphere within the decision-making arena.

The challenges in metropolitan regions require public managers to implement policies and planning that consider social inequalities and climate change to ensure the quality of life for their socially vulnerable residents. Highlights include the expansion of urban resilience through the implementation of programs to recover environmental functions based on ecosystems, changing patterns of land use and occupation, reviewing construction standards and construction codes, implementing and expanding green infrastructure, improving air quality and urban mobility, strengthening the resilience of infrastructure and the universalization and qualification of sanitation services. However, local governments are also required to reinforce preventive management to deal with climate risks and promote adaptation measures considering social and environmental vulnerabilities, with transparency and social participation, emphasizing actions guided by climate justice. This demands management practices that consider strengthening dialogues essential for implementing adaptation policies. The role of NGOs that already operate in the local sphere and in communities where public authorities are often not present strengthens capillarity and rapprochement with the population for the elaboration of participatory diagnoses and for the execution of projects in a more horizontal and locally legitimized way.

This requires thinking about the climate crisis and adopting new practices in a context of networked, transversal, and democratic governance with interdisciplinary and aggregating approaches that incorporate into climate governance and intersectional and transformational policies that should be included in plans, policies, and programs for climate justice.

References

APAC. Agência Pernambucana de Águas e Clima (2022) https://www.apac.pe.gov.br/. Accessed on 2 Feb 2023

Barbi F, Rei FCF (2021) Mudanças climáticas e agenda de adaptação nas cidades brasileiras. Rev Catalana Dret Ambient 12(1)

Beck U (2010) Sociedade de risco. São Paulo: Ed 34:49–53

CAF/WAYCARBON/ICLEI (2019) Análise de riscos e vulnerabilidades climáticas e estratégia de adaptação do município do Recife—PE—Resumo para tomadores de decisão. http://www2.recife.pe.gov.br/sites/default/files/sumario_clima_recife_portugues.pdf

Camarinha P (2019) Status e Limites do Conhecimento em Mudança do Clima no Nível Municipal. Índice de Vulnerabilidade dos Municípios Brasileiros. CEMADEN (PPT)
Chou SC et al (2014) Evaluation of the eta simulations nested in three global climate models. Am J Clim Chang 3:438–454
Coutinho SM, Santos DV, Bursztyn M, Marengo JA, Lucena AF, Andres D, Maia SM (2019) A adaptação sob uma perspectiva integrada no Brasil: abordagem Nexus+. Rev Sustentabilidade em Debate, Brasília. https://periodicos.unb.br/index.php/sust/issue/view/2174/520
Coutinho SM, Maglio IC, carbone AS, Neder EA (2021) Adaptação às mudanças climáticas no Brasil: complexidade, incertezas e estratégias existentes. ClimaCom—Coexistências e cocriações [Online], Campinas, Year 8, Num. 20, April, 2021. https://climacom.mudancasclimaticas.net.br/adaptacao-as-mudancas-climaticas/
Debortoli NS, Camarinha PIM, Marengo J, Rodrigues RR (2017) An index of Brazil's vulnerability to expected increases in natural flash flooding and landslide disasters in the context of climate change. Nat Hazards
De Souza TO et al (2020) Bibliometria das publicações da revista 'Saúde em Debate': uma fotografia da saúde mental. Saúde Debate [Online] 44:305–319. spe 3, Accessed on 1 June 2022. https://doi.org/10.1590/0103-11042020E325. Epub: 12 Aug 2021. ISSN 2358-2898
González Arellano S (2022) The future scenarios of cities: an analysis of their institutional construction. Int J Urban Reg Res 46(5):865–873. https://doi.org/10.1111/1468-2427.13103
Google (s.f) Hydrography of Recife, Pernambuco, Brazil. Accessed on 15 Mar 2022
Huntjens P, Zhang T (2016) Climate justice: equitable and inclusive governance of climate action. Hague Inst Global Justice 1(1):1–29
ICLEI—Local Governments for Sustainability; CEPS—Centre for European Policy Studies (2013) Climate change adaptation: empowerment of local and regional authorities, with a focus on their involvement in monitoring and policy design. https://cor.europa.eu/en/documentation/studies/Documents/climate-change-adaptation.pdf. Accessed on 18 Oct 2018
IPCC (2022) AR6 climate change 2022: impacts, adaptation and vulnerability. impacts, adaptation and vulnerability. https://www.ipcc.ch/report/sixth-assessment-report-working-group-3/. Accessed on 11 Apr 2022
Jacobi PR, Sulaiman SN (2016) Governança ambiental urbana em face das mudanças climáticas. Rev USP 109:133–142
Jacobi et al (2019) Lacunas de conhecimento em adaptação às mudanças climáticas. Relatório Diagnóstico Brasil. Red Regional de Cambio Climático y Toma de Decisiones. Programa UNITWIN de UNESCO, Proyecto LatinoAdapta. https://saras-institute.org/wp-content/uploads/2020/05/Informe-Brasil_2019.pdf
Lopez FG, Abreu RA (2014) Participação das ongs nas políticas públicas: o ponto de vista de gestores federais. IPEA—Instituto de Pesquisa Econômica Aplicada—1949. Text for discussion, Brasília, April 2014. https://sbsa.com.br/wp-content/uploads/2020/08/Estudo-a-participa%C3%A7%C3%A3o-das-ONGS.pdf. Accessed on 20 Sep 2022
Margulis S (2017) Guia de Adaptação às Mudanças do Clima para Entes Federativos. Relatório completo WWF, Brasília-DF
de Moraes DA (2020) REVISITANDO AS ZONAS ESPECIAIS DE INTERESSE SOCIAL NO RECIFE: LIMITES E CONTRADIÇÕES PARA A GESTÃOE REGULAÇÃO DE FAVELAS. Oculum Ensaios 17
Oliveira DD, Souza-Santos LP, Silva HKP, Macedo SJ (2014) Toxicity of sediments from a mangrove forest patch in an urban area in Pernambuco (Brazil). Ecotoxicol Environ Saf 104:373–378. https://doi.org/10.1016/j.ecoenv.2014.02.004
Osorio G (2016) Planejamento Público: Desafios para a integração do tema adaptação. Rev Adaptação Mudança Clima/Ministério Meio Ambiente 1(1), June, 2016, Brasília: MMA
Recife, Prefeitura do (2023) Prefeitura do Recife. https://www2.recife.pe.gov.br/. Accessed on 16 Feb 2023
Rede Clima (2022) Mudanças Climáticas Globais no Brasil. http://redeclima.ccst.inpe.br/. Accessed on 2 Feb 2023

Romero-Lankao P, Gnatz D (2019) Risk inequality and the food-energy-water (FEW) nexus: a study of 43 city adaptation plans. Front Sociol 4

Sievers-Glotzbach S, Tschersich J (2019) Overcoming the process-structure divide in conceptions of social-ecological transformation: assessing the transformative character and impact of change processes. Ecol Econ 164(106361). https://doi.org/10.1016/j.ecolecon.2019.106361

Sultana F (2021) Critical climate justice. Geogr J [S.L.] 188(1):118–124. https://doi.org/10.1111/geoj.12417

Teixeira RLP, Pessoa ZS, Di Giulio GM (2020) Cidades, mudanças climáticas e adaptação: um estudo de caso de Natal/RN, Brasil. Desenvolvimento Meio Ambiente 54:468–483

Travassos L et al (2021) Why do extreme events still kill in the São Paulo Macro Metropolis Region? Chronicle of a death foretold in the global south. Int J Urban Sustain Dev 13(1):1–16

Chapter 13
Anticipatory Action and Future Urban Living in a Context of Increasing Temperatures: An Analysis from the Barcelona Climate Shelter Network

Brais Estévez, Julia Hidalgo, and Marion Bonhomme

Abstract In recent years, the world has undergone a significant transformation from a realm of certainties to one filled with uncertainties and looming disasters. These catastrophes, which are both ecological and social, are primarily attributed to the climate crisis caused by human activities. As a result, the future of cities and the planet is at stake, with the prospect of becoming uninhabitable. To address this urgent issue, local governments worldwide have taken action to mitigate and adapt to climate change. One exemplary initiative is the Barcelona City Council's Network of Climate Shelters, which was established in 2019. This network aims to prepare the city for the anticipated rise in temperatures by transforming existing public spaces, such as libraries and schools, into climate shelters. What sets the Barcelona Network of Climate Shelters apart, at least in one of its modes of assemblage, is its collaborative approach. The citizens actively participate in designing and implementing some of these infrastructures. That ensures the shelters are efficient and effective and fosters a sense of collective responsibility toward the climate crisis. By involving the public, the initiative prepares the city for climate change and builds a more resilient citizenry capable of facing an uncertain future together.

B. Estévez (✉)
Departamento de Xeografía, Universidade de Santiago de Compostela, Praza da Universidade, 1, 15703 Santiago de Compostela, Spain
e-mail: brais.estevez.villarino@usc.es

J. Hidalgo
National Centre for Scientific Research (CNRS), Laboratoire Interdisciplinaire Solidarités, Societés, Territoires (LISST), Toulouse II University, Toulouse, France
e-mail: julia.hidalgo@univ-tlse2.fr

M. Bonhomme
National Institute on Applied Sciences (INSA), Laboratoire Matériaux et Durabilité des Constructions (LMDC), Toulouse III University Toulouse, Toulouse, France
e-mail: marion.bonhomme@insa-toulouse.fr

S. Gonzalez-Arellano and B. Gandlgruber (eds.), *Cities as Anticipatory Systems*, The Urban Book Series, https://doi.org/10.1007/978-3-031-78162-9_13

Keywords Anthropocene · Cities · Climate shelters · Climate change adaptation · Heatwaves

13.1 Introduction: Why Do We Need to Talk About Climate Shelters?

Global warming is an unequivocally anthropogenic phenomenon that has triggered widespread and irremediable impacts that threaten life-support ecological systems and human and nonhuman well-being (Arias et al. 2021). In this sense, the IPCC reports are increasingly transparent about the direct link between global warming and the increase in extreme weather events such as heat waves, heavy precipitation, floods, droughts, wildfires, and others. The accumulation of scientific evidence on climate change and the awareness of its damaging effects has led governments and policymakers worldwide to include climate change mitigation and adaptation projects in their agendas—especially since the Paris Agreement at COP21 in 2015.

However, despite scientific evidence alerting to the risks for the viability of contemporary human societies of global warming exceeding 1.5 °C or 2 °C above pre-industrial levels, greenhouse gas emissions continue to rise. Neither mitigation measures aimed at reducing emissions nor anticipatory actions aimed at preparing territories and adapting our social systems appear robustly deployed. The disconnection between the knowledge and data on climate change and the hesitant and inconsistent action of governments and citizens, as analyzed by Latour (2015), is at the heart of the disorientation and hesitation surrounding climate action today, with the incommensurability of the climate crisis representing one of its greatest challenges. Some authors (Stengers 2015; Bonneuil and Fressoz 2016; Danowski and Viveiros de Castro 2016 or Latour 2017) consider that this disorientation is yet another consequence of the profound transformations caused by the so-called modern model of development, based on the unlimited use of natural resources and indifferent to the planet's environmental limits. The Anthropocene, referred to by these authors as an epoch characterized by a warmer and more populated world with increased risks and hazards, represents a new geological era distinct from the Holocene, where the relationships between humans and the Earth are no longer guaranteed due to the transformations caused by centuries of human activity based on fossil energies. This concept challenges the traditional notion of nature as a static backdrop and instead portrays Earth as an uncertain and threatening reality that can only be understood through the study of its concrete and localized manifestations. We consider this reflection to also apply to climate emergency governance in the context of cities adapting to extreme weather events such as heat waves.

In this sense, in dialogue with theories of resilient adaptation of communities and territories (Meerow et al. 2016; Reeghezza-Zitt and Rufat 2015; Ruíz-Mallén et al. 2022), urban spaces are particularly relevant environments for testing climate-resilient policies. First, because more than half of the world's population lives in cities and, according to the UN-Habitat World Cities Report 2022, this figure is expected to rise to 68% by 2050 (UN-Habitat 2022). Second, cities are spaces particularly vulnerable to heat due to the heat island effect: the rise in temperature caused by the massive presence of materials that absorb temperatures during the day and emit heat at night (Cuerdo-Vilches et al. 2023). Finally, cities serve as diverse and innovative spaces that encourage citizen participation and bring together a heterogeneous mix of individuals, institutions, and knowledge, fostering the potential for collaborative grassroots responses and the creation of shared knowledge.

This chapter focuses on the study of climate shelters, which are extreme heat adaptation devices that are being tested in cities around the world, consisting of the use of pre-existing infrastructures and public equipment, such as libraries, schools, or parks, to accommodate vulnerable people—babies, the elderly, or people with chronic illnesses—during the so-called extreme weather events. While these spaces maintain their previous uses and functions, adapting their infrastructures to the heat by introducing ventilation elements, water points, or vegetation provides thermal, environmental, and social conditions that are welcoming and healthy for people.

Climate shelter networks can be a powerful climate resilience initiative for two reasons. First, they provide heat adaptation for people while optimizing the use of air conditioning—if needed—and can, therefore, be considered a virtuous device from the point of view of climate change mitigation objectives. Second, they are an initiative that can make the climate emergency commensurable. Rather than assuming that climate change is a matter of fact, climate shelters can approach the climate crisis as a reality whose conditions of intelligibility and evidence also depend on community involvement in shared worlds where that evidence is collectively produced and can be meaningful to all actors involved.

The act of transforming an incomprehensible phenomenon into a comprehensible one aligns with Maria Puig de la Bellacasa's perspective on care practices, which emphasizes the importance of giving ethical and political significance to socio-material practices, generating care for overlooked issues, and actively engaging with them to create transformative effects (Puig de la Bellacasa 2017). In this sense, we argue that climate shelters can be both collective infrastructural adaptation equipment to prevent the effects of extreme heat and instances of urban community resilience capable of collectively elucidating the socio-technical terms of such adaptation.

13.2 What Are We Talking About When We Talk About Anticipatory Action and Future Urban Living in the Context of Increasing Temperatures?

13.2.1 Heatwaves in the City: Living the (Future) Climate Change Now

Climate change is causing a general increase in temperatures and an intensification of heatwaves that in urban areas interact with the urban heat island phenomenon, creating a new composite risk characterized by periods of extreme temperatures. Its negative impacts are expected to increase in the future, affecting the habitability of cities and the health of many of their residents, especially the most vulnerable groups (Jones et al. 2015). Indeed, the impacts of exposure to extreme heat are numerous: (i) public health problems such as increased mortality and morbidity, (ii) lack of thermal comfort in indoor and outdoor spaces, (iii) excessive energy consumption and exacerbated greenhouse gas emissions, (iv) infrastructure damage caused by temperature itself or shrinkage and swelling of clay soils, and (v) weakening of biodiversity.

In the context discussed here, anticipatory action is formulated through climatic adaptation policies (Arias et al. 2021).

Those policies are typically implemented at three levels: national, municipal, and building. At the national level, reactive measures such as heatwave alert systems are implemented. Municipal interventions focus on longer-term strategies related to urban planning and design, including green (nature-based), grey (technical infrastructure), and soft (behavioral) solutions. At the building level, improving indoor thermal comfort involves building design, window characteristics, insulation, and occupants' behaviors.

The so-called climate shelters we discuss in this chapter are hybrid devices that consider some of the heat adaptation strategies mentioned above. Although they may involve urban-scale and building-scale transformations, we argue that such strategies have great potential to connect a building's need for thermal comfort with a series of collaborative processes for community and territorial resilience that can be replicated in a wide range of spaces worldwide.

13.3 What Are Climate Shelters, What May They Be, and What is Their Genealogy?

One of the first bibliographic references to climate shelters is the so-called cooling centers (Changnon et al. 1996; Palecki et al. 2001; Widerynski et al. 2017; Nayak et al. 2017; Sehgal and Sehgal 2023). They emerged in the US and are an important resource in extreme heat, especially for people who are homeless or lack adequate cooling in their homes. According to Widerynski et al. (2017), cooling centers are typically air-conditioned or cooled buildings designated as sites for respite and safety during extreme heat events. They can utilize existing infrastructures and staff and be relatively easily implemented. Also, in terms of their management and typology: "they may be a government-owned building such as a library or school, an existing community center, religious center, recreation center, or a private business such as a coffee shop, shopping mall, or movie theatre" (p. 4). In this regard, Black-Ingersoll et al. (2022) distinguish between formal heat shelter buildings or spaces designated by the city council and informal/volunteer air-conditioned shelter spaces open to the public, such as malls, museums, churches, or cinemas not listed by local authorities.

The severe heatwave in Chicago in 1995, resulting in many deaths, was a pivotal moment in shaping heatwave prevention policies in the US, leading to improvements in adaptation and preparedness plans, including the establishment of cooling centers. These measures, along with enhancements to the city's power supply system and the implementation of heat action plans, contributed to a significant reduction in the death rate during the 1999 heatwave compared to 1995 (Klinenberg 1999, 2015; Whitman et al. 1997; Palecki et al. 2001).

The urban heat island is considered an aggravating factor in the Chicago disaster, but climate change is barely mentioned. In the second edition of the book *Heat Wave: A Social Autopsy of Disaster in Chicago*, Klinenberg (2015) added a new preface reflecting how climate change has made extreme weather events in urban centers a global challenge. According to Klinenberg, addressing this challenge requires a commitment to climate-resilient infrastructure changes rather than just relief responses. Additionally, he highlights how social ties, cohesion, and urban dynamism in some working-class neighborhoods in Chicago played a crucial role in saving lives during the 1995 heatwave. That was made possible by a robust social infrastructure of local commerce, public facilities, dense community ties, and organizations that fostered mutual support and intense community resilience.

In this sense, increasingly frequent and intense heatwaves have led local governments worldwide to develop heat adaptation, extreme heat preparedness, and response actions. These actions include, for example, the setup of the so-called cooling centers.

More than 1402 cooling centers in the US have been identified from publicly available sources (Adams et al. 2023). There are also many in Canada and a few in Europe and Australia. Cooling centers are also being built in South Asian countries such as India (Knowlton et al. 2014). However, the existing cooling centers are expensive and depend on a stable electricity supply for the air conditioners to function correctly. Other limitations identified by the literature include the following: (a) some people are not aware of the existence of cooling centers in their city (Alberini et al. 2011); (b) accessibility to cooling centers can be difficult for some citizens: city plans include cooling centers but not transportation to reach them (Bedsworth 2009); (c) some people refuse to go in a cooling center in the event of a heatwave because they do not like the idea of sleeping in a dormitory with nothing to do (Kosatsky et al. 2009; Sampson et al. 2013); and (d) most people do not see themselves vulnerable enough for cooling centers, considering they are not ill (Kosatsky et al. 2009; Sampson et al. 2013).

Despite the need to adapt the city to heat waves, current cooling centers do not meet population needs. Studies show that outdoor cool sites like public pools and parks are used more frequently during heatwaves than indoor cooling centers (Nayak et al. 2017). Public reluctance toward climate shelters may stem from the fact that they are not associated with spaces of well-being, sociability, learning, and community life. Instead, they evoke images of exceptionality and urgency, like flood, fallout, and air raid shelters.

In other regions, noteworthy initiatives, such as those promoted by some local governments in partnership with stakeholders and citizen movements, seek resilient adaptation of small-scale urban services and infrastructures in the context of climate change. One of these initiatives is the Climate Adapted People Shelters (CAPS), a project aimed to reimagine Sydney's bus shelters as Climate Adapted People Shelters through tenets of co-design and co-production that adapt bus shelters to heat (Jacobs et al. 2018).

Nevertheless, since 2019, the notion of climate shelter connects to a specific type of urban initiative for resilient adaptation to heatwaves, explored in Barcelona. This initiative aimed to transform and adapt some facilities, such as public schools, to extreme heat and thus provide thermal comfort spaces—as well as protection, rest, and safety spaces accessible to all—to citizens during extreme weather events (Vilar and Barnada 2020; Baró et al. 2021; Satorras-Grau et al. 2023).

While cooling centers serve as emergency shelters for vulnerable citizens during extreme weather events, the climate shelters in Barcelona are everyday living spaces that adapt their infrastructures and communities to a new world threatened by extreme heat.

13.4 Barcelona Climate Shelters Network

13.4.1 Characteristics and Limitations

The Barcelona Climate Shelter Network initiative launched by the Barcelona City Council in 2019 began to be deployed in the summer of 2020. It consists of a set of pre-existing municipal facilities and public spaces—such as collective social centers, municipal sports facilities, museums, public libraries, schools, parks, and gardens—which, while maintaining their regular uses or functionalities, can be activated, and used otherwise as places to shelter from extreme heat. Barcelona climate shelters can be indoors or outdoors; the former does not necessarily have to be air-conditioned, but they must guarantee a setpoint temperature of 27 °C in summer. The latter should provide greenery, shade, benches for seating, and fountains.

Barcelona is a Mediterranean city with hot and humid weather in summer and is particularly exposed to the heat island effect (Martín-Vide and Moreno-García

2020). Some research indicates that, in the period 1990–2004, mortality in Barcelona soared on extreme heat days by up to 27% (Tobías et al. 2014). Indeed, according to data from the Public Health Agency of Barcelona—ASPB, according to its initials in Catalan—in the 1992–2015 period, extreme heat caused more than 3500 deaths. When temperatures were high, the mortality risk increased by 14% in women and 4% in men (Marí-Dell'Olmo et al. 2022). More recent research indicates that mortality risk in Barcelona on the hottest days also varies unevenly within the city according to age, gender, social class, and neighborhood biases (Ingole et al. 2020). Moreover, climate predictions foresee increasingly intense, prolonged, and recurrent heatwaves in Barcelona (Altava-Ortiz et al. 2017; Pereira et al. 2017; Arellano Ramos et al. 2022).

However, although climate shelters were initially conceived by the City Council as an urban adaptation measure to heatwaves, since 2023, the network has also been attentive to cold waves to protect vulnerable citizens from low temperatures during the winter season. This innovation has not been without controversy, and one of the informants we interviewed for this research, a technician involved in the deployment of the network, shared with us his doubts about it:

> I am very critical of winter shelters; I think they have a radically different logic. Winter season is a different world; those who can die in winter in a city are the people who are homeless. And in this sense, I think there is a gap to be bridged between the reality of winter and the devices that deal with that reality. Summer and winter are different realities and cannot be treated as equivalent (Miquel, interviewed 1).

In this sense, even though initially the climate shelters were only activated in heat-related episodes—there was an activation protocol that operated during the so-called preventive phase of heat waves—today the City Council considers that the indoor spaces are permanently activated—all year round—while the outdoor spaces are activated only during the heat prevention phase, from June 15 to September 15.

According to the Climate Change and Sustainability Office of the Barcelona City Council website, a climate shelter must fulfill the following characteristics:

(a) The indoor spaces—climate-controlled or not—must provide a setpoint temperature of 27 °C in the summer and 19 °C in the winter.
(b) They are designed for people vulnerable to the heat and the cold (children, elderly people over 75 years old, and people with chronic illnesses).
(c) They are not designed for people who need medical care.
(d) They must be easily accessible and safe, providing comfortable rest areas (chairs and benches) and free water.
(e) The urban parks, which form part of this network, have abundant urban greenery (surface area > 0.5 ha) and an NDVI index (which indicates the quality of the vegetation) > 0.4. They are accessible to people with reduced mobility and have fountains and seats.
(f) All climate shelter spaces are adequately signposted.

The Barcelona Climate Shelter Network is one of the responses envisaged in the *Climate Plan* (Ajuntament de Barcelona 2018), the strategic document with which the Barcelona City Council establishes a roadmap for its climate change adaptation and mitigation policies for 2018–2030. This document focuses on four strategic axes. First, mitigation policies are understood as the set of actions aimed at reducing greenhouse gas emissions. Second is resilient adaptation, i.e., actions aimed at preparing the city for the effects of climate change, reducing the vulnerability of citizens and urban infrastructure. Third, climate justice places the most vulnerable groups at the heart of climate action. Fourth is promoting citizen involvement in climate action and governance by committing to democratic and participatory climate policy governance. Indeed, action line 3 of the *Climate Plan* is dedicated to preventing excessive heat. It proposes "improving thermal comfort in the city's public spaces and buildings and creating a network of places to take shelter in high temperatures, to protect people's health and provide special care for the most vulnerable groups" (p. 72).

Through this initiative, the Barcelona City Council aimed to provide all city residents with a climate shelter space within 10 min of walking distance from their homes. Indeed, according to data presented by Barcelona City Council at a press conference on 30 June 2023, 97% of Barcelona's citizens have a climate shelter less than 10 min walk from their home, while 58% have one even closer, just five minute walk from their home (see Fig. 13.1).

Methodologically, the network was designed firstly through benchmarking work inspired by similar projects already in existence in Paris—the *Paris Climate Action Plan* (City of Paris, 2018) set a goal that by 2020, all Parisians could be within a

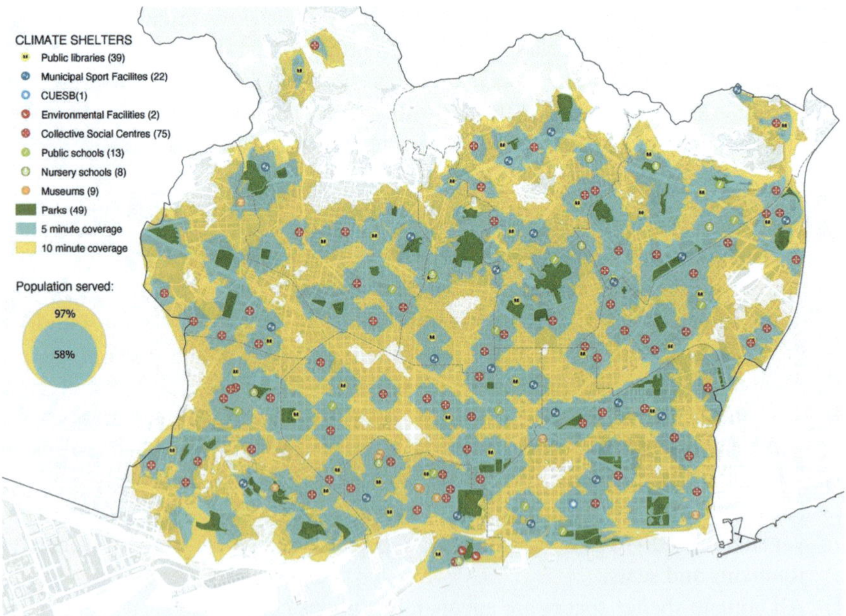

Fig. 13.1 Climate shelter map in Barcelona for the summer of 2023. *Source* Barcelona Regional Agency

seven-minute walk of a cool island (here understood as a "fresh" outdoor space)—and in cities in the US and Australia, and secondly through fieldwork carried out to identify and map potential shelter spaces.

The network has accelerated pace quickly, during which the COVID-19 pandemic also interfered with its development. In 2020, the network was deployed for the first time in the city, making 70 climate shelters available to the public. In 2021, the network reached 155. Moreover, in the summer of 2023, Barcelona expanded its network of climate shelters to more than 200 spaces: 227 climate shelters for the summer and 141 for the winter.

However, in the words of another informant—also a technician involved in the network design—the municipality's obstinacy with the quantitative growth of the network has its ups and downs:

> The local authorities maintain a discourse on the proliferation of the network; however, in the summer of 2021, the City Council carried out a survey asking the level of public awareness of the network. The results showed that 20% of citizens were informed about the network, but only 8% had used it. Therefore, I believe that the City Council should emphasize on raising awareness of the network and once it has become a recognized resource for citizens, focus on expanding it (Aina, Interviewed 2).

According to the fieldwork we have carried out—direct observations of some shelter spaces and interviews with actors involved in the project—some of the principles that inspire the Climate Plan are only half fulfilled in this initiative. Below, to illustrate these shortcomings, we share some testimonies from key informants obtained in our fieldwork:

> A climate shelter should be where a person arrives and can stay in a cool environment where someone offers them a glass of water and asks them how they are feeling. But they are not working like that. Barcelona has 210 shelters right now, but when someone asks about the annual budget allocated to these spaces, the City Council has two answers. Either they answer that there is no specific budget or that the budget is that of the city's facilities, parks, and gardens (Josep, Interviewed 3).

> In the beginning, we were very inspired by some experiences tried in Paris, such as the opening of some parks and gardens until midnight or even allnight during heat waves. What happens in Barcelona is that logistical problems arise quickly: either there is no staff, or there is no staff at that time, so some actions that may be necessary are not carried out (...). Unlike, for example, the American cooling centers, Barcelona's network of climate shelters is made up exclusively of public infrastructures. Our spaces already have a timetable and a staff and, ultimately, what often changes in these spaces is that they gain an identifier and a service of which their staff often have no clear idea, apart from knowing that they have to welcome the people who come here to shelter from the heat. However, our research shows that most people who go to public libraries to take shelter from the heat do so because they are refreshing and welcoming spaces, not because citizens know they are climate shelters. The network has been growing, and the City Council has a discourse of increasing the number of shelters, which, in my opinion, can distort the initiative. What happens? You build up the network, but many of these spaces close or limit their opening hours when August arrives. There should be a network that offers territorial coverage, but perhaps instead of being obsessed with creating many, it is more important to cover densely populated areas of the city. The municipality is sometimes hostage to its own words. There has been a certain fascination with the idea of a climate shelter five minutes away from everyone's home, but have you

> asked yourself about the conditions of such shelters? Some spaces identified as shelters are halls of sports centers, sometimes hardly welcoming (Miquel, Interviewed 1).

> Public libraries have always been a natural climatic refuge. During heat waves, we have always received users looking for a cool place to connect to the internet, consult magazines, and be comfortable. But, at the same time, adapting our spaces as shelters has some deficiencies. According to the network principles, a climate shelter should have a water point that allows users to hydrate themselves. In many of our libraries, the water point is the toilet tap; we do not have proper fountains; only the new libraries have a proper water point (Lara, Interviewed 4).

The testimonies gathered in these interviews highlight the logistical and budgetary weaknesses of the network, which, while showing its value, highlight the complexity that managing such a network can present in the long term. However, some of the challenges identified have been addressed otherwise through a European project on school-based climate shelters recently launched in Barcelona. We discuss some aspects of this project in the next section of the chapter.

13.4.2 Focus on a Participatory Project: The "Blue, Green, and Grey: Adapting Schools to Climate Change Project"

In 2018, the City of Barcelona submitted the project "Blue, Green & Grey: Adapting Schools to Climate Change" to the third call of Urban Innovative Actions (UIA)—an initiative of the European Commission that provides support for innovative measures in the field of sustainable urban development to test new and unproven solutions to address urban challenges. The proposal was selected, and the project was funded with €5 million (€4 M with European ERDF funds and €1 M with funds from Barcelona City Council) to be implemented over three years. Barcelona's schools house more than 120,000 children considered vulnerable, and their buildings and schoolyards have a low level of heat adaptation. According to the technical document "Good practices: climate shelter projects" (Gerència d'Àrea d'Ecologia Urbana & UIA 2022), Barcelona's proposal was presented as an opportunity to improve the thermal comfort conditions of the city's schools and transform them into healthier spaces for students and teachers, taking advantage of the potential of this transformation to bring benefits to all citizens throughout the year, outside school hours, during and outside the school calendar. The project started in 2019 and finished in 2022.

As a pilot test, the project set out to transform 11[1] public schools to adapt them to heat through three types of climate-resilient measures: (i) Green measures, naturalizing the schoolyards incorporating trees and plants; (ii) Blue measures through the provision of water points, fountains and unique places for children to play with

[1] The schools were chosen from a list of all the city schools that applied to participate in the project. For this purpose, different parameters were analyzed: from climatic vulnerability and the presence of vegetation in the urban environment, or the energy efficiency of the buildings (e.g., number of classrooms exposed to the sun), to the type of paving of schoolyards, and others.

water; and (iii) Grey interventions: i.e., architectural solutions on the buildings such as thermal insulation, shading, protection for façades, cross ventilation, and energy generation.

Although the project was led by Barcelona City Council—under the coordination of the Urban Ecology Area—it also included the following partners: Public Health Agency of Barcelona, Barcelona Consortium of Education, Barcelona Cycle of Water—Public Service Provider—Barcelona Institute for Global Health (ISGlobal), Institute for Environmental Science and Technology (ICTA-UAB) and Vila Olimpica School—the only school that participates in the project since its initial design.

The project turned 11 public schools into a prototype space to implement Barcelona's climate shelter initiative in a profound and committed way. Indeed, the main difference between the adaptation of schools and other climate shelter spaces in the city—such as libraries or museums—is that, in the case of schools, there is a relevant physical transformation process and, at the same time, a participatory and pedagogical process that allows us to imagine and enact a community concerned by the climate crisis.

13.4.3 Co-design and Collective Learning

The transformation of the 11 public schools into so-called climate shelters was done through co-design. However, it should be noted that although the project transforms the school buildings and the schoolyards, only the schoolyards are integrated into Barcelona's network of climate shelters. Only schoolyards are considered spaces open to all citizens during heat waves.

A participatory process made it possible to reach a consensus on the design of the new spaces between the educational community—students, families, school staff, and members of neighborhood associations concerned with the school—and the experts and technicians mobilized by the city council.

First, each school chose a steering group that led the project in the school itself, working on the concept of climate change and imagining possible actions in the school related to heat adaptation. Indeed, in Can Fabra School—one of the schools we visited as part of our fieldwork—the steering group diagnosed on a map the school's (see Fig. 13.2) weaknesses in terms of thermal comfort and proposed ideas on how to solve them.

Four participatory sessions were held over two school years. The specific objectives of these participation sessions were threefold: (a) to involve the educational community in the project, (b) to collectively define the measures to be implemented, and (c) to collectively evaluate the impact of the measures carried out. The first two sessions were held in the 2019–2020 academic year. In the first session (September 2019), they discussed and prioritized the challenges that each school had identified in terms of heat adaptation based on a catalog of available solutions drawn up by the City Council. In the second session (January 2020), the educational community helped the team of architects assigned to each school specify the chosen measures

Fig. 13.2 Details of the adaptation process of Can Fabra School. *Source* photo by the authors

and draw up the executive project for the works. In April 2020, the outbreak of the COVID-19 pandemic made it necessary to hold an extra session—held virtually—to finalize some design issues. In October 2020, the third session was held virtually due to the pandemic. In this case, the works were already executed, and the initial evaluation of the measures implemented was discussed. In June 2021, the fourth session (also virtual) was held to evaluate and verify whether the implemented actions were reducing the temperature of the schools. The total number of participants was 1.264 people.

According to Vilar and Barnada (2020), at the end of the participatory sessions, the chosen adaptations were the following: (i) at the building level, to favor cross ventilation; (ii) at the schoolyard level, to incorporate vegetation, water points, and surface treatment—less reflective and more permeable materials and colors. However, above all, what stood out were the mixed solutions that incorporated elements of shade, greenery, and water in the same layout (i.e., pergola with greenery and water diffusers). It should be noted that while most cooling centers in the world operate through air conditioning, the schools converted into climate shelters in Barcelona operate exclusively through passive measures. However, this decision was not free of doubts and questions. According to one of our informants, passive measures would not be sufficient to cool the school buildings:

> Through passive modes, you can go down by 4-6° C, we have sensorised and checked it, but if you are at 33-34° C, you can reach 27-28° C, with the humidity in Barcelona, maybe the thermal sensation is 32° C (Josep, interviewed 3).

In parallel to the participatory process, the adaptation of the schools as climate shelters carried out a pedagogical process aimed at awareness-raising among the educational community in the face of the climate crisis. In the same way that the participation process mentioned above involves the educational community in the school's adaptation to heat, the pedagogical process aims to involve the same educational community in the fight against climate change, involving in climate action a type of actors—for example, children, but also their families—who do not usually inhabit the forums where these issues are discussed.

During the same two school years (2019–2020, 2020–2021), three training seminars were held in the 11 schools for teachers, as well as educational support for 5th-grade students—the educational level leading the project in each school—on the issue of climate emergencies. This training took the form of a teaching guide entitled "The Challenge of the Climate Emergency: An Information Guide and Proposal for Educational Activities"—a tool for teachers to help them communicate the challenge of climate change—a so-called pedagogical suitcase—and a series of classroom activities and workshops with the project's scientific partners (ISGlobal, ICTA-UAB, ASPB). The pedagogical suitcase to face the climate emergency is aimed at students between 3 and 12. It includes materials and educational resources in physical and digital format on how to deal with the climate emergency in the classroom, accompanied by a teacher's handbook. The activities are organized according to age and with the creation of a learning sequence based on three thematic sections: (a) "Let's talk about climate change," (b) "The climate crisis is already affecting us," and (c) "Let's react to the emergency"—activation dynamics. In the first thematic axis (a), some activities present problems associated with climate change, identifying its origin. In the second block (b), there are activities aimed at discovering in an experimental or sensorial way the consequences of global warming, emphasizing how they will affect our lives. The third block (c) aims to motivate students to act against the climate emergency.

The first seminar for the teachers took place in September 2019. It shared scientific information on climate change and provided teachers with educational resources on climate change.

In September 2020, the second seminar took place to deepen the pedagogical use of the teaching initiatives promoted so far and to socialize the concept of climate refuge among the educational community. In October 2021, it was discussed how to give continuity to the project over time, and the pedagogical suitcase was handed over.

Nevertheless, all this explanation would only be incomplete with some empirical register that would allow us to identify critically the transformations carried out in the schools.

On 22 and 23 May 2023, the authors of this text traveled to Barcelona to visit three schools adapted to the heat to find out the extent of the transformations and interview teachers from each school. The schools visited were Escola Can Fabra—which opened in 2017 on the grounds of Fabra i Coats, a former textile factory that closed in 2005 and converted by Barcelona City Council into a robust arts and culture factory in 2012—(see Fig. 13.2), Escola Cervantes—located in Ciutat Vella, the old district of Barcelona and housed in the Antic Palau Dou, a historic building with baroque and neoclassical architecture—and Escola de les Aigües—located in the district of Baix Guinardó and with an institutional legacy linked to the Spanish army. The first two are part of the climate shelters project financed with European resources, while the last one is part of *Transformem els patis,* a municipal project inherited from the former consisting of transforming schoolyards through participatory methodologies.

According to one of Can Fabra's teachers:

> We already had the idea of transforming the schoolyard, and the Climate Shelter's project fitted perfectly with our aims; we could transform some spaces of the school and at the same time raise awareness of how we can respond to the climate emergency. We love the result, but we really appreciate the participatory process. A technical team accompanied us throughout the process, and the children played a vital role, as did the families. We believe participation has been one of the project's strong points. Before attending the participatory sessions, we had done much work listening to the children, the families, and the technical team that accompanied us, and this work, indeed, allowed us to socialize on the issue of the climate emergency. We modified the schoolyard, replacing the sports court with a Mediterranean garden: a sandy area with trees, shrubs, aromatic plants, etc. We also installed a wooden grandstand —which we call the *agora*— a pergola and a shaded area covered by a sunshade. We installed two multifunctional fountains for cooling, and in the building, we installed solar shading between facades —fixed aluminum solar shades— and a mechanical solar shading system: external screens that are activated mechanically.
>
> We have gained shaded areas, more pleasant, more habitable, and naturalized spaces, and we have also ensured that the summer classroom temperature is not as high as it used to be. This project has been an opportunity to transform the space of our school and the mental framework of the entire educational community; we have experienced how, in the face of a serious problem, such as the climate emergency, there can be a real response that involves a transformation of spaces and that also has a direct consequence in increasing our well-being. The physical transformation needed to make all the participatory and pedagogical work visible was very important, and, in this sense, I would like to highlight the relevance of the physical transformations (Anna, Interviewed 5).

The headmaster of Cervantes School (see Fig. 13.3) also insisted on the positive assessment of the participatory process while showing us some of the most successful infrastructural adaptations of his school:

> We are in a relatively disadvantaged area, a very dense neighborhood with very few green areas, and in this sense, we fit in with the criteria of territorial vulnerability that the project specified.
>
> Our schoolyard used to be a basketball court and has been transformed into a greener space with a multifunctional fountain. We have gained in cross-ventilation in the building thanks to straightforward interventions such as making casement windows and installing solar panels on the roof. Both the participatory and pedagogical processes have been precious; the technicians and mediators sent by the municipality helped us to ground the children's wishes and, in this way, we went from an initial imaginary that immediately thought of a swimming pool to a very subtle type of action that I believe has improved the habitability of the school

Fig. 13.3 Blue, green, and grey solutions at Cervantes School. *Source* photo by the authors

> during heat waves. However, I must also point out that discussing the climate with the children triggered ecological anxiety among many. It, therefore, remains an open question that we must continue to work on (Marta, Interviewed 6).

In short, the main difference between most of the spaces that compose the network of climate shelters in Barcelona and the schools adapted to climate change through the European project is that, in the latter, the issue of climate emergency can be elaborated collectively as shared evidence. We will call this modality of co-construction of climate shelters participatory climate shelters.

13.5 The Notion of Participatory Climate Shelters Regarding Anticipatory Action and Urban Futures

13.5.1 Beyond the Limitations, to What Extent is Barcelona's Network of Climate Shelters an Example of Urban and Community Resilience?

In a context where scientific evidence is not strong enough to guide a collective response to climate threats and conspiracy theories about global warming abound, is it sufficient to adapt urban infrastructures without providing spaces for collective rebuilding in the new climatic regime? According to the anonymous text attributed to the Invisible Committee (2009), the evidence goes beyond logic and reasoning; it resides "on the side of the sensible, on the side of the worlds" and is what is shared (p. 31). How can climate shelters counter the crisis of sensibility that hinders collective responses to the climate crisis?

We use the concept of the "crisis of sensibility," like Morizot (2022), who describes it as both a crisis of our productive relations with living environments and a crisis of collective and existential relations (p. 17). Morizot highlights a Kantian sensibility that separates humans from non-humans, treating the latter as subjective projections of the former. This concept helps us understand the rupture of ties between humans and between humans and territories. To address this, we must understand territories otherwise, no longer as transcendent scenarios in which human life takes place but as a living world with which we are bound by deep relations of interdependence and with which we are obliged to compose our ways of life for habitability.

Morizot's idea connects with the work of the late Latour (2022), particularly with his appeal to the vital role that affect plays in responding to the climate crisis. In Latour's words, "If you do not have the affects capable of metabolizing the ecological situation, everything becomes too hard (…) and the work is overwhelming" (p. 98).

Therefore, in our view, heat adaptation policies require both an infrastructural shift and a new sensibility that allows us citizens to reconnect with other human and non-human entities with whom we maintain relationships of interdependence—often imperceptible—but fundamental for the sustainability of our lives. Moreover, for this, we understand that we need spaces of collective experimentation where climate threats are made sensible in two necessarily complementary ways: (i) as measurable threats that we—an alliance of academic, professional, and social expertise—can recognize and identify together; and (ii) as a new climate regime that oblige us to act in our daily living environments to adapt our territories to the consequences of climate change and thus limit its impact.

In our view, some of the challenges we have outlined have already been tested in Barcelona through the mentioned project that adapted 11 public schools as climate shelters.

13.5.2 An Anticipatory Community?

Section 13.2.1 of this chapter discusses how climate shelters can serve as an urban anticipatory device in response to climate change. Two issues are highlighted. On the one hand, Barcelona City Council's Network of Climate Shelters addresses the climate crisis from the present, when many citizens and organizations still imagine it as a distant issue: it would indeed be a probable future, but above all, a future. On the other hand, the network itself, especially in its school composition, positively contributes to build a more resilient citizen community that can be considered as an anticipatory community. As explained in González Arellano's chapter in this book, an anticipatory community is a group of actors who share a common anticipatory practice: they combine a shared perspective on the future with a specific enactment of that perspective. But what kind of anticipatory community resonates with Barcelona's network of climate shelters? González Arellano (2025) proposes five types of anticipatory communities: (i) repairers—those who respond quasi-random phenomena by designing and implementing protocols to deal with events with known patterns of behavior; (ii) managers—a community that seeks to introduce modest changes that will improve the functioning of the system in the face of events based on regular patterns; (iii) anticipators-strategists—actors with a very open perspective on predictive models and interested in the future scenario building; (iv) visionary-advisors—collectives that share a techno-scientific rhetoric to construct medium and long-term scenarios; and (v) utopian-prophets—those who share concerns about the long-term future, with positions between utopian and collapse theories. Of course, these typologies are ideal models and doesn't exist in a purely discrete urban form. But, at least, in our view, each typology has certainly played a role in the historicity of the Barcelona City Council's Climate Shelter Network. For example, initial Barcelona's school individual-based initiatives to transform schoolyards and adapt them to the heat and sunshine could be thought of as a community of repairers dealing with local increase of temperature due to climate change and urbanization. The phase in which eleven public schools in Barcelona participate in the European project "climate shelters in schools," adapting to the heat through subtle actions aimed at improving thermal comfort can be thought of as the so-called community of managers. Moreover, the technicians and politicians who, based on different climate scenarios, have opted to build a network of climate shelters in Barcelona could be thought of as visionary-advisors. Specially since this type of public policy was completely new in Spain. Finally, On the other hand, the current spokespersons of the Barcelona's network of climate shelters who meet on a regular basis to discuss and imagine the terms of the Barcelona's public facilities adaptation to climate change can be considered anticipators-strategists as they have a good understanding and vision of the network complexity and are able to identify relevant interactions between components, redundancies and institutional or social weaknesses which make it difficult to implement, consolidate and develop such a public policy.

13.6 Final Remarks: Climate Shelters Are Open to All People

On July 10, 2023, as we were writing these lines, the world's media echoed a news item published by the World Meteorological Organization (WMO), which reported that, according to preliminary data, during the first days of July, the world had experienced the hottest week on record. In the words of Christopher Hewitt, Director of Climate Services at WMO, "We are in uncharted territory, and more records can be expected to fall (…) this is worrying news for the planet" (WMO 2023).

As we argued in the previous section, apart from the specific threats that extreme heat poses to the health and well-being of the most vulnerable groups—as well as to the thermal comfort of the population as a whole—we consider that a significant obstacle in the field of anticipatory actions regarding heatwaves has to do with a certain collective inability to make climate change concern us and not just worry us.

Although most institutions, as well as some authors, consider climate shelters as a service aimed at hosting vulnerable populations during specific meteorological events—such as heat waves—based on our research focused on the case of Barcelona, we believe that climate shelters can be much more powerful and decisive spaces both in the fight against climate change and in the adaptation of cities to heat.

We argue that climate shelters have the virtue of operating on the logic of anticipatory action by assuming that climate change is occurring in the present and not just in a still uncertain future that we can continue to wait and not act upon. In this sense, the anticipatory and almost visionary character of climate shelters allows us to think of them as future infrastructures operating in the present.

Nevertheless, why do we consider that a public network of climate shelters such as Barcelona's is relevant to deal with such a sensitive issue as adapting cities to heat?

On the one hand, in a context of uncertainty, where the climate crisis may be driven inequitably, through private solutions that leave critical sectors of the citizenry behind, a public network may bring collective, democratic, and inclusive solutions that have the potential to strengthen the resilient adaptation of communities and infrastructures about the climate crisis.

On the other hand, climate shelters, as a scalable adaptation device, pool energy resources in a crisis that calls for sobriety, makes climate action commensurable, and can involve the population in issues traditionally restricted to the field of experts. However, it is also a device that is easily replicable and adaptable to other types of territories—even non-urban—and latitudes—dealing with the cold or the heat.

Considering the European project that made it possible to adapt 11 schools in Barcelona through participatory methods involving heterogeneous actors from the educational community and the neighborhood, we argue that spaces linked to cultural and educational practices are particularly suitable environments for adaptation as climatic shelters. Thus, despite the difficulties and budgetary or logistical obstacles detected (highlighting problems with maintenance or the difficulty of arbitrating practices with radically heterogeneous opening hours in the same space, which can often strain the calendars and forms of governance of the facilities), we consider climate shelters to be a powerful initiative to anticipate the future. In this sense, we want to conclude this chapter by asking a speculative question that is also a wish: Will climate shelters be able to become a public infrastructure, analogous to public transport, capable of accommodating most citizens during heat waves regardless of their vulnerability?

References

Adams QH et al (2023) Examining the optimal placement of cooling centers to serve populations at high risk of extreme heat exposure in 81 US cities. Public Health Rep 138(6):955–962. https://doi.org/10.1177/00333549221148174

Ajuntament de Barcelona (2018) Pla Clima 2018–2030. https://www.barcelona.cat/barcelona-pel clima/sites/default/files/pla_clima_cat_maig_ok.pdf

Alberini A, Gans W, Alhassan M (2011) Individual and public-program adaptation: coping with heat waves in five cities in Canada. Int J Environ Res Public Health 8(12):4679–4701. MDPI AG. Retrieved from https://doi.org/10.3390/ijerph8124679

Altava-Ortiz V, Barrera-Escoda AA, Amaro J (2017) Escenaris climàtics regionalitzats a l'Àrea Metropolitana de Barcelona (Projecte ESAMB). Projeccions estadístiques regionalitzades, Fase II. Servei Meteorològic de Catalunya & Departament de Territori i Sostenibilitat, Generalitat de Catalunya

Arellano Ramos B, Roca Cladera J, Serra de Larrocha C, Martínez Santafé MD, Biere Arenas RM, Lana Pons FJ (2022) Olas de Calor en la ciudad de Barcelona: 1971–2020. ACE: Archit City Environ 17(50):1–24. https://doi.org/10.5821/ace.17.50.11684

Arias P, Bellouin N, Coppola E, Jones R, Krinner G, Marotzke J, Zickfeld K (2021) Climate change 2021: the physical science basis. Contribution of working group I to the Sixth Assessment Report of the Intergovernmental Panel on Climate Change; technical summary

Baró F, Camacho DA, Del Pulgar CP, Triguero-Mas M, Anguelovski I (2021) School greening: right or privilege? Examining urban nature within and around primary schools through an equity lens. Landsc Urban Plan 208:104019. https://doi.org/10.1016/j.landurbplan.2020.104019

Bedsworth L (2009) Preparing for climate change: a perspective from local public health officers in California. Environ Health Perspect 117(4):617–623. https://doi.org/10.1289/ehp.0800114

Black-Ingersoll F, De Lange J, Heidari L, Negassa A, Botana P, Fabian MP, Scammell MK (2022) A literature review of cooling center, misting station, cool pavement, and cool roof intervention evaluations. Atmosphere 13(7):1103. MDPI AG. Retrieved from https://doi.org/10.3390/atmos13071103

Bonneuil C, Fressoz JB (2016) The shock of the Anthropocene: the earth, history and Us. Verso Books

Changnon SA, Kunkel KE, Reinke BC (1996) Impacts and responses to the 1995 heat wave: a call to action. Bull Am Meteor Soc 77(7):1497–1506. https://doi.org/10.1175/1520-0477(1996)077%3c1497:IARTTH%3e2.0.CO;2

Cuerdo-Vilches T, Díaz J, López-Bueno JA, Luna MY, Navas MA, Mirón IJ, Linares C (2023) Impact of urban heat islands on morbidity and mortality in heat waves: observational time series analysis of Spain's five cities. Sci Total Environ 890:16441. https://doi.org/10.1016/j.scitotenv.2023.164412

Danowski D, Viveiros de Castro E (2016) The ends of the world. Polity Press

Gerència d'Àrea d'Ecologia Urbana & Urban Innovative Actions (2022) Bona pràctica: projecte refugis climàtics. https://bcnroc.ajuntament.barcelona.cat/jspui/handle/11703/127541

González Orellano S (2025) Cities as anticipatory systems. In: González Orellano S, Gandlgruber B (eds) Cities as anticipatory systems. Springer

Ingole V, Marí-Dell'Olmo M, Deluca A, Quijal M, Borrell C, Rodríguez-Sanz M, Achebak H, Lauwaet D, Gilabert J, Murage P, Hajat S, Basagaña X, Ballester J (2020) Spatial variability of heat-related mortality in Barcelona from 1992–2015: a case crossover study design. Int J Environ Res Public Health 17(7):2553. https://doi.org/10.3390/ijerph17072553

Jacobs B, Schweitzer J, Wallace L, Dunford S, Barns S (2018) Climate adapted people shelters: a transdisciplinary reimagining of public infrastructure through open, design-led innovation. In: Fam D, Neuhauser L, Gibbs P (eds) Transdisciplinary theory, practice and education. Springer, Cham. https://doi.org/10.1007/978-3-319-93743-4_17

Jones B, O'Neill BC, McDaniel L, McGinnis S, Mearns LO, Tebaldi C (2015) Future population exposure to US heat extremes. Nat Clim Chang 5(7):652–655. https://doi.org/10.1038/nclimate2631

Klinenberg E (1999) Denaturalizing disaster: a social autopsy of the 1995 Chicago heat wave. Theory Soc 28(2):239–295. http://www.jstor.org/stable/3108472

Klinenberg E (2015) Heat wave: a social autopsy of disaster in Chicago. University of Chicago press, Chicago

Knowlton K, Kulkarni SP, Azhar GS, Mavalankar D, Jaiswal A, Connolly M, Ahmedabad Heat and Climate Study Group (2014) Development and implementation of South Asia's first heat-health action plan in Ahmedabad (Gujarat, India). Int J Environ Res Public Health 11(4):3473–3492.https://doi.org/10.3390/ijerph110403473

Kosatsky T, Dufresne J, Richard L, Renouf A, Giannetti N, Bourbeau J, Julien M, Braidy J, Sauvé C (2009) Heat awareness and response among Montreal residents with chronic cardiac and pulmonary disease. Can J Public Health 100(3):237–240. https://doi.org/10.1007/BF03405548

Latour B (2015) Waiting for Gaia: composing the common world through arts and politics. In: Yaneva A, Zaera-Polo A (eds) What is cosmopolitical design? Design, nature and the built environment. Routledge, pp 21–33

Latour B (2017) Facing Gaia: eight lectures on the new climatic regime. Polity

Latour B (2022) Habiter la terre. Entretiens avec Nicolas Truong. Les Liens qui Libèrent et Arte éditions

Marí-Dell'Olmo M, Oliveras L, Arechavala T, Ariza C, Borrell C, Font-Ribera L, Gómez-Gutiérrez A, González-Marín P, Grau E, Mercuriali L, Montalvo T, Olabarría M, Pérez C, Portaña S, Realp E, Rico M, Sánchez-Martínez F, Valero N, Villalbí JR (2022) Canvi climàtic i salut a la ciutat de Barcelona. Agència de Salut Pública de Barcelona

Martín-Vide J, Moreno-Garcia MC (2020) Probability values for the intensity of Barcelona's urban heat island (Spain). Atmos Res 240:104877. https://doi.org/10.1016/j.atmosres.2020.104877

Meerow S, Newell JP, Stults M (2016) Defining urban resilience: a review. Landsc Urban Plan 147:38–49. https://doi.org/10.1016/j.landurbplan.2015.11.011

Morizot B (2022) Ways of being alive. Polity Press

Nayak SG, Lin S, Sheridan SC, Lu Y, Graber N, Primeau M, Rafferty CJ, Hwang SA (2017) Surveying local health departments and county emergency management offices on cooling centers as a heat adaptation resource in New York state. J Community Health 42(1):43–50. https://doi.org/10.1007/s10900-016-0224-4

Palecki MA, Changnon SA, Kunkel KE (2001) The nature and impacts of the July 1999 heat wave in the midwestern United States: learning from the lessons of 1995. Bull Am Meteor Soc 82(7):1353–1368

Pereira SC, Marta-Almeida M, Carvalho AC, Rocha A (2017) Heatwave and cold spell changes in Iberia for a future climate scenario. Int J Climatol 37(15):5192–5205. https://doi.org/10.1002/joc.5158

Puig de la Bellacasa M (2017) Matters of care: speculative ethics in more than human worlds. University of Minnesota Press

Reeghezza-Zitt M, Rufat S (2015) Resilience imperative: uncertainty, risks, and disasters. ISTE Press—Elsevier

Ruiz-Mallén I, Satorras M, March H, Baró F (2022) Community climate resilience and environmental education: opportunities and challenges for transformative learning. Environ Educ Res 28(7):1088–1107. https://doi.org/10.1080/13504622.2022.2070602

Sampson NR, Gronlund CJ, Buxton MA, Catalano L, White-Newsome JL, Conlon KC, O'Neill MS, McCormick S, Parker EA (2013) Staying cool in a changing climate: reaching vulnerable populations during heat events. Global Environ Change Hum Policy Dimensions 23(2):475–484. https://doi.org/10.1016/j.gloenvcha.2012.12.011

Satorras Grau M, Lara-García Á, Ruiz-Mallén I, Del Moral L, Berraquero-Díaz L, Oteros-Rozas E, March H (2023) Social involvement on urban climate action and governance: lessons from Barcelona and Seville. BAGE (96). https://doi.org/10.21138/bage.3241

Sehgal NK, Sehgal AR (2023) Spatial access to cooling centers in the city of Boston. J Clim Change Health 11:100231. https://doi.org/10.1016/j.joclim.2023.100231

Stengers I (2015) In catastrophic times: resisting the coming Barbarism. Open Humanities Press/Meson Press

Tobías A, Armstrong B, Gasparrini A, Diaz J (2014) Effects of high summer temperatures on mortality in 50 Spanish cities. Environ Health 13(1):1–6. https://doi.org/10.1186/1476-069X-13-48

UN-Habitat (2022) Envisaging the future of cities—world cities report 2022. UN-Habitat, Nairobi, Kenya

Vilar M, Barnada J (2020) We are transforming Barcelona's schools to adapt them to the effects of climate change. Barcelona Soc Rev Coneixement Anàlisis Soc 27:135–147. https://ajuntament.barcelona.cat/dretssocials/sites/default/files/revista-ingles/12_exp_vilar_i_barnada_bcn27_ang.pdf

Whitman S, Good G, Donoghue ER, Benbow N, Shou W, Mou S (1997) Mortality in Chicago attributed to the July 1995 heatwave. Am J Public Health 87(9):1515–1518. https://doi.org/10.2105/ajph.87.9.1515

Widerynski S, Schramm P, Conlon K, Noe R, Grossman E, Hawkins M (2017) The use of cooling centers to prevent heat-related illness: summary of evidence and strategies for implementation. Climate and Health Technical Report Series. Climate and Health Program, Centers for Disease Control and Prevention

World Meteorological Organization (2023) Preliminary data shows the hottest week on record. Unprecedented sea surface temperatures and Antarctic sea ice loss (July 10, 2023). World Meteorological Organization. https://public-old.wmo.int/en/media/news/preliminary-data-shows-hottest-week-record-unprecedented-sea-surface-temperatures-and

Chapter 14
Conclusion

Bruno Gandlgruber, Salomón González-Arellano, Nora A. Morales Zaragoza, and Marco Jaso

Abstract Building a research agenda of cities as anticipatory systems starts with the recognition of the complexity of dynamic processes of transformation in contemporary urban settings that reflect problems of local interrelatedness and global embeddedness. The discipline of anticipation as a new approach to future studies introduces a central role for fundamental uncertainty that shows the importance to visualize existential risks and emerging opportunities associated to long-term horizons. By widening the possibility spaces for future scenarios, different pathways for transformation can be visualized and socially constructed. All relevant stakeholders are included with new forms of agency and governance of the future, including anticipatory communities. The cases in the book provide different kinds of insight based on combinations of conceptualization of futures with varying forms of participation and rule-building. Predictive models are constantly adapted depending on capacities to imagine futures and to introduce these images in the processes of transformation. Anticipatory systems are nested in different levels and types of organizations, connecting thereby images, locations and temporalities. Practices of anticipation are institutionalized formally and informally with heterogeneous levels of futures literacy. Cities as anticipatory systems as a new field of interdisciplinary research still face challenges in topics like anthropocentrism or decolonization.

Keywords Cities · Anticipation theory · Urban futures · Futures literacy

B. Gandlgruber (✉) · S. González-Arellano · N. A. Morales Zaragoza · M. Jaso
Universidad Autonoma Metropolitana, Mexico City, Mexico
e-mail: bgandl@cua.uam.mx

S. González-Arellano
e-mail: sgonzalez@cua.uam.mx

N. A. Morales Zaragoza
e-mail: nmorales@cua.uam.mx

M. Jaso
e-mail: mjaso@cua.uam.mx

S. Gonzalez-Arellano and B. Gandlgruber (eds.), *Cities as Anticipatory Systems*, The Urban Book Series, https://doi.org/10.1007/978-3-031-78162-9_14

14.1 Rethinking Cities and Anticipation Systems

In this book, we have explored different perspectives and modes of the "use of the future" in the urban context, with applications to various sectors and subsystems in other parts of the world, particularly in the Global South. All these examples reinforce the idea that the current challenges of urban transformation are phenomena with high levels of systemic complexity in terms of their local interrelatedness and global embeddedness. Extending explicitly the role of the future in these issues creates additional difficulties related to the consideration of fundamental uncertainty of all expressions of the future and the prevalence of growing existential risks, but also significant opportunities to understand the diverse options of transformation and to build socially and environmentally sustainable pathways towards preferable scenarios in the future. That is why we introduced the proposal of conceiving and studying cities as anticipation systems.

The "discipline" of anticipation distinguishes itself from earlier approaches in future studies for several relevant reasons. Firstly, it recognizes the epistemic change in (social) sciences as a shift from theory building that is exclusively history-based toward inquiries that systematically consider the emerging character of representations of the future, embracing uncertainty and change. In the second place, contemporary future studies have shifted from the traditional focus on quantitative forecasts of specific developments and scenarios toward enriched forms of multi-disciplinary foresight and anticipation exercises with diverse methods and perspectives. In the third place, anticipatory systems and their particular characteristics are analyzed "normatively" without implying standardized solutions towards a (non-existing) optimal outcome. In the fourth place, anticipation theory recognizes the need to integrate all stakeholders and their dynamic mental models to build alternative futures. In fifth place, strengthening anticipatory capacities means challenging established forms of governing the future by including all actors in different participatory and interactive exercises beyond the traditional "expert futures."

The cases presented in this book provide several insights to help us understand urban anticipatory systems better. An analysis of these cases oriented on actors, levels, and scales of anticipation, institutionalization, and futures literacy offers a picture of how city anticipatory systems are developed through open combinations of formal and informal conceptualization, participation, and rule-building processes. Indeed, anticipation can be found at every level, in all sectors and structures of cities, in a more or less organized way. In this sense, we should not confuse the inherent attribute of anticipation as a pre-configurated planning mode. The more we understand the details of the anticipation process, the better the cities' anticipatory systems will be able to guide the transition.

As explained in the introduction, the chapters in this book do not follow a closed methodology of anticipation theory but explore different aspects and subsystems of other cities as anticipatory systems. Seeing the city as an anticipation system is a way to understand how diverse representations of the future guide urban transformation. It provides insight into how citizens and institutions could actively use the future, strengthening anticipatory capacities on different levels. Following the general idea of anticipation systems, all cities as complex systems have the inherent attribute of anticipation as the process of generating images of themselves in the future and using these images for decisions in the present. Urban anticipatory systems are not stable or controlled by specific government entities. Unlike other types of more vertical organizations, cities in particular, and urban systems in general, have developed decentralized, open, diverse, and multilevel anticipatory systems that are continuously changing. Some of the main findings of the different chapters regarding the characteristics of urban anticipatory systems relate to anticipatory actors, levels of anticipation, forms of institutionalization, and futures literacy.

In the first place, the chapters show a great diversity of urban actors involved in different mundane situations where patterns of conceptualization from either individual or collective entities are invariably guided toward building specific narratives of the future that create images of themselves in the future in crucial areas like climate change, the quality life of groups of citizens, resource efficiency, environmental risks, or promoting sustainable development. All these intentions lead to an effort to translate future images into present behaviors. All actors consider aspects of an anticipatory model of a city and adopt their predictive model of themselves and the environment, allowing the system to change. Still, that change is governed by its rhythms and peculiarities. We are especially interested in revealing what type of images preceded specific actions considering the two anticipatory dimensions; the first one is the willingness to imagine a future scenario within a particular disposition of arrangements of actions, and the second one is how they integrate these visions to take action, in other words, their agency or transformative capacity to change the system in function to the future vision that they build and share.

Several chapters have provided an alternative approach to analyze and connect the role of multiple actors, their partnerships, narratives, and citizen participation from a perspective that does not necessarily respond to a geographic denomination but is instead identified as an anticipatory community with complex problems and emergent contexts. We consider it essential to use a mixed multi-perspective focus that involves an epistemological one and the inherited tacit knowledge structure that affects certain actors and practices embedded in specific socio-ecological contexts toward a more long-term visioning that corresponds to the anticipatory literacy model suggested.

Some new topics have emerged in this work, revealing insights toward a deeper analysis of how those visions led to actions. This is the case of the "anticipatory communities" concept, which signals a certain group of actors with a specific anticipatory practice within a regime. Furthermore, we could reflect on hope as a "special force" that catalyzes the actors' impulse to channel their energies toward co-creating a desired future. We could identify different types of urban actors in an anticipatory taxonomy. We see "repairers" and "managers" responding with urban infrastructure works for climate adaptation in Recife, Brazil, or who use the future or "Anticipators" who believe in mushrooms as an alternative to food; locate "Visionaries" that pursue an image of a Healthy City by integrating new indicators of quality of the physical and social environment, stress levels, educational and opportunities in India.

14.2 Anticipatory Actors

We have found a great diversity in how each chapter portrays urban actors using the future in action, and it doesn't matter if we talk about individuals or collective entities; all of them are guided toward particular intentions that might be either avoiding climate change problems, enhance the quality life of cities residents, improve resource efficiency or eliminate toxic materials, or even promote sustainable development. All these intentions involved translating future images into the present behaviors, which needed to deal with what is offered in its environment and specific spatial and temporal contexts and scales.

All actors take into account an anticipatory model of a city and adopt their predictive model of themselves and the environment, allowing the system to change. Still, that change is governed by its rhythms and peculiarities. We are particularly interested in revealing what type of images preceded specific actions considering the two anticipatory dimensions following Poli; the first one is the willingness to imagine a future scenario within a certain disposition of arrangements of actions (which most of the chapters take into account in this book). The second one is how they integrate these visions to take action and their agency or transformative capacity to change the system.

All Chapters have provided an alternative approach to analyze and elucidate the role of multiple actors, their partnerships, narratives, and citizen participation from a perspective that does not necessarily respond to a geographic denomination but is instead identified as an anticipatory community that complex problems and emergent contexts have impacted. We consider it important to use a mixed multi-perspective focus that involves an epistemological one and the inherited tacit knowledge structure that affects certain actors and practices embedded in specific socio-ecological contexts toward a more long-term visioning corresponding to the anticipatory literacy model suggested by this book.

14.3 Levels and Scales of Anticipation

The contributions in the previous chapters have shown how anticipatory systems are nested at different levels of organizations and geographic scales. As the literature has documented, anticipation is an inherent attribute at the cellular level and the level of individuals, organizations, and more complex social systems such as cities. This multilevel and multiscale game of anticipation was demonstrated in several of the contributions in this book, showing its complex nature and a kind of fractal logic of anticipatory systems. The case studies presented in the previous chapters have demonstrated various ways of forming anticipatory communities by sharing images of futures and translating those images into present actions at the level of small neighborhood groups and urban collectives. Thus, a collective of neighbors in Huelva, or small entrepreneurs for mushroom growing in Mexico City whose future project takes shape locally, contrasts with the anticipatory systems nested in large cities articulated with global-scale processes. The cases related to climate change in Barcelona and Quito are examples that articulate local anticipatory policies with global-scale future scenarios.

On the other hand, it is important to note that the two components of anticipation, the imagination of futures and the capacity to transform the present, can function at multiple levels and geographic scales. As the case of Buenos Aires illustrates, the project of a future without patriarchy draws on images on a global scale, while mobilizations to change a law have a local or national focus. When a successful anticipatory policy is established locally, it can trigger future imaginaries in nearby or distant places, causing a multiplier effect at other levels or geographic scales.

14.4 Institutionalization of Anticipation

The chapters clearly illustrated that anticipatory communities are heterogeneous collectives, each with a different response model to conceptions of the future; however, they raise questions regarding the feedback between the routinization of anticipatory practices, their standardization, and their stabilization through formal or informal norms. In this sense, the chapters raise concerns in the reader regarding the systems under which the anticipatory model of each community acquires legitimacy and authority.

In this line of thought, it will be beneficial to offer further on a more comprehensive explanation of the process in which actors and their actions acquire subjective meaning from a more formal or institutionalized process of crafting future images through actionable steps encompassing various forms of institutionalization such as cultural-cognitive, normative, and regulatory.

14.5 Futures Literacy

The review of the cases analyzed in this book reveals that futures literacy—consisting of both the capacity to conceive alternative futures and to employ such information and images to impact the present—is distributed in a highly heterogeneous manner among groups of actors interacting around the same theme or problem, even within the same city. In its simplest expression, we find groups of actors with different future images and horizons, not always possessing the characteristics of a mature anticipatory community as understood in this book. In these cases, the future images of certain actors contrast with a situation opposite to a pressing issue in the present. In these instances, a different future emerges from criticism of a current situation (for example, a non-patriarchal system, an inclusive healthcare system, a waste-free community, etc.). In other cases, groups of actors have managed to go further and characterize alternative futures quantitatively or qualitatively based on a particular anticipatory logic or model. Interestingly, in these cases, futures have been disaggregated into horizons with different timeframes and establish the groundwork for greater transformative capacity (for example, through establishing goals in transitional policies and programs).

As observed in previous sections, the capacity for transformation has been observed at various levels. At a micro level, emerging organizations from individual enterprises have found niches for experimenting with new social practices, addressing uncertainty from a market logic (as in the case of mushroom restaurants). Meanwhile, at a meso level, implementing new practices has required the involvement of governmental and state actors, who have been crucial in ensuring the necessary levels of investment and corresponding normative and institutional stability (climate refugees, security services, transportation systems, material substitution). In these cases, the ability to respond to unwanted inertia has been valued as an essential legitimacy capital for citizens.

At a subsequent level of systemic complexity, the identification of unwanted futures at a planetary or macro level, the heterogeneity of interpretation systems by actors with dissimilar interests has hindered the construction of consensus and the implementation of effective transformational policies, programs, or actions. Thus, the developing capacity for change has been announced in the laws or an ideology of change without materializing into social practices with a higher level of coverage or acceptance. However, the introduction of incipient practices and routines could be documented.

For any of the levels and cases studied, it would be highly enriching to understand how developing exploration and use capacities associated with forecasting, foresight, or other creative exploration practices mutually reinforces the evolution and maturation of inherent anticipatory capacities in urban systems.

14.6 The Pursuit of a Research Agenda

Examining anticipatory systems across various cities has consistently highlighted the pervasive nature of crafting narratives and visualizations of the future, particularly about the pressing issues or topics of interest concerning urban planning, governance, or integral components of a global system. Authors have drawn upon various future urban concepts, identifying the most pertinent or defining visions, narratives, or scenarios. They have reflected on their implications for current plan implementation, programs, or targeted actions.

Nevertheless, it becomes evident that one of the foremost challenges in comprehending the development of urban anticipatory systems lies in the comprehensive and precise identification of actors, agents, or stakeholders and in understanding their specific modes of operation in response to future-oriented information (such as data, concepts, images, or narratives). Further elucidation is required regarding the mechanisms facilitating the development of anticipatory capabilities within urban actors and institutions, which is an accumulative process grounded in organizational, group, and citizen capacities. Specifically, there is a need for more research to comprehend the interconnectedness between micro, meso, and macro levels. Authors concur on the necessity for further research into the factors impeding the development of anticipatory capabilities. In this line, it would be beneficial to offer a more comprehensive explanation of the concept of "enacting," wherein future visions gain significance through actionable steps, encompassing various forms of institutionalization such as cultural-cognitive, normative, and regulatory.

It is essential to have a deeper understanding of the interplay between fundamental future literacy capabilities inherent in anticipatory systems and traditional approaches in future studies, such as forecasting and foresight. Additionally, exploring connections with other creative tools utilized for future-oriented thinking would enrich our comprehension.

Finally, all the cases we have studied in this book are anticipatory systems of human social systems. This is an anthropocentric approach that, if we want to overcome, we will have to adopt an approach from socio-ecological systems and the actor-network theory to incorporate non-human actants such as other species (animals, plants) in the analysis. Finally, we consider it essential that in times of uncertainty and unprecedented socio-ecological transitions, it is necessary to systematically and explicitly incorporate a justice perspective in the study and pragmatics of anticipatory systems. Finally, the diverse contributions in this book show the importance to understand the role of hegemonic processes of futures building and the possibilities for decolonization of the future from the Global South.

MIX
Papier aus verantwortungsvollen Quellen
Paper from responsible sources
FSC® C105338
www.fsc.org

If you have any concerns about our products,
you can contact us on
ProductSafety@springernature.com

In case Publisher is established outside the EU,
the EU authorized representative is:
Springer Nature Customer Service Center GmbH
Europaplatz 3, 69115 Heidelberg, Germany

Printed by Libri Plureos GmbH
in Hamburg, Germany

Financing Innovations of Small and Medium-Sized Enterprises
in a Systemic Perspective

Sozio-ökonomische Perspektiven in Südosteuropa

Socio-Economic Perspectives in South Eastern Europe

Herausgegeben von
Siniša Kušić / Slavko Kulić / Vladimir Cvijanović

Band 3

PETER LANG
Frankfurt am Main · Berlin · Bern · Bruxelles · New York · Oxford · Wien

Vladimir Cvijanović

Financing Innovations of Small and Medium-Sized Enterprises in a Systemic Perspective

Theoretical Foundations and Cases of Hungary, Slovenia and Croatia

PETER LANG
Internationaler Verlag der Wissenschaften

Bibliographic Information published by the Deutsche Nationalbibliothek
The Deutsche Nationalbibliothek lists this publication in the Deutsche Nationalbibliografie; detailed bibliographic data is available in the internet at http://dnb.d-nb.de.

Zugl.: Frankfurt (Main), Univ., Diss., 2010

Cover Design:
Olaf Gloeckler, Atelier Platen, Friedberg

D 30
ISSN 1864-483X
ISBN 978-3-631-61324-5

www.peterlang.de

Table of Contents

List of Figures

List of Tables

List of Abbreviations

AT	Agency theory
BA	Business angels
BAN	Business angel network
BERD	Business expenditures on research and development
BICRO	Business Innovation Center of Croatia
CEE	Central and Eastern Europe
CEEC	Central and Eastern European Country
CEO	Chief Executive Officer
CG	Corporate governance
CIS	Community Innovation Survey
ECOP	Economic Competitiveness Development Operational Programme
EPO	European Patent Office
EU	European Union
EU27	European Union with 27 member states
EUR	Euro
FDI	Foreign direct investment
FP7	Framework Programme 7 (of the European Union)
GDP	Gross domestic product
GEM	Global Entrepreneurship Monitor
GNP	Gross national product
GVOP	Economic Competitiveness Operative Programme of the Hungarian Government
HAMAG	Croatian Agency for Small Economy
HBOR	Croatian Bank for Reconstruction and Development
HR	Croatia
HU	Hungary
ICT	Information and communication technology
JPO	Japanese Patent Office
NACE	Statistical classification of economic activities in the European Community
NGO	Non-governmental organisation
NSI	National system of innovation
NUTS	The Nomenclature of Territorial Units for Statistics
OECD	Organisation for Economic Co-operation and Development
ORTT	National Radio and Television Commission
PE	Private equity
PPP	Purchasing power parity
R&D	Research and development
SII	Summary Innovation Index

SITC Rev. 3	Standard International Trade Classification, Revision 3
SLO	Slovenia
SME	Small and medium-sized enterprise
STI	Science, technology and innovation
TCE	Transaction cost economics
TEA	Total entrepreneurial activity
TPP	Technological product and process innovation
UK	United Kingdom
US	United States (of America)
USA	United States of America
USD	United States Dollar
USPTO	United States Patent and Trademark Office
VC	Venture capital
VC/PE	Venture capital/private equity

Preface

This work is a dissertation which I defended successfully at the Faculty of Economics and Business Administration of the Johann Wolfgang Goethe University Frankfurt am Main, Germany.

Christensen's 1992 study greatly inspired me to develop an approach to financing innovations.

I thank my colleagues from whom I have learned a lot before and during the process of writing this dissertation: Zoran Aralica, Slavko Kulić, Đuro Medić, Domagoj Račić. My biggest thank you goes to my supervisor, Professor Tamás Bauer – for trusting my work and always being ready to talk to me. Regardless of their suggestions, all errors remain my own.

I am very grateful to all of my interviewees who made the empirical field research possible.

I would like to thank several persons who helped me with some of the interviews and with gathering data on some firms. Stevo Cvijanović helped me with several interviews in relation to Croatia. Several interviews in relation to Slovenia were done by Mario Filipan. In the case of Hungary, I was assisted by Réka Csernátonyi (for a few interviews), Csaba Grünblatt and Zsolt Varga (for databases and websites) and Mónika Tóth (for translating).

Last but not least I thank my wife Petra and daughter Una for keeping up with me during the "destructive stage" of my work.

1. Introduction

There have not been many studies on financing innovation activities (financing innovations). The literature is usually divided into two parts, with few exceptions – one that deals with questions regarding innovations, and the other that deals with financing. One of the reasons for this situation is certainly the data on which such studies would be based. Namely, there are no standardised data on financing innovations. They are not even gathered by the most famous instrument for collecting innovation-related data: the Community Innovation Survey (CIS). Statistics on financing innovation usually only incorporates expenditures on research and development[1] (cf. Iliev, 2001), which is only one of the innovation activities. The other reason for this situation is probably the complex nature of financing innovations: innovation activities involve considerable uncertainty and risk, and last for an extended period of time. The process requires financing during its entire duration. It is not finance theory alone that is uppermost here, since it is the real sector that produces (product) innovations. It is the quality of innovations produced, coupled with the needs and demand of the real sector, that constitutes a necessary condition for innovation activities to be financed. Whether it is a sufficient condition depends on the financial system. The aforementioned reasons for the lack of studies in this area of research apply even more to the case of small and medium-sized enterprises (SMEs), which is what we shall be focussing on. SMEs are particularly prone to experiencing financial barriers to innovation activities, but can be more agile than larger enterprises when it comes to adapting to market conditions.

Innovation is crucial for economic progress (Nelson and Winter, 1977; Freeman, 1982) and economic growth (see e.g. Freeman, 1988 and Grupp, 1998 for discussion). This has undoubtedly been recognised by both mainstream and heterodox economists. Financing innovation activities is therefore an essential precondition that allows the above-mentioned processes to unfold. There is hence a need to understand ways of financing innovations and their relationships with other elements of an innovation system. As these linkages have largely been neglected by the academic literature, this study attempts to illuminate them and suggest a possible approach for their analysis. More precisely, this study aims at contributing to the literature on financing innovations from a systemic perspective. In essence, the financing of innovations for SMEs is seen in a framework defined by the national system of innovation (NSI), corporate governance (CG), and the financial system. They are all influenced by culture and social capital, as shown in Figure 1 below. The concepts and their actual representations partially overlap and are mutually interdependent. Hence, the predominant approach that we follow is systemic, since we see the financing of innovations for SMEs as

1 Abbreviated as R&D.

decisively determined at the macro level. The approach is also institutionalist, since we recognise the role and importance of institutions for financing innovations.

Figure 1: Framework for analysing financing innovations

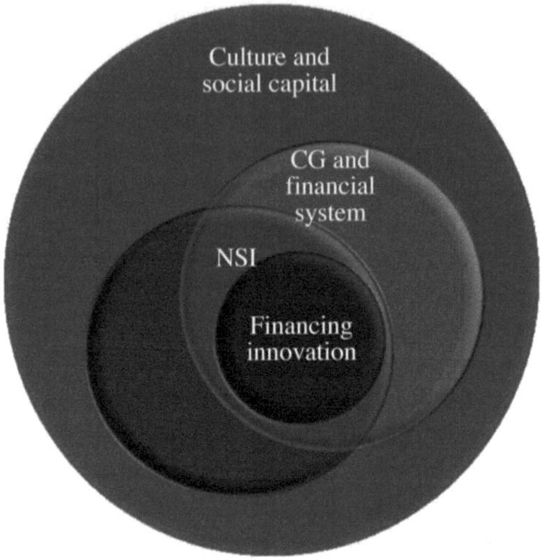

Note: CG stands for "corporate governance"; NSI stands for "national system of innovation". Source: author.

In the empirical components, we analyse the financing of innovation activities for SMEs from the high-technology sectors of three post-transition countries of Eastern or South-Eastern Europe: Hungary, Slovenia and Croatia. The present study has two empirical components: the first is based on desk research of the existing literature and focused on the NSIs; the second contains an analysis of case studies, which is based on interviews conducted with qualified representatives of the SMEs (but also on other sources like firms' websites). Each country shares a border with both of the others, and all three are comparatively small. Slovenia and Hungary are already in the EU, while Croatia is only a candidate country. While Slovenia and Croatia were both part of the former Yugoslavia (hence, their common historical legacy), Hungary belonged to the former Eastern Bloc. Nevertheless, all three have similar socio-economic characteristics, although there are noticeable differences between them in terms of innovative performance and financing innovations. As there is a marked shortage of studies on financing innovations covering these countries, the present study aspires to contribute to that segment of research as well. Furthermore, we have so far witnessed no studies on financing innovations that compare these three countries. We study the innovation activities and innovative performance of SMEs from these countries in the context of their NSIs, and discuss the implications for financing their innovations. Contrary to what one would expect from some earlier studies, we do not find financial problems to be the greatest barrier to innovation activities. Although all of these countries have bank-based financial systems, state funds are the preferential source for financing innovations, while venture capital (private equity) plays a marginal role. This indicates a possibly important role that the State could play in stimulating (financing) innovations, and thereby circumventing the proneness of bank-based financial systems to incremental innovations.

Following the introduction, the present study comprises five chapters. In the second chapter, we describe its framework. This part contains crucial definitions and concepts that we shall be using later on. In this part, we explain how corporate governance and the financial system, the national system of innovation, culture, social capital, competitiveness and entrepreneurship can be brought together under a single framework and employed in a study of financing innovations. The relevant hypotheses and the methodological approach are set out as well. The third part is theoretical, comprising important building blocks for establishing the principles of financing innovations set out at the end of this chapter. These building blocks are the economics of innovation, the theory of the firm, and financial theory. We highlight important contributions from each of these research areas and work towards a synthesis. The next two chapters contain the empirics. It needs to be stressed that the list of references and other sources is followed by important Appendices that should be viewed in conjunction with this analysis. Appendix I contains a case study database comprising most pertinent results of the data collection, and Appendix II contains case study reports in which our findings are specified in greater detail. An interested reader would be well-advised to check Appendix III as well, in which innovation indicators from the European Innovation Scoreboard, as well as selected indicators on competitiveness in the studied countries, are presented. The fourth chapter elaborates on the national systems of innovation in Hungary, Slovenia and Croatia. We review their historical legacies, which we consider pertinent to an understanding of the present state of their NSIs (*that being as far as our data goes*). The results of our empirical field research are described and analysed in Chapter Five: Financing innovations of SMEs in Hungary, Slovenia and Croatia. After a brief introduction and methodological notes to the survey, we proceed by analysing SMEs based on the samples for each country, followed by a summary that completes the chapter. Finally, the conclusion is presented in the last part of the study.

We hope that the methodology and findings of this study will encourage future research on financing innovations, and motivate researchers to produce more studies in which the financial sector and the real sector are observed as interdependent.

2. Framework of the Study

In this part of the study, we shall elaborate on the basic concepts and definitions that we have used in this work. We first explain why we have chosen to study small and medium-sized enterprises, and go on to stress the distinctive features of high-technology manufacturing sectors (ch. 2.1.). The subsequent chapter (2.2.) builds on that by defining and describing innovation (activities) and its (their) characteristics. Financing innovations is different from financing already known products, which is examined in Chapter 2.3. The next chapter (2.4.) deals with the conceptual framework for studying financing innovations, and is essential for our approach. It comprises sections on corporate governance and financial systems (2.4.1.), national systems of innovation approach (2.4.2.), culture and social capital (2.4.3.), competitiveness and entrepreneurship (2.4.4.) and finally, the hypotheses of the study (2.4.5.). The methodological approach (ch. 2.5.) concludes the chapter on the framework of the study by explaining the extent of our inquiry (2.5.1.) and provides its institutional foundations (2.5.2.).

2.1. Why High-Tech Small and Medium-Sized Enterprises?

In this study we shall call all enterprises that employ fewer than 250 employees SMEs, which is a widely accepted criterion[2]. Micro enterprises are encompassed by this definition, and they employ fewer than 10 employees. Small enterprises are those that employ 10-49 employees, and medium enterprises have 50-249 employees (see Eurostat, 2009). It is challenging and exciting to study SMEs, particularly in high-tech industries. First of all, there are many more SMEs than large enterprises, and their importance is crucial in terms of production and employment. For example, in the EU share of SMEs in the total number of non-financial enterprises is more than 99,75%. They have a huge importance for the economy, since their contribution in terms of value added amounted to 57,6% in 2005[3], and they accounted for 67,1% of total employees[4] (Eurostat, 2008: 115-116). In the countries we compare in the empirical part, the SMEs have a similar importance for their economies. Namely, in Hungary, 99,9% of all enterprises in 2004 were SMEs[5], whereas in Slovenia, in the same year, they added up to 99,7% (Eurostat, 2007). In Croatia, the SMEs accounted for 99,4% of all enterprises[6] in 2007 (FINA 2008, and own calculation). For similar reasons, it is im-

2 Besides the value of the balance sheet and total revenues.

3 Aggregate for EU27, calculated for 2005.

4 See previous footnote.

5 These SMEs contributed with 51,9% to the value added and with 72,2% to the number of persons employed (Eurostat, 2007).

6 They employed 64,2% of total employees and had a share of 51,7% in total revenues (FINA, 2008 and own calculation).

portant to study high-tech industries, since they produce the highest value added. The relevance of both SMEs and high-tech industries can be seen in relation to large enterprises (to which the SMEs often maintain close business relations) and technological activities and industries of a lower level of sophistication (which are essential for the normal functioning of any economy). In addition to the importance they have for any national economy, the SMEs are much more exposed to market conditions than large enterprises, and data on them are harder to obtain.

High-technology enterprises that shall be analysed in the empirical part belong to the manufacturing industries, as listed below in Table 1. The classification of manufacturing enterprises based on their levels of technology was performed by the OECD (2005), distinguishing between four levels of technology on the basis of expenditures on research and development (R&D) in relative terms[7]. The four kinds of industries in terms of the technology level are the following, listed from the relatively highest to the relatively lowest expenditures on R&D: 1) high-technology industries, 2) medium-high-technology industries, 3) medium-low-technology industries, and 4) low-technology industries. Investment into innovation activities – of which R&D is a part – is crucial for raising the competitiveness level (cf. OECD, 2005), while raising competitiveness is positively correlated with the GDP (cf. Sala-I-Martin et al., 2008). The rationale for considering high-technology industries also lies in the fact that they have invested most into R&D and hence need finance for their innovations. Aralica (2007) stresses the need for high-tech industries to cooperate with universities and research institutes.

Table 1: High-technology industries

High-technology industries	**ISIC Rev. 3***
Aircraft and spacecraft	353
Pharmaceuticals	2423
Office, accounting, and computing machinery	30
Radio, TV, and communications equipment	32
Medical, precision, and optical instruments	33

* International Standard Industrial Classification, Revision 3.
Source: OECD (2005: 182).

The OECD methodology of classification of industries according to technological intensity used here (low-technology, middle-low-technology, middle-high-

7 For details see OECD (2005).

technology and high-technology) is the most widespread, but it is not perfect. Although there is product taxonomy as well, it would not make sense to use it without the detailed and precise data that we do not have access to in this study. In addition, there is no similar classification of services because of the lack of adequate data (Srholec 2006). However we may presume that the knowledge content of services directly connected with high-technology branches is similarly high. Srholec (2006) problematizes the classification of manufacturing industries herein adopted on the ground of products produced, as well as activities actually performed in those industries. Firstly, judging by the share of R&D in its total revenues, not all products produced by a high-technology company are high-technology products. Secondly, it is not a foregone conclusion that workers in industries classified as high-technology will necessarily be engaged in demanding knowledge-based activities. For instance, they may be engaged in assembly, and not in R&D. In other words, the classification of an industry as high-tech does not guarantee that any activities performed there are actually high-tech. It would, indeed, be important to understand what is the knowledge content of the innovation activities actually performed inside a firm, since we would expect the demand for financial sources to be bigger in high-tech firms in which there is a higher knowledge content. Namely, knowledge-demanding innovation activities (like R&D) are necessarily more sophisticated than e.g. assembly, and require a specialised workforce and/or specialised equipment, which entails more costs.

Small and medium-sized enterprises (SMEs) differ from large enterprises in many aspects (cf. Ratz, 1988: 13-17; Kaufmann and Kokalj, 1996: 14-15; Modified). The agglomeration of resources in large enterprises makes it possible for them to more easily achieve synergies visible through the economies of scale and economies of scope. In particular, the SMEs cannot sustain such a specialized division of labour as that found in large enterprises. For example, they often lack personnel specialized for R&D, which is especially crucial for high-technology activities (Rothwell and Zegvelt, 1982, as referenced in Major and Cordey-Hayes, 2003: 668). In addition, large enterprises, as opposed to SMEs, have bigger opportunities for diversification. SMEs have to specialize in the production of a limited range of products. Innovations in some industries (such as the automobile and aircraft industries) are usually developed within large firms with big R&D departments, which are integrated with their manufacturing and marketing departments. Because of the economies of scale in these innovation activities, the corresponding market structure tends to be concentrated (Schertler and Stolpe, 2000: 6). "It is when knowledge and ideas are the most important inputs that the innovation process is more likely to be dominated by independent new firms which are, at least partly, owned by the creative talent itself" (Schertler and Stolpe, 2000: 6). Therefore, the importance of SMEs and large enterprises, as far as innovation is concerned, differs depending on the industry an enterprise is active in. SMEs have certain advantages over large enterprises because their size makes them more adaptable to rapidly changing market conditions (e.g. more direct contact with the customers and, therefore, faster reaction to their needs, more flexible organisation, etc.). Some empirical research shows that SMEs create more jobs than large enterprises (for some sources see Ratz, 1988: 16-17). Furthermore, since an SME is likely to be owner-managed, the principal-agent conflicts between shareholders and managers are often non-existent (Berger and Udell, 1998).

However, the financial situation of SMEs is weaker than that of large enterprises[8]. Financing problems for SMEs look different from financing problems for large enterprises. Because large enterprises have existed much longer on the market, they have much bigger possibilities for internal financing and financing on the capital market. On the contrary, SMEs (especially newly founded enterprises) have smaller internal financing possibilities and are limited in financing at the capital market or by the banks. Financing from retained earnings can only be applied by firms which have already existed on the market for several years (cf. Niederöcker, 2002). Moreover, SMEs have smaller equity-to-assets ratios than large enterprises, signalling a higher risk for the banks, which are, therefore, not willing to supply them with credits. Hence, sometimes not even profitable investments (especially in innovation activities, because of the risk inherent in such projects) will be financed (Stratmann, 1998: 25). Because of SMEs' size, some SMEs will be rationed by the bank, since the fixed costs for a banking screening procedure are as high as with larger enterprises (which could get larger credits), but the credits are smaller (cf. Tschach, 2003). Furthermore, as banks often look for collateral, and not a quality project, SMEs (especially younger ones) are not as fully able to provide valuable collateral as larger enterprises are. From that perspective, SMEs are caught in a vicious circle. In order to get out of this circle, SMEs primarily need equity investments or credits under more favourable conditions than are available on the market (e.g. credit guarantees or subsidised credits from government agencies). However, SME owners/managers could be reluctant to relinquish an equity share, which means SME growth and innovation prospects could be limited, at least in the short run. SMEs are somewhat limited in issuing shares comparable to large enterprises, since they do not have a comparable reputation, and neither can they issue a large enough number of shares to minimize fixed costs per share. These and other characteristics of SMEs will be revisited when we discuss financing innovations in more detail.

8 This is because of various reasons, such as weak bargaining position, lack of information about financing possibilities, high proportion of fixed costs in relation to financing needs, etc.

2.2. Defining Innovation

In this work, we will be looking at innovations from at least two perspectives. The first one will imply a view of innovations as novel products and services, as well as processes. It will be applied when we look at the literature on (financing) innovations, as well as when we review previous studies on (financing) innovations. The second one considers innovations as incrementally or radically new (physical) products. It will be applied in the empirical field study of Hungarian, Slovenian and Croatian high-tech companies (see below).

In defining innovation for the purposes of this study, we turn to the OECD's definition, which calls it technological product and process innovation (TPP). It can be defined as "new and significantly improved **products** (goods and services) **and processes**" (OECD, 1997: 8, emphasis original) at the firm level.

"*A technological product innovation is the implementation/commercialisation of a product with improved performance characteristics such as to deliver objectively new or improved services to the consumer. A technological process innovation is the implementation/adoption of new or significantly improved production or delivery methods. It may involve changes in equipment, human resources, working methods or a combination of these.*" (OECD, 1997: 9; emphasis original).

Hence, technological innovation "requires an objective improvement in the performance of a product" (OECD, 1997: 28), and not just a "cosmetical" improvement, which will satisfy the customer only on a subjective level. However, the product innovation of one firm (which produces, say, machines) could be a process innovation of the other (e.g. which employs such machines in its production process). The OECD (1997) makes a distinction between 1) a technologically new product[9], and 2) a technologically improved product[10]. We shall call them radical and incremental innovations respectively. Notwithstanding these quite precise definitions, Goldsmith and Foxall (2003) highlight the problem of different usage of the concept of innovation – in different contexts and

9 "**A technologically new product** is a product whose technological characteristics or intended uses differ significantly from those of previously produced products. Such innovations can involve radically new technologies, can be based on combining existing technologies in new uses, or can be derived from the use of new knowledge" (OECD, 1997: 32; Emphasis original).

10 "**A technologically improved product** is an existing product whose performance has been significantly enhanced or upgraded. A simple product may be improved (in terms of better performance or lower cost) through use of higher-performance components or materials, or a complex product which consists of a number of integrated technical sub-systems may be improved by partial changes to one of the sub-systems" (OECD, 1997: 32; emphasis original).

different scientific fields – and its meaning when it comes to products, services or production processes that are innovated.

Innovations are created as an output of innovation activities. "TPP innovation activities are all those scientific, technological, organisational, financial and commercial steps, including investment in new knowledge, which actually, or are intended to, lead to the implementation of technologically new or improved products or processes" (OECD, 1997: 39). Hence, (TPP) innovation activities encompass many different activities, like intramural research and experimental development (R&D), acquisition of extramural R&D, acquisition of machinery and equipment for innovation activities, acquisition of other external knowledge for innovation activities, training connected with innovation activities, market introduction of innovations and design, other preparations for production/deliveries (Kurik et al., 2002: 95)[11]. Hence, research and development (R&D) represents just one of the innovation activities. However, R&D plays an especially important role in innovation activities because it could result not only in novel products and processes, but also in enhanced capacity of an enterprise to create such innovations. This capacity is enhanced through ability to absorb external information and through learning (Cohen and Levinthal, 1989). Innovation always presumes creativity. Carayannis and Gonzales mention several characteristics of creativity at the individual level, relating them to thinking "laterally, to perceive, conceive, and construct ideas, models, and constructs that exceed or supersede established items and ways of thinking and perceiving" (Carayannis and Gonzales (2003: 587). Innovation activities are characterized by a high degree of uncertainty, longevity, mostly intangible investment and output (primarily knowledge) as well as asset specificity of the production factors. Due to an increased knowledge content of high-technology products and fast-changing market conditions, they require more complex innovation activities and a wider range of financial instruments than firms in other industries.

Creating innovations cannot be left to the market *in toto*, since there are research and development (R&D) spillovers that cause the public return on R&D to be higher than private returns on R&D (Arrow, 1962 as in Christensen, 1992: 252). Aside from the head start advantages and trade secrets[12] that firms use in order to protect their innovations, there is another legal tool to achieve that – patent protection. However, innovations can only be protected (and extra returns from

11 This research on innovation in Estonian enterprises 1998-2000 (Kurik et al., 2002) represents the Community Innovation Survey (CIS). The CIS is methodologically based on the Oslo manual (OECD, 1997) and provides an internationally comparable statistical basis for an analysis of innovations.

12 Trade secret is information known to a company and hidden from the public, from which economic value is or can be derived (cf. Bouchoux, 2001: 193-194).

them appropriated) for a limited period of time. During this time period, innovators can appropriate extra returns. As innovations by definition end up on the market, knowledge incorporated in them eventually becomes public (cf. Nelson, 1988). Ideally, there ought to be a balance between private and public benefits from innovation. Namely, too much legal protection for a firm's innovations could create an overly rigid system, which stops others from utilising an innovators' knowledge. On the other hand, a too liberal system, which does not protect the innovators' knowledge, would also diminish the incentives to innovate (cf. Boynton, 2004). The appropriability of profit from innovation could be a problem because, with fewer possibilities to appropriate part of the profit from innovation, a possible investor is more likely to withdraw from financing a project (Scherer, 1999: 54). Especially troublesome is the financing of fundamental research, since its benefits are the hardest to appropriate. Hence, government financing of fundamental research is especially important (Scherer, 1999: 56).

In order to sum up this section regarding product innovations, it ought to be stressed that the latter presuppose the commercialisation of a product, which does not have to be entirely new. Product innovation therefore stands in contrast with invention, which is the creation of new knowledge.

2.3. The Nature of Financing Innovations

Financing innovations or investing into innovation activities in practice looks different from other real investments, which we are about to explain in detail below, as well as to summarize here in three points. First of all, it comprises not only real investments (investment into physical assets), but also investments into intangibles like training and intellectual property rights. Secondly, it differs from some other investments on the grounds of their longevity, as these investments take a considerable amount of time. Thirdly, it is often not a predetermined object that these investments are going to create, but a process involving a notably higher level of uncertainty and risk than any other frequent or "regular" investment. In addition, uncertainty and risk also follow from the first two properties. The aforementioned characteristics have repercussions on the ability of both large enterprises and SMEs to finance their innovations. Moreover, they feature even more prominently when observing high-tech industrial activities, for which specific financial instruments, institutions, and markets are more suitable than others.

The financing of innovations is ingrained in culture, social capital and the institutional environment, as well as the economic variables pertinent to the financing of innovations. Before extensively elaborating on each of the concepts considered in this work, at this point we would like to offer a brief sketch of the logic that we follow. Obviously, "financing" is in the domain of the financial sector, and "innovation" (in products) in the domain of the real sector. Hence, on the one hand, we should observe financial markets, financial institutions and financial instruments for financing innovations. On the other hand, we need to analyse firms in their environment. Finally, it is the relevant socio-economic characteristics of both sides involved in financing innovations (investor and investee) and of the innovation itself that matter. Therefore, it does not suffice to have abundant financial sources for the financing of innovations to take place. In the sense that applies here, financing is an action that is always performed in relation to the agent requiring financing, and *vis-à-vis* the object meant to be financed, i.e. innovation. Innovations can differ in their quality, much like the managers/owners of the enterprises differ with respect to whether they prefer this way of financing (f.i. banking credits) or the other (f.i. venture capital), because they have different corporate governance implications. In this respect, enterprises are far from homogenous. Furthermore, both (potential) investors and (potential) investees have limited information, which is asymmetrically distributed. The former are never fully informed about the actions of the investees' management, and the latter need to plan their innovation activities in advance, not knowing what difficulties they could face in the process. The actors are neither fully rational nor perfectly informed. Because of the aforementioned "imperfections", it follows that there must be something causing them, which is outside the scope of "purely" economic transactions. Accordingly, there is a need to encompass culture, social capital and the institutions, along with economic variables important for financing innovations. This, in turn, has repercussions on both the theoretical and/or the conceptual approach, as well as the empirical research that we are dealing with. We shall be looking at the financing of innovations through the concept of the national systems of innovation, which – roughly speaking – encompasses institutions pertinent to innovations and their linkages, which contribute to the creation and diffusion of innovations[13].

Technological investment or investment into innovation activities encompasses any investment into technological products and process innovation. There are four different ways of looking at technological investment (OECD, 1995): 1) from the perspective of the management theory, 2) from the functional perspective, 3) from the perspective of the "physical content", and 4) from the product cycle perspective. The management theory claims that there are two basic forms of investment: product-portfolio investment and investment in resources. The former concerns investment related "to particular production programmes" (OECD, 1995: 22) and comprises investment in equipment and, to a lesser extent, investment into R&D. Investment in resources comprises investment in

13 Diffusion of innovations is a process whereby innovations are spread at a societal level.

most of R&D, training, commercialisation, and organisation, and is “designed to enable the firm to incorporate ‘vectors of technological competitiveness’ into products and manufacturing processes” (OECD, 1995: 22)[14]. “Modern management theory defines ‘resource investment’ as the assets generated by the acquisition or expansion of know-how that can be used in the manufacture of more than one good or service” (OECD, 1995: 22). Vectors of technological competitiveness may be incorporated into products and processes by incurring three kinds of costs (OECD, 1995: 22-23): the costs of tangibles, the visible costs of intangibles, and the hidden costs of intangibles. The functional perspective distinguishes between core technological investment and complementary investment, “which guarantees the efficiency of the core investment” (OECD, 1995: 26). The physical content perspective distinguishes between tangible and intangible investment. The product cycle perspective differentiates R&D investment, investment for production preparation, and investment for production launch.

There are certain market failures which hinder investment into innovation (cf. OECD, 1995: 56-57). Because of these, government intervention is needed. First, there is uncertainty, which inhibits the financial sector from investing into innovation. If the government did not intervene, particularly in basic and pre-competitive research, underinvestment into innovation would be a fact of life. On the other hand, of course, public sector funds might not be as efficiently invested as private sector funds (cf. Rosenberg, 1992). Secondly, the innovator cannot appropriate all the benefits from innovation, or, more precisely, the social benefits derived from innovation are larger than the private benefits derived from innovation because of spillovers. The selection of innovation projects by the market is biased against small and medium-sized enterprises (SMEs), product innovations, and medium-high-technology innovations (OECD, 1995: 60-61). In order to increase private returns from innovation, intellectual property rights[15] can be used. However, excessive intellectual property rights can have an adverse effect on innovation activities in a society (Drahos, 2005). As far as SMEs are concerned, the financial obstacles they face are higher than those faced by large enterprises. As for the product innovations, they involve a higher share of (commercial) uncertainty than do process innovations. Regarding medium-high-technology innovations, they “combine three drawbacks – are too risky for banks, promise returns that are too slim for ‘venture backers’, and are too lacklustre to attract government aid” (OECD, 1995: 61).

---

14 Vectors of technological competitiveness are: high-tech capital goods, technology-intensive product components and advanced substances (OECD, 1995: 22).

15 In the case of inventions patents are primary and most obvious form of protection. However, there are other forms of intellectual property rights, like copyrights and trade marks.

Furthermore, because of two specific characteristics of the product life-cycle, the nature of technological investment requires the involvement of specialised financiers and/or government backing: the average costs of R&D are gradually increasing, and products that reach the market stay there for a shorter period of time than they used to. The consequence of this is that investors have to be more patient for their investment to pay back. Besides, it is a characteristic of new innovation projects that they have a higher variance of cash flow, so cash flow requirements are far more irregular and bigger than they used to be (OECD, 1995: 65-67).

The financing of innovations is chiefly influenced by uncertainty, informational asymmetry, the intangibility of a firm's assets, and by the conditions prevailing on the financial and product markets (Lerner, 2007: 406). Uncertainty, in the sense of not knowing what the future outcome will be, is exacerbated by the duration of already complex innovation activities. Even when information is known, it is not equally distributed, as an investee knows more about his or her projects than does the investor. Furthermore, it is easier to evaluate – and, therefore, to finance – projects or firms that predominantly involve tangible resources or assets, rather than those that predominantly involve intangible ones.

Instead of trying to explain the financing of innovations through determinants that are strictly at the firm level – although this perspective as well will be applied later in this work –, financing in general is to a large extent determined by the characteristics of national financial systems and corporate governance (see below). The characteristics of capital market-based and bank-based financial systems have implications for financing in general and the financing of innovations in particular. These financial systems conceptually and practically overlap with different corporate governance arrangements and other concepts and phenomena, which we explain in more detail below. In parallel to the development of concepts, we shall also develop our research framework in the next chapter.

2.4. Conceptual Framework for Studying the Financing of Innovations

Knowing that the phenomenon at hand (the financing of innovations) is multilayered and complex we need to be able to observe it and explain it in terms of its connections with other phenomena. In our view, it is necessary to analyse complementarities and inconsistencies in socio-economic systems we intend to observe. This approach is similar to the one applied in cultural studies that look for a cultural fit (see Zagoršek and Štembergar, 2007). The institutions – as defined here – are seen as complementary among themselves, and as shaping a particular industrial structure and developmental patterns either due to their coherence (Amable, 2000), or to the lack of it. In this chapter, we explain several concepts that are pertinent to an understanding of the financing of innovations, having in mind Figure 1, in which we have depicted their relationships. Socio-economic systems are embedded in a certain cultural and social environment, so we need to understand the impact of culture and social capital on innovation activities and their financing. The financing of innovations is then strongly influenced by corporate governance and the financial system on the one hand, and the national system of innovation on the other. In addition, we study two other related concepts – competitiveness and entrepreneurship. Having reviewed these concepts, we elaborate our hypotheses.

2.4.1. Corporate Governance and Financial Systems

National financial systems are to a significant extent determined by corporate governance systems (cf. Račić, Cvijanović and Aralica, 2008). However, corporate governance thus perceived transcends its interpretation as a system of management and control (as defined by Bajuk et al., 2005) confined to companies, as the mainstream understanding of this concept suggests. Corporate governance should be seen as a process in a social context, not as a static state of affairs (Račić and Podrug, 2004)[16]. A national financial system comprises much more than just the financial sector, since it also includes a demand for finance from the real sector, as well as the flow of relevant information and control between the real sector and the financial sector (cf. Hackethal and Schmidt, 2000). Because of the linkages within and between the two sectors going beyond the flow of goods, services and money, corporate governance and the financial system are intertwined and cannot be observed separately. Namely, the institutional environment of enterprises like accounting standards, disclosure requirements at the stock exchange, and the protection of minority shareholders, are all corporate governance elements providing information (e.g. to financial analysts, financial institutions, and the general public) about the financial viability of an enterprise, and influencing managers' or owners' financing decisions.

Corporate governance (CG) includes "the structures, process, culture and systems that engender the successful operation of the organisations" (Keasey and Wright, 1993, as quoted in Keasey, Thompson and Wright, 1997: 2). A broad enough definition of corporate governance, which still allows for contextualisation in terms of innovation activities and their financing, is given by Hirschey

16 International standards of corporate governance (OECD, 2004) are focused on issues of corporate governance pertinent to a company and its main stakeholders – in continental Europe these are supervisory board, management board, shareholders and stakeholders in general. However, the Preamble of the aforementioned document clearly frames corporate governance in social fabric of society.

(2003: 67): a "system of controls that helps corporations and other organizations effectively manage, administer, and direct economic resources". CG influences the setting and fulfilment of strategic objectives of a company (OECD, 2004). There are two contending views of corporations – the shareholder model and the stakeholder model. The shareholder-centred model of the company focuses on the corporation as an entity legally and legitimately pursuing the goal of profit maximisation, whereby managers need to be properly incentivised and monitored for compliance more or less independently of the societal context. The stakeholder model of the company is contextualised within a social environment, which is crucial for innovation activities. Besides the management and supervisory boards, the shareholders and workers, stakeholders in a company are also suppliers, buyers, members of the local community, etc. These actors can contribute to a company's innovativeness, as recognised in the Community Innovation Survey (CIS). Nevertheless, both the shareholder theory and the stakeholder theory of corporate governance neglect innovation activities, because they focus on optimal allocation or existing resources. Because innovation is necessary for firms to be able to compete in the market, and corporate governance is essentially about control of corporate resources, the concept of corporate governance has to incorporate innovation (O'Sullivan, 2000). Furthermore, the concept of corporate governance has been predominantly applied to (large) corporations, that is, companies that are publicly traded (cf. Račić, Cvijanović and Aralica, 2008), which has limited insight into the CG of SMEs.

Corporations and other business organisations are strategically guided[17], managed and controlled through a number of internal and external factors (mechanisms or processes). In keeping with a broad definition of corporate governance, we can discern institutional factors (i.e. legal and nonlegal institutions), as well as socio-cultural and economic factors, which corporate governance depends on, in addition to its dependence on public policy (cf. Račić and Podrug, 2004). Key (legal) institutions influencing corporate governance are the Constitution and the laws and by-laws regulating the business of companies (e.g. company law and accounting laws and by-laws), principles of corporate governance (e.g. the National Code of corporate governance or a code valid for companies listed at the stock exchange) and company statute. Non-legal institutions (such as the coupling of business with corrupt practices or informal ways of negotiating) interact with legal institutions, cultural dimensions (such as collectively shared modes of behaviour, social norms), and those of social capital, like trust. The conduct of companies is also influenced by economic factors, like ownership and organisation of companies, markets for products and services, as well as the capital and labour market (cf. Račić, Cvijanović and Aralica, 2008). While markets for

17 Strategy can be understood as a process needed to position a company in relation to other companies so that a competitive position is attained (cf. Porter, 1996).

products and services, or the labour market, create competitive pressures that management needs to cope with, the capital market serves as a mechanism through which the actions of management are reflected in the share price, and which, therefore, acts as a constraining mechanism for management's actions. Corporate governance of SMEs has distinct characteristics, particularly if they are owned by a single person or a family, whereby the ownership function is not separated from the control function. This can cause conflicts of interest resulting in lack of efficiency. For example, when equity investment is needed in a company, the owner (being, at the same time, the manager of the business) may reject it on the grounds of retaining control over the company (Wilson and Bates, 2003: 6). Even if SMEs look different from large enterprises in terms of ownership structure and organisational structure, the CG of SMEs is still influenced by external factors like products and services markets (even if the shares of SMEs are unlisted at the stock exchange), accounting standards, the community in which an enterprise is located, etc.

What the literature leaves unresolved is the debate on which system of corporate governance is better, the Anglo-American model or the continental European model (such as that applied in Germany). There is a view that stresses market myopia – a widespread focus on the short-run company performance and neglect of the long-term company performance. Since managers in market-based systems pay close attention to the share price, they can disregard the long-run value of the company[18]. This happens because the capital market undervalues capital investment and investment into R&D (Keasey, Thompson and Wright, 1997: 5). Allen and Gale (2000) believe that markets are better than financial intermediaries in financing industries for whose activities little information is produced, and for which there are different opinions about an investment project. Countering the latter view, Christensen (1992: 143-144) believes that the intermediaries and the enterprise itself can use internal financing to deal with uncertainty much better than the market, when it comes to financing innovations. He believes that these financing instruments enhance learning and making of better contracts *ex post*, as well as allow for better screening and monitoring than the market. However, Hall and Soskice (2001) remind us that the role of intermediaries and/or markets should be seen in connection with other institutions in an economy, that is, whether they are complementary or not. Using the language of the "varieties of capitalism", they assert that liberal market economies (i.e. like those in the USA and the UK, in which firms mostly rely on market forms of coordination)

18 There are instruments to combat management myopia, such as tying a certain percentage of compensation for managers to the performance of a company's share in the long run, employee stock ownership plans (Hirschey, 2003: 75), and others. However, these instruments can only be applied at the level of individual companies, which is not enough to achieve widespread, systemic effects.

have an institutional structure that enables firms to quickly adapt to new market conditions (e.g.: in a recession, managers can quickly lay off workers, and financial markets will deny funding if a firm has profitability problems in the short run; this will create conditions for restructuring). Coordinated market economies (like Germany – i.e. those that rely more heavily on non-market forms of governance) have corporate governance systems facilitating long-run financial arrangements irrespective of the short-run profitability of firms. They believe that liberal market economies tend to be more supportive of radical innovations, while coordinated market economies better support incremental innovations[19].

Although the framework for financial and corporate governance systems here is primarily national, that is not to deny the importance of international factors. As globalisation blurs national boundaries, a firm does not depend on national financing sources the way it did before. The owners of the firm may also be foreigners, with an authentic corporate culture and the principle financing source with the parent company. An example of (high-tech) firms bypassing the national financial system could be found in Israeli high-tech companies that do initial public offerings and subsequently list their shares on the NASDAQ Stock Exchange in the USA. However, as is well established in the literature, financial systems are still predominantly national, and the usual distinction is made between bank-based and capital market-based financial systems[20] - see table 2 below.

Table 2: Capital market-based and bank-based financial systems

Type of the financial system	**Capital market-based (like UK and USA)**	**Bank-based (like Germany, France and Sweden)**
Financing patterns	Debt-equity ratio is relatively low; Firms are usually financed through retained earnings and issuance of new equity and debt securities.	Debt-equity ratio is relatively high; Financing is dependent on retained earnings and loans.
Characteristics of ownership	Ownership concentration is very low to medium; Market capitalization to GDP is high and shares and other	Ownership is very concentrated; Market capitalization to GDP is low with resultant ownership stabil-

19 One of the conclusions that can be reached through studying the varieties of capitalism literature is that financial systems (or corporate governance systems, depending on the perspective taken) are embedded in the wider institutional structure of an economy (see e.g. Vitols (2004) for a study on Germany's bank-based system).

20 They are alternatively named bank-oriented and capital market-oriented financial systems.

	securities are actively traded at the stock exchange. Hence, ownership stability is quite low.	ity.
Characteristics of monitoring (corporate governance)	Main financial actors are either anonymous agents or non-bank financial institutions; Since capital markets are strong, the mechanism for controlling management is in the market.	Main financial actors are either universal banks or investment banks; Insider control is stronger than outsider control, because of ownership concentration and relatively small capital markets.
Coping with problems, risk or uncertainty	There is a correctional mechanism by the capital market, e.g. a company can be taken over by another company; Venture capital is a developed form of financing for highly uncertain and risky projects; Government intervenes if problems escalate, and supports the defence industry.	Companies are very dependent on banks; Government intervenes or supports enterprises if problems escalate.

Source: author, based on OECD (1995: 69) – original source OECD Secretariat drawing upon M. Aoki, M. Aoki & G. Dosi, J. Franks & C. Mayer, M. Roe, R. Gilson & R. Kraakman, R. Gilson & M. Roe, C. Sabel & J. Griffin, J. Zysman.

What is implicit in the above exposition of the characteristics of financial systems (and CG systems) is that financial systems have certain implications on firms' financing, ownership characteristics, monitoring, and dealing with uncertainty and risk. All of these factors naturally vary from one country (system) to the other. As for firms' financing or firms' capital structure, it is not only dependent on factors specific to the firm, but also on the overall characteristics of the economic and institutional system (cf. Antoniou, Guney and Paudyal, 2008). We shall devote more space below to financial theories on factors specific to the firm or its immediate environment that have an influence on the financing of the firm (or the firm's capital structure). Although it is indisputable that specific factors at the micro-level can be crucial for the financing of firms, seeing things from a systemic perspective enables us to more easily notice complementarities between the constituents of the system (e.g. the financial sector and the real sector) and explain innovation-financing patterns (discussion about this follows below). In addition to the characteristics of the bank-based and market-based financial systems from the above table[21], the former are characterised by the control-orientation of investors and small financial markets, while in the latter investors are portfolio-oriented, and financial markets are large (Berglöf, 1997). Seen together, these characteristics make up a logical and coherent system for each of the two financial systems – e.g. in a market-based system, investors can have a portfolio orientation because of the large financial markets with many securities to chose from. Strong financial markets provide a guaranty that management will try hard to produce value for shareholders, etc. On the other hand, bank-based financial systems enable more control for insiders, who have concentrated ownership in order to secure their interests.

2.4.2. National Systems of Innovation Approach

Broadly speaking, the national system of innovation is a set of institutions and organisations – as well as linkages between them – that directly or indirectly influence the innovation performance of a national economy. Organisation and governance of an innovation system is crucial for economic growth (OECD, 1992, as referenced in Švarc, 2009: 13). Below we provide only the essence of an NSI approach, and we revisit this concept in the chapter on the economics of innovation[22].

What is lacking so far is a perspective on innovation, particularly one that is capable of combining the aforementioned discussion on financial/CG systems with innovations. In that respect, the national systems of innovation (NSI) perspective seems appropriate as a starting concept to develop our understanding of financing innovations. Although far from being a theory, the NSI is holistic, interdisciplinary, and emphasizes the historical perspective, while studying innovation as a learning process which involves feedback and interaction (Edquist and Hommen, 1999: 65-66). The NSI keeps us focused on the issues of innovation and is broad enough to encompass financing issues as well. However, the literature on the NSI has not to any significant extent dealt with financing innovation so far. The system of innovation approach is not conceptually complete, which leaves it open for application in particular cases. It does, however, put into focus enterprise and the results of innovation processes and allows them to interact with other institutions relevant to innovation activities and other framework conditions. Especially important is an analysis of the interaction of firms with other institutions that results in some information and knowledge flows, financial flows, flows of skilled workers, etc. (Fischer, 2001: 209). A researcher's role is to identify and analyse interactions that are crucial in terms of new

21 Berglöf (1997) calls the former control-oriented financial systems, and the latter arm's length financial systems.

22 See chapter 3.1.

knowledge in each of the systems of innovation (Johnson, 1997, as referenced in Fischer, 2001: 209). This is because national frameworks for stimulating innovation differ, and there is no consensus in the literature as to how exactly to proceed in each case. Besides, it is not entirely uncontroversial to observe systems of innovation strictly within national borders (cf. Quéré, 2004). Lundvall (2007) does, however, suggest a four-step procedure for analysing the NSIs:

"[T]he *first step* would be to analyse what takes place inside firms in terms of innovation and competence building. A *second step* would be to analyse the interaction among firms including competition, cooperation and networking, and how firms interact with knowledge infrastructure. A *third step* would be to explain international differences in these respects with a reference to the specificities of national education, labour markets, financial markets, welfare regimes and intellectual property regimes. As a *fourth step* firm organization and network positioning may be used to "explain" the specialization, competitiveness and growth performance of the innovation system." (Lundvall, 2007: 102, emphasis original).

The most important part ("the core") of the innovation system consists of firms in interaction with their immediate environment, which is visible as a result of application of the first two steps. Of course, application of the NSI concept is subject to availability of data and such societal specificities as the distribution of power and need for institution building (Lundvall, 2007). Such a comprehensive analysis of the NSIs at hand is beyond the scope of this work, in particular because we have to focus on an analysis of the financing of innovations (hence, some elements and interactions in the NSIs are more important than others). The data that we have gathered in the case studies (interviews) and from other sources limit the picture that we can obtain by analysing these data. On the other hand, Edquist (2004) proposes that a special emphasis be put on the "activities" that are most directly connected with the creation and diffusion of innovations, while considering individuals and organisations (as "players"), as well as institutions. Activities are "[t]he factors influencing the development and diffusion of innovations" (Edquist, 2004: 487). These activities are, for example, the financing of innovative activities, provision of knowledge for innovation activities, making changes to the legal institutions that directly influence innovation activities, etc.

The concept of the NSI essentially claims that innovation is determined by a number of socio-economic factors, predominantly on the level of nation states. A version of the NSI concept called the Aalborg-version stresses the importance of the non-market forms of governance, i.e. non-price relationships (Lundvall, 1985, as referenced in Lundvall et al., 2002: 218), coordination and cooperation resulting in facilitating interactive learning. These relationships (and, therefore, innovation) can be facilitated in a cultural and social environment characterised by complementarity, an environment marked by social cohesion, equality and trust. In addition, the concept of the NSI includes corporate governance, in the sense that short-termism and long-termism produce different outcomes. These institutional and social structures should exist side by side with the legal institutions. The advantage of such a concept that integrates legal and non-legal institutions embedded in the social fabric of a society is visible when it comes to explaining innovation activities in which trust and tacit knowledge play a predominant role (Lundvall et al., 2002)[23] – as in the case of high-tech manufacturing.

Complementing previously mentioned research on the NSI, Souitaris (2003) distinguishes quite a broad spectrum of factors that determine the rate of technological innovation. Having identified the common determinants of innovations in the literature, Souitaris (2003) establishes a portfolio model by synthesising the determinants of innovation. He divides the determinants into four categories: 1) contextual variables, 2) strategy-related variables, 3) external communications, and 4) variables related to the "organisational context". Contextual variables comprise a firm's profile (consisting of some objective variables from financial statements and other) and the competitive environment. Strategy-related variables are those connected with business strategy, innovation budget, management attitudes, and CEO's profile. External communications in this model consist of communication with the stakeholders, networking-scanning external information, and co-operation with external organisations. Souitaris (2003) underlines the fact that (management) culture could be a factor exerting influence on others (Souitaris, 2003: 517). Besides, the external communications part of this model in particular could largely be denoted as social capital.

The concept of the NSI is connected to some other concepts that shed light on innovation activities and their financing. We have already explained the need and intention to incorporate concepts of the financial system and corporate governance in the NSI. Both of them comprise institutions, in addition to the flows of money and information. First of all, it should be pointed out that it is institutions that matter, as Borrás (2004) explains. Formal and informal institutions[24] interact with the production structure, as well as the environment comprising politics, culture, the market, science, technology. It is important to analyse institutions, as they have specific functions in the NSI - e.g. in connection with the production, diffusion and appropriation of knowledge, as well as with risk reduction (Borrás, 2004). Entrepreneurial activity in general (comprising intrapre-

23 Tacit knowledge is normally opposed to codified knowledge. While the former is inherent to a person from whom it cannot be separated, the latter is specified in detail and articulated.

24 As explained below, we shall call them legal and non-legal, respectively.

neurial activity[25]) is important for the creation of new firms as possible carriers of innovative activity, as well as for the expansion of the innovative capabilities of existing firms. In this sense, entrepreneurship is an important building block of any NSI (Edquist, 2004). This is connected with the question of the "creator" of innovations: how does an agent (entrepreneur) influence structure (NSI) (Lundvall, 2007)? In addition, the NSI can stimulate innovations which contribute to the competitiveness of a country (Freeman, 1987, as referenced in Quéré, 2004).

In accordance with its structure, the NSI can be influenced by a set of public policies respecting its complexities and complementarities.

> "A broad concept of innovation system implies a new perspective on a wide set of policies including social policy, labour market policy, education policy, industrial policy, energy policy, environmental policy and science and technology policy. Specifically, the concept calls for new national development strategies with co-ordination across these policy areas." (Lundvall et al., 2002: 227).

Public policies should stimulate innovation processes by enabling them to develop in the long-run, as is inherent in their nature. On the other hand, the short-termism typical of finance should be limited in size and scope (Lundvall et al., 2002: 227).

2.4.3. Culture and Social Capital

The socio-economic and cultural conditions of an economy influence all (economic) activities in a given country. The heuristic model of socio-cultural factors of developmental performance devised by Adam et al. (2005) suggests as much. The model has two main constituent blocks, civilisational competence and developmental factors, which in their turn influence the third – developmental performance. Civilisational competence is represented by historical conditions that have influenced both the material aspects of a society and cultural (intangible) ones. Developmental factors are a complex set of phenomena[26] covering various spheres of socio-economic life on which social capital exerts a crucial influence. Although we still lack a uniform definition of social capital, the literature on social capital is vast and provides enough material to derive an operational concept of social capital relevant in the context of this book. Social capital comprises forms of trust and norms (cf. Landry, Amara and Lamari,

25 Entrepreneurial activity inside business organisations.

26 Although we will not discuss that in detail here, these are the interrelated phenomena of social cohesion, cognitive mobilisation, quality of governance, and entrepreneurial spirit as internal factors. Openness (internationalisation) is considered an external factor (Adam et al., 2005: 8).

2001), and networks (cf. Adam et al., 2005) in social interactions. The importance of social capital in general lies in its positive correlation with the GDP (Svendsen, 2005) and, in particular, its impact on innovation performance (cf. European Commission, 2008). There are two basic strands in the literature on social capital (Grootaert et al., 2004: 3). One perceives social capital as immaterial resources (such as ideas) that one can obtain through his or her position in the network. The other refers to the framework conditions of a society. We will be focusing on the latter strand found in the literature. Although culture has many manifestations (and definitions), we are interested in seeing how and to what extent culture – as a set of shared patterns of thought and behaviour – influences innovation activities and their financing.

In terms of metaphor, one can view social capital as a glue and/or a lubricant. In the case of the former, it ties individuals and groups together, while from the perspective of the latter as applied in economics, it facilitates business transactions. However, although members of the same group enjoy advantages, members of other groups (outsiders) can find themselves at a disadvantage. The powerfully bonding social capital of e.g. interest groups could be detrimental to society as a whole (Sabatini, 2006). A widespread classification of social capital in the literature (Grootaert et al., 2004: 4) distinguishes between the following: 1) bonding social capital, 2) bridging social capital, and 3) linking social capital. The first form of social capital refers to ties among individuals who share some sociological characteristics. Its second form covers situations in which ties are built between groups of different sociological characteristics. The bonding and bridging types of social capital are horizontal because the individuals or groups in its possession are on the same or similar level of the social hierarchy. When it comes to ties with individuals or groups occupying a position of power or authority, we are dealing with linking social capital, which is, therefore, vertical. Trust, as a dimension of social capital, can be divided into three main groups (Raiser, 1999: 4-5): 1) between family members, i.e. a kinship group; 2) trust between parties to a transaction that has been repeated; and 3) extended or generalised trust reflecting the quality of legal institutions, the moral characteristics of parties to a transaction, and their experiences. The latter kind of trust is the most important in modern (Western) economies. The role of the state and its institutions can be crucial for the establishing this kind of trust. The role of the state in stimulating extended trust can, according to Raiser (1999: 15), be divided into three different instruments: 1) constitutional safeguards (that make possible the division of political power and its proper functioning); 2) distributional policies (understood in the broad sense of socio-economic policies that contribute to social justice); and 3) policy interventions. Low trust can be detrimental to innovation (Landry, Amara and Lamari, 2000: 8). In the literature, norms, as another dimension of social capital, usually refer to individuals' behaviour, like reputation for trust-worthiness. Finally, networks, as a dimension of social capital, are considered as communication channels with relevant actors (Landry, Amara and Lamari, 2000: 8).

On a more operational level, which is important for our study, one has to observe the importance of social capital on the activities of firms. Westlund and Nilsson (2003: 1) suggest that the social capital of a firm can be defined "as social and economic networks in which enterprises intentionally and unintentionally invest and interact in". The social capital of a firm can be broken down into two main categories, as Westlund and Nilsson (2003: 4) have done: 1) internal, and 2) external. Internal social capital is reflected in the company spirit, cooperation climate, and methods such as those used for knowledge codification. External social capital can be related to production, the environment, or the market and incorporates links or relations to actors from the company's environment. When investigating innovation processes in firms, Landry, Amara and Lamari (2001) find that the quality of social capital influences business decision making and has a bearing on whether the innovation will be incremental or radical. In their study, the observed level of social capital varies from industry to industry. Social capital is not just "out there": rather, it can be influenced by e.g. firms' investments, even if only partially and not at all in the short run. Hence, we could say that it is through its impact on innovation performance that social capital can influence the financing of innovations.

Westlund (2006) puts forward a concept of culture as related to institutions and social capital, claiming that culture is actually a broader concept than social capital, and that it intersects with the concepts of institutions. There are many conceptualizations of culture[27], and we will start from the immaterialistically oriented definition[28], which is "that complex whole which includes knowledge, belief, art, morals, law, custom, and any other capabilities and habits acquired by man as a member of society" (Ritzer, 2005: 181 quoting Tyler, 1871: 1). However, thus defined, the notion of culture intersects with the notion of institutions as we understand it. As defined and used here, institution is a socially embedded rule that may, or may not, be established in legal documents. It follows that it does not have to be written down, but definitely has to be shared and widely accepted among individuals in a society. Culture can also be understood as socially transmitted information comprising beliefs, preferences, values, knowledge, etc. (Boyd and Richerson, 2005). The concept of culture as an im-

---

27 "Culture" is a very interesting notion, rich in meaning and (possible) uses. For a comprehensive discussion on culture an interested reader is invited to refer to Eagleton (2000).

28 Culture can also be defined in the sense of the artefacts created by the human kind. We will not be using that concept of culture here.

material sphere is also shared by Westlund (2006), who quotes the definition of culture by Casson and Godley (2000: 2), which suggests that culture comprises "shared values and beliefs" (Westlund, 2006: 8). Culture can be framed as e.g. a national and organisational/corporate culture (Zagoršek and Štembergar, 2007; Brown and Ulijn, 2004), a professional culture peculiar to a given profession (Brown and Ulijn, 2004).

Entrepreneurial culture exerts a more direct influence on innovation (financing). Sztompka's concept of entrepreneurial culture comprises "innovativeness, competence, rational calculation" (Adam et al., 2005: 21, referring to Sztompka, 1993)[29], and this is the concept of culture that we shall employ in this study. However, we shall do so in a less restrictive manner, understanding (entrepreneurial) culture as cultural variants that (may) influence innovations and their financing. Tidd, Bessant and Pavitt (2001) discuss culture in connection with the innovative capabilities of individuals and companies. In that sense culture (or "climate") can be either innovative or not conducive to innovation, impossible to change in the short run. Hampden-Turner and Trompenaars (1995) conceptualize culture in terms of valuing processes of wealth creation, whereby cultures differ in alternative ways needed to fulfil them[30]. Kutlača et al. (2006) undertake a thorough examination of the innovation culture in Serbia, scrutinizing it from the level of the firm, through the sectoral level, to the national level. In so doing, they view culture as determining the national system of innovation, which is our approach as well.

---

29 Sztompka (1993 – as referenced in Adam et al., 2005: 21) actually discusses entrepreneurial culture as one of the building blocks of civilisational competence. The other ones are: "2) civic culture (political activism, participation); 3) discursive culture, in terms of an intellectual exchange of ideas based on mutual tolerance, pluralism and scepticism; and 4) everyday culture (a sense of punctuality, hygiene, precision, orderliness, technical competence)" (Sztompka, 1993 – as referenced in Adam et al., 2005: 21).

30 The seven countries that they analyse include the United States, Great Britain, Japan, Germany, France, Sweden, and the Netherlands. The seven valuing processes they study involve (the opposing approaches offering a solution are given in the brackets): 1. making rules and discovering exceptions (universalism vs. particularism), 2. constructing and deconstructing (analysis vs. integration), 3. managing communities of individuals (individualism vs. communitarianism), 4. internalizing the outside world (inner-directed vs. outer-directed orientation), 5. synchronizing fast processes (time as sequence vs. time as synchronization), 6. choosing among achievers (achieved status vs. ascribed status), and 7. sponsoring equal opportunities to excel (equality vs. hierarchy) (Hampden-Turner and Trompenaars, 1995).

2.4.4. Competitiveness and Entrepreneurship

Entrepreneurship can be defined as "a process of exploiting opportunities that exist in the environment or that are created through innovation in an attempt to create value. It often includes the creation and management of new business ventures by an individual or a team" (Ulijn and Brown, 2004: 5). It is important to note from this definition that there are opportunities to be exploited that may be exogenous to an enterprise. Later on (in the empirical part), we shall be discussing the importance of the entrepreneurial environment in connection with innovation activities. Notwithstanding the fact that entrepreneurs are important as creators of value added, we have already briefly touched upon our position that it is the system (NSI) that is crucial for innovations and their financing. By the same token, competitiveness is no less under a systemic influence[31]. It has to do with attaining a better position in relation to competition. Although this position can be attained through an increase in efficiency (e.g. reduction in production costs), it is through innovations (e.g. in products and/or services) that a firm attains an advantage over the competitors, whereby rents can be extracted.

Porter's "diamond of national advantage" (cf. Porter, 1990) is a useful basis on which to build a conceptual framework for studying the interconnected phenomena of entrepreneurship, innovation and competitiveness. Porter's diamond is a model for studying the competitive advantage of a country and/or a firm in it with a view to the world market(s). There are four factors influencing the national competitive advantage (Porter, 1990: 77): 1) factor conditions, 2) related and supporting industries, 3) demand conditions, and 4) firm strategy, structure, and rivalry. In Porter's model, these factors comprise "the playing field that each nation establishes and operates for its industries" (Porter, 1990: 77). Despite not doing so explicitly in the model itself, Porter underlines that "[d]ifferences in national values, culture, economic structures, institutions, and histories all contribute to competitive success" (Porter, 1990: 74). Porter's model of competitive advantage upgrades David Ricardo's static principle of comparative advantage, and emphasizes the fact that competition is a necessary precondition for competitiveness. Porter writes: "Companies gain advantage against the world's best competitors because of pressure and challenge. They benefit from having strong domestic rivals, aggressive home-based suppliers, and demanding local customers" (Porter, 1990: 73). Another message transmitted by the model is that a lot depends on domestic (that is, local) conditions. A critique of the model could be based on the aforementioned fact that it does not include "soft factors" (like cul-

31 This is especially visible in the concept of systemic competitiveness, which includes micro, meso, macro and meta levels, whereby the latter includes social and cultural factors that contribute to competitiveness. However, the concept is not fully operational (Nielsen, 2000).

ture and institutions), and that it does not account for differences in the size of a country. The latter would be especially important if we wanted to distinguish national from international processes influencing competitiveness. However, this model underlines the importance of factors that influence innovative activities as well.

Entrepreneurship and its offspring – innovation – is a multi-layered phenomenon that ought to be analysed extensively in terms of its determinants and in a historical perspective. The complexity of the entrepreneurship phenomenon is perhaps best articulated by the following quotation from the Global Entrepreneurship Monitor (GEM):

> "[E]ntrepreneurship is an important mechanism on which the economic growth depends, but the rank of countries in terms of the level of entrepreneurial activity is very stable. This suggests that the level of entrepreneurship depends on institutional, social and cultural factors that are difficult to change at short notice. Therefore, the evaluation of different programs and policies implemented by many governments in their countries with the aim of improving the entrepreneurial activity is not always possible because the effects cannot be seen in a short period of time" (Singer et al., 2003: 51).

The GEM framework makes it explicit that the "social, cultural and political context" counts when studying entrepreneurship. The role of these factors is central and is split into two further impact blocks: that of the national framework conditions and that of the entrepreneurial framework conditions. Finally, there is the impact on GDP growth (for details see Singer et al., 2006: 12). We shall be using findings from the GEM below, when analysing the national systems of innovation of the studied countries.

2.4.5. Hypotheses

First and foremost, we are interested in seeing what economic theory, and especially finance theory, has to say on financing in general and financing innovation in particular. Ardalan (2004) notes that modern finance is practiced from the point of view of the functionalist paradigm, which is to say that academic finance theoreticians try to explain and predict phenomena visible (only) at the surface, in a value-free and objective manner. Current mainstream financial theory/theories do not view financing as a complex social process, and do not consider fundamental uncertainty, but only risk. These concepts and theories (i.e. the neoclassical economic theory and the new institutional economics theory) have the problem of capturing the complex and long-term nature of innovation activities. Mainstream finance theories especially neglect institutions (in a broader sense, and not just legal institutions), as well as social capital and culture. In fact, they are not specifically designed for observing financing innovations. However, we believe there are some useful concepts and theories in mainstream finance that can be utilised when discussing financing innovations, especially those that come from the "economics of information" literature. Furthermore, heterodox economics theories (especially the post-Keynesian theory) are much more realistic, but fall short of developing concepts of financing innovations. Their main advantage (in comparison with mainstream theories) is the stronger emphasis on the side of the real sector.

At this point, it would be useful to briefly describe what we mean by "mainstream" economics theories. These are in particular the neoclassical economics and the new institutional economics. Arnsperger and Varoufakis (2006) claim that the three distinguishing characteristics of neoclassical economics are the following axioms: 1) methodological individualism[32], 2) methodological instrumentalism[33] and 3) methodological equilibration[34]. Aspects of neoclassical economics and the new institutional economics will be elaborated in the chapter on the theory of the firm (3.2.) and the chapter on financial theory (3.3.). The new institutional economics is very closely aligned with neoclassical economics in that they share methodological individualism and methodological instrumentalism. Franičević (2003) asserts that it is characterised by bounded rationality and the fact that it recognises the factor of time. We may, therefore, say it does not necessarily observe equilibrium states. We shall call the other approaches mentioned here "heterodox". When we make reference to "institutional economics", we usually mean the branch of institutional economics called the "old" institutional economics or the "original" institutional economics. Its methodology is the opposite of the three aforementioned axioms of neoclassical economics. It allows for interdisciplinarity and qualitative research practice, and it observes processes, rather than states (cf. Foster, 1991). It essentially claims that economic activities are embedded in a wider institutional, cultural and social environment. The other heterodox school of economic thought we mention is post-Keynesian economics, which is also at odds with the axioms of neoclassical economics. More importantly, it follows its own methodological practice, of which we shall mention only some aspects (Dow, 1991 as in Medić and Cvijanović, 2001): it pursues an interdisciplinary approach and offers an integrative view of the economy (i.e. it is not in favour of a separated micro and macro analysis). The theories and/or concepts that we present in this work should be

32 Methodological approach which tries to explain socio-economic phenomena by focusing on actions of individuals.

33 Methodological approach that treats people as maximisers of their preferences.

34 Methodological approach that imposes observation of socio-economic situations as if they are in equilibrium and as if they need special events to get them out of equlibrium.

seen in relation to the above-mentioned characteristics of their schools of economic thought.

We argue that the current mainstream economics theories are not capable of explaining the financing of innovations in a systemic perspective because of their inherent reductionist reasoning. Heterodox economics theories do not explain the financing of innovations, since they have not sufficiently tackled this topic. But there are important theories from both the orthodox and the heterodox economic thought that can be used to describe and explain the financing of innovations. We need a concept of financing innovations that is able to encompass the dialectic between the real sector (which produces innovations and demands financing) and the financial sector (which can provide financing). The intertwinement of the real and financial sectors is crucial here, since both have to interact – not in a one-off way, but continuously. Because the economic system is embedded in the social reality, this concept needs to incorporate not only economic variables, but also institutions, social capital, as well as culture, since the financing of innovations is likely to reflect those characteristics. Governance through and by markets is not the predominant mode of governance, particularly not in those economies and national systems of innovation in which institutions are not fully functional – as in such post-transition countries as Hungary, Slovenia and Croatia. Furthermore, it is not just market imperfections (that is, supply-side obstacles to financing) that we have to study, but also obstacles inside firms (on the demand side). For example, the enterprises from the post-transitional countries that we have studied are likely to be closed to the venture capital/private equity (VC/PE)[35] option because of the predominance of the banking culture, insufficient knowledge of VC/PE, and/or the entrepreneurs' fear that they might lose grip of the firm. In addition, some forms of financing are more available than others so firms will predominantly use those that are available and/or those that they are used to (e.g. they will use banking credits, since they have always done so).

---

35 In this work the author concentrates on venture capital as "[i]ndependently managed, dedicated pools of capital that focus on equity or equity-linked investments in privately held, high-growth companies" (Lerner, 2000: 522). Venture capitalists actively participate in the development of a firm they have invested in. They can do this by following two different strategies: 1) by helping develop products and processes, and 2) by doing financial engineering. They can also combine these two strategies (cf. Günter and Frommann, 1998: 4, 7). Lerner (2000: 522) notes that "[m]any venture capital funds, however, occasionally make other types of private equity investments" and that "[o]utside of the United States, this phrase is often used as a synonym for private equity". Although the author uses the term "venture capital" in this broader sense that encompasses private equity, we are primarily interested in its function as an innovation financing source.

The second point is going to be tackled in the empirical parts of the study. Although firms often complain about financial problems as the worst problems they encounter in the innovative process, – especially regarding the lack of finance and financial sources that are too expensive –, in reality there are deeper and/or other problems causing them even greater concern. In Chapter 5, we shall see that many firms: 1) either do not utilise own resources and competences that would enable them to be sufficiently innovative (or are not innovative at all), or 2) do not fully commit to securing possible financial sources for financing innovations. The former problem can be manifested in the poorer quality of innovations – incremental innovations, as opposed to radical innovations[36]. Reasons for this can lie in a firm's strategy not being conducive to innovation, for example because a firm does not cooperate with other firms or organisations, or because of insufficient ambition for further development of the business and/or lack of business plans. Naturally, this problem may be aggravated by conditions external to the firm. We say "aggravated", since they neither uniformly influence each firm, nor are they necessarily predominant: there are successful and less successful firms. The institutional environment may be constraining factor hampering a firm's innovation activities, while culture and social capital may not be conducive to innovations of a higher quality. Even in the case of a stimulative institutional environment, inherited cultural patterns, which are much more resistant to change than f.i. legal institutions, may pose certain limitations in this regard. As to problem no. 2, the lack of commitment could be seen from suspicion and/or mistrust towards some forms of external finance, especially those with larger governance capabilities (i.e. equity). It can be manifested in a reluctance to use venture capital. This reluctance may be culturally contingent and/or influenced by lack of social capital. However, as finance (capital) in its many forms can only augment successful enterprises (those with innovations), we could say finance has a complementary, not a primary, role in financing innovations. In other words, finance is only auxiliary and it can only support propulsive ventures (products/innovations). But it is the quality of a project that largely determines financing options. Therefore, incremental innovations are more likely to be financed by banking credit, and radical innovations by venture capital.

In addition, in innovation studies one often comes across observations about the severity of financial problems for innovation activities. Although there is no doubt that they could have negative effects on innovation activities, it seems that

36 It goes without saying that not all innovations are – or can be – radical. There is nothing wrong with incrementally innovating products, services, processes, etc. On the contrary, it may be a viable strategy for firms coping with the increasingly limited life-cycles of their products. Hence, in the absence of better indicators, whether an innovation is incremental or radical in this context serves only as a proxy for the quality of innovations.

the problem is that of the viewpoint: who is it that answers the question on problems regarding financial factors. In the Community Innovation Survey (CIS), it is the key personnel in the enterprises. So, instead of this problem lying on the supply side (that is, the market side or the side of the potential financier), it might as well lie on the demand side (that is, the side of the enterprise).

2.5. Methodological Approach

2.5.1. Extent of Our Inquiry

Because of the insufficient treatment of the financing of innovations in the literature, we seek to develop both theoretical and empirical explanations that can be used to elaborate this topic. The multifaceted nature of the financing of innovations leads us to develop a concept that should be realistic and enable us to conduct empirical research. Likewise, we shall be using empirical research to try and uncover not just the functional relationships pertinent to the financing of innovations, but also structural relationships. Hence, we follow a comparative method and highlight the peculiarities of each of the countries that we analyse. Before we proceed, it is necessary to explain our methodological approach, particularly in terms of defining the theory (concept) and in order to see what the purpose of inquiry in economics is, and what it is that we should observe.

Theory "is based on research of phenomena, logical generalization of cause-and-effect relations between phenomena, setting hypotheses on their relationships, verification of set hypotheses, as well as positing general principle of the explanation" (Proleksis enciklopedija, 2009). In essence, it was once a hypothesis (or hypotheses) that has (or have) been proven (using scientific methods), and not yet falsified, using Karl Popper's language. However, it could also consist of statements and axioms one can use to explain reality. The role of theories is both theoretical/cognitive and applied (cf. Žugaj, 1989: 146). Two or more theories can coexist and compete for dominance – just as various schools of economic thought do. A theory needs to be coherent, correspond with reality, and be clear and simple. However, the two latter criteria should only be considered secondary. That is, they are applied only after the criteria of coherence and correspondence have been applied (Eichner, 1983, as in Fox, 1997: 35-36). A concept may involve all the elements of a theory (hypotheses, causal explanation, and general principles), but its cognitive value is lower than that of theory. This is because its hypotheses and explanations have not been as extensively tested.

There is a discussion going on among economists as to what the purpose of their inquiry is: (simple) explanation and prediction on the one hand[37], or understand-

37 This methodological approach is also known as instrumentalism (Torrance, 1991).

ing phenomena on the other? While quantitative measurement is underlined in the first tradition, the other tradition lends a greater emphasis to the qualitative approach (Fox, 1997), maintaining that phenomena cannot be measured with precision. Related to the latter is the issue of deductivist vs. inductivist reasoning. The former is more likely to be found in (currently) mainstream economics. Neoclassical economics presupposes reality can be described (explained, approximated) as a closed system[38], whereas the opposite is more realistic. Namely, although closed-system reasoning can indeed be very helpful when applied to certain phenomena, such as approximation (e.g. stock market prices in developed capital markets), it applies under certain preconditions (presumptions) which need not be fulfilled. Because of inherent uncertainty about future events, factors previously unaccounted for might come about. This is the open-system complexity a social science researcher faces, not least because he/she cannot use experiments.

In addition, social reality is not only an open system, but its explanation calls for an understanding of the many different layers it comprises. Visible phenomena are just the "tip of the iceberg". Social reality "[c]omprises not only actual events and states of affairs, some of which we may directly experience, but also deeper structures, powers, mechanisms and tendencies, etc., which produce, facilitate or otherwise condition these events and states of affairs" (Lawson, 2003: 79). In order to understand the causal relationships behind phenomena ("at surface"), it is necessary to be able to observe and understand the layers of the social reality below the surface. In other words, it is necessary to understand the ontology of the social sphere (cf. Lawson, 2001, 2002 and 2003; cf. Dow, 2005). This is critical realism applied to economics, and we believe it should be immanent to economic theory.

We, therefore, believe it is feasible to employ not only functional analysis, but also historical and structural analysis. Moreover, the social scientist's conclusions are insofar more reliable as the analysis is comparative (that is, by using comparative methodology). According to Ragin (2004), "[t]he primary theoretical objective of comparative research is not theory testing, but concept formation, elaboration, and refinement, and theory development". Comparative methodology requires what Lawson has termed the "contrast space". "A contrast space is merely a domain over which it is meaningful, given our current understandings (…), to draw comparisons" (Lawson, 2003: 89). The contrast space needs to be sufficiently broad, and the observed phenomenon or contrast should stretch over a considerable space and/or time (cf. Lawson, 2003). Consequently,

---

38 A closed system is "[o]ne in which regularities of the form 'whenever event (or state of affairs) *x* then event (or state of affairs) *y*' (or stochastic near equivalents) occur" (Lawson, 2002: 3).

there is a need to broaden the inference beyond induction and deduction to include the method of retroduction (see Lawson, 2003). "Retroduction involves first a theoretical-abstract re-description of the events based on existing qualitative and quantitative material, this is followed by explaining the events through postulating and identifying the structures and causal mechanism producing them; and it ends with the theory being checked empirically" (Lee, 2002: 792).

No single theory that explains everything has yet been devised – not in the ("hard") natural sciences, nor in the ("soft") social sciences. That would not be realistic, due to an enormous amount of unknowns. Even regarding the capitalist system itself, it seems realistic to presume that one theory cannot explain everything[39]. The validity of a single theory in social sciences can vary when applied to different countries/systems or even different point in time[40]. This conclusion follows from the fact that individuals and societies are ever-changing. The capitalist system has changed from its early beginnings (so called mercantilist phase in the 16th century) to modern global capitalism characterised by a high content of knowledge in production, and fast and huge financial flows. The criterion for adopting a certain theory would be its validity and realism, as compared to the advantages and disadvantages of earlier (or competing) theories. However, unlike theories in the natural sciences, in which one theory is more realistic and prevails over the other[41] (older theory), theories in economics (and other social sciences) are much more rigid in nature. Namely, shifts in economic theories (and paradigms[42]) are not as swift and clear as those in natural sciences. Rather, different economic theories often coexist and compete with each other. A realist's view regarding economic theories is that reality is not predictable, that is, event regularities are not prevalent. Moreover, that is apparently not the case only with economics, but also with natural sciences (Lawson, 2001).

39 Such is e.g. the standpoint of the followers of the post-Keynesian school of economic thought (see Dow, 1991).

40 This could be illustrated by the labour theory of value, which was probably more applicable to the mercantilist period of capitalism (when there were no robots, and economies were mostly producing material goods) than to modern capitalism (with robotized production processes, as well as production of knowledge).

41 Perhaps the most famous example of this in history is the replacement of the geocentric theory by the heliocentric theory, although the latter has only been gradually replaced in science.

42 The notion of paradigm was applied to science by Thomas S. Kuhn in his 1962 book "The Structure of Scientific Revolutions", in which he showed that progress in science is rather non-linear. Paradigm can be defined as a fundamental explanation of a phenomenon or process valid at a certain time and in a certain scientific community (Cvijanović, 1999). As such, it is considered a cornerstone of further scientific practice.

In a debate on whether the purpose of a scientific theory is simply understanding or also explaining and predicting social phenomena (cf. Fox, 1997), the author is more inclined to adopt the former view[43]. Namely, the former view is closer to a definition of (economics as) a social science that does not imply the existence of rational individuals, whose behaviour is predictable. On the contrary, it presumes that the object of observation of a social science (economics) is ever-changing, and recognises the fact that the socio-economic systems we observe are many (and that each of them has its peculiarities)[44]. In line with the Lawsonian[45] critical realism approach as briefly outlined above, and with the aforementioned methodological notes, we will try to derive the principles and a concept of the financing of innovations that will be fit to be used in this study. Before we proceed with the theory, we need to tackle the institutional foundations of our methodological approach.

2.5.2. Institutional Foundations

Economy as a system is embedded not only in a materialistic, but also in a social (institutional) context[46]. This is the basic supposition of institutional economics, be it the "old" (or "original") institutional economics or the new institutional economics. Institutional economists are typically concerned with questions like: a) how is a socio-economic system organised, and b) who controls it (cf. Foster, 1991)? In this section, we are looking to gain an insight into what institutions are, and what implications an institutional structure has for economic activities in general. We are doing this on a more fundamental level that will help lead our inquiry on the NSI and the financing of innovations in the later chapters. Namely, a particular institutional structure results in a certain structure of incentives, which is "*embedded in the institutional framework*" (Mantzavinos, 2001: 169; Emphasis original). An incentive is something that gives a person a motive to act in a certain way. It is either financial or non-financial factors that could trigger an incentive. To a certain extent, institutions can be intentionally designed (and eventually adopted), and incentives can direct people's behaviour in

43 This view is more likely to be found among representatives of heterodox schools of economic thought, such as the post-Keynesians.

44 The aforementioned questions are connected with understanding whether the scientific theory in economics and other social sciences should be appraised by using the same criteria as those in "hard" or exact sciences, and whether they should be structured in similar ways by using similar methods (cf. Fox, 1997).

45 Named after its most prominent representative Tony Lawson (cf. Stanford Encyclopedia of Philosophy, 2003).

46 Defined broadly to encompass everything created by humans (the human society), which includes culture (in the sense of "the sum total of ways of living built up by a group of human beings and transmitted from one generation to another" (Webster's, 1996: 353)).

a certain direction[47]. Furthermore, an institutional structure conveys information (Melody, 1987), determines cognition, motivation, and emotions which influence economic actions (Dequech, 2003).

Institutions can be defined as "systems of established and prevalent social rules that structure social interactions" (Hodgson, 2006: 2)[48]. Institutions have an impact on costs of production and on transaction costs[49] (cf. North, 1990). Institutions can be said to comprise conventions and rules. Convention is "a particular instance of an institutional rule" (Hodgson (2006: 2), referring to Sugden (1986) and Searle (1995))[50]. "The term *rule* is broadly understood as a socially transmitted and customary normative injunction or immanently normative disposition, that in circumstances *X* do *Y*." (Hodgson, 2006: 3, emphasis original). Institutions or rules can be either legal or nonlegal (Hodgson, 2006)[51]. Interesting for our subject of analysis among the nonlegal institutions are moral rules and social norms (cf. Mantzavinos, 2001). Government, as well as private companies, industrial associations, civil society, church, political parties and even individuals exert an influence on both moral rules and social norms[52].

"The moral rules are, to a great degree, culture-independent since they provide solutions to the more or less identical problem of cooperation that is bound to exist whenever individuals interact. Social norms, on the other hand, being also social rules, are less fundamental than moral rules in that they regulate settings appearing mainly in specific times and places. In

47 For more on incentives see Laffont and Martimort (2002).

48 Institutional rules can be codified (e.g. constitution) or not (e.g. afterwork parties). Also, they can be known by an individual or by a community tacitly (e.g. Hodgson (2006: 4) mentions grammar, which is likely not to be known by everyone where a language is spoken, although everyone can speak the language) or explicitly (e.g. chair umpires in tennis know all tennis rules by heart).

49 Transaction costs could be broadly defined as the explicit or implicit costs of exchange of goods and services. Explicit costs are all the costs of an economic activity which can be specified based on the market value of goods and services involved in a transaction, and partly on the opportunity costs. Implicit costs encompass the part of the opportunity costs that cannot be specified, such as information gathering.

50 Hodgson (2006: 2) takes the convention to drive on the left hand side (e.g. as it applies in England) or on the right hand side (for example, in continental Europe) as an instance of traffic rules.

51 As Hodgson (2006) argues, there is a variety of terms in the realm of institutions that could be best described as legal and nonlegal institutions. This typology differs from e.g. Douglass North (1990) or Mantzavinos (2001).

52 For example, the Croatian government has financed commercials reminding citizens that taxes should be paid; a private company producing software promoted licenced software because it wanted to raise people's awareness of illegal software (both are examples of moral rules being imposed). If an individual tries to jump a queue, in some countries he is going to be warned by other people in the queue that his place is at the back of the queue (an example of a social norm being imposed).

addition, social norms are enforced by an enforcement agency external to the agent, which is usually the other individuals in the group" (Mantzavinos, 2001: 118).

Institutions both constrain and enable economic action. Their constraining function is fulfilled because they indicate what is allowed and what is not, whether explicitly or implicitly. On the other hand, the existence of institutions enables (economic) action (even predominantly enables such action – see Searle, 2005) because it implies a relatively predictable social environment where uncertainty is minimized, and where (economic) actors do not have to renegotiate the basic "rules of the game"[53]. One can distinguish between three dimensions of institutions: regulatory, normative, and cognitive (cf. Scott, 1995). The regulatory dimension is the most obvious one, as it consists of enactment of institutions and their implementation (e.g. a parliament enacts laws, whose implementation is monitored by police and the judiciary). The normative dimension of institutions presupposes the understanding that they exist, and that sanctions may be imposed when they are violated. The third aspect of institutions is a cognitive one, and it regards acceptance (awareness) of institutions. It follows from these elaborations that, in order for individuals, particular groups in a society, or society as a whole to adopt them (if the regulatory level of the institutions is already in place), one has to learn how to deal with them. Learning inevitably takes time. If any of the three dimensions is missing, institutions cannot fulfil their function: that is, they cannot be adopted.

Institutions can be differentiated based on how quickly they change. There are slow-moving institutions and fast-moving institutions (e.g. laws may change faster than social norms) (Roland, 2004). Understanding how policies can affect the institutional structure requires an understanding of the permanency and stability of institutions (Hollingsworth, 2000). Since legal rules as a whole are necessarily incomplete, since they cannot predict and specify every state of nature, nonlegal institutions are there to fill the gap. Nonlegal institutions are far more pervasive than legal institutions. These can be either in conflict or complementary, thereby enhancing the efficiency of the socio-economic system (cf. North, 1990). Nevertheless, in order for the socio-economic system to function properly, legal institutions must be enforced. Because of the character of the legal institutions as a public good, enforcement must be left to a third-party. "Third-party enforcement means the development of the state as a coercive force able to monitor property rights and enforce contracts effectively" (North, 1990: 59).

53 As Searle (1996: 94) puts it, "[i]nstitutional power – massive, pervasive and typically invisible – permeates every nook and cranny of our social lives, and as such it is not a threat to liberal values but rather the precondition of their existence".

In order to understand how institutions emerge and how they work, one has to understand the interplay of agency and institutional structure, as Hodgson (2006) reminds us. If, according to the aforementioned elaborations, we define institutional structure as a system of institutions at a given time in a given socio-economic environment, then it remains to be explained how the (human) agency in question acts. Institutional structure could be a product of actions of individuals (as in the new institutional economics), but what remains to be explained is how institutions determine individuals' behaviour (Setterfield, 1993). One of the crucial answers lies in culture – loosely defined as a set of shared patterns of thought and behaviour, it constitutes and guides economic actors and their action (Dequech, 2003: 467). As both culture and social capital share these characteristics, we may conclude that they form a robust and long-lasting social structure, along with legal and nonlegal institutions. Furthermore, market action does not necessarily have to be guided by institutions alone, but also by culture and social capital. Hence, history matters, since the institutional structure is intertwined with culture and social capital.

Given the fact that production and innovation activities must be planned in advance and the fact that they take time, informational and cognitive functions of institutions are very important, but they cannot eliminate risk, ambiguity and uncertainty that entrepreneurs encounter in the market. While risk is a situation in which the probability of an event can be ("objectively") calculated, the notions of ambiguity and fundamental uncertainty denote situations in which the actors involved have less knowledge or no knowledge at all about future events. "Ambiguity is uncertainty about probability, created by missing information that is relevant and could be known" (Camerer and Weber, 1992: 330, as quoted in Dequech, 2000: 45). Consequently, ambiguity can be decreased (or even reduced to risk) when relevant information is revealed (with the passage of time). As opposed to ambiguity, fundamental uncertainty is a situation when actor(s) have no knowledge about future events that may happen[54]. In other words, they are not even aware of potential events that may make an impact on the observed phenomenon (transaction, object, etc.). However, the degree of fundamental uncertainty is variable[55] (as Keynes claimed, too). It is dependent on the individual (his or her cognitive capacity) and on the socio-economic context that he or she lives in (Dequech, 2000)[56]. Our cognitive capabilities – including what Keynes

---

54 And therefore cannot attach to them any probability, objective or subjective.

55 Hence, it is not necessarily equal to complete ignorance (this kind of theoretic approach could be termed nihilistic – cf. Dequech, 2000).

56 When the aforementioned notions are defined in this way, it becomes clear that both Knight (who was the first to refer to the difference between risk and uncertainty) and Keynes (who attributed a great importance to uncertainty), when writing about uncertainty, actually wrote about ambiguity and fundamental uncertainty. In the words of

termed “animal spirits” –, as well as legal and nonlegal institutions, guide us through a world of risk, ambiguity and uncertainty. Their stability (as opposed to frequent changes), complementarity with other elements of the system, and quality will constitute a stimulative environment for innovations, provided the relevant institutions are widely adopted.

Knight, “[i]t is a world of change in which we live, and a world of uncertainty. We live only by knowing *something* about the future; while the problems of life or of conduct at least, arise from the fact that we know so little. This is as true of business as of other spheres of activity. The essence of the situation is action according to *opinion,* of greater or less foundation and value, neither entire ignorance nor complete and perfect information, but partial knowledge. If we are to understand the workings of the economic system we must examine the meaning and significance of uncertainty; and to this end some inquiry into the nature and function of knowledge itself is necessary” (Knight, 1921: III VII 6).

3. Financing Innovations in Theory

The principle goal of this chapter is to conceptualise the financing of innovations, which is done in its last part (ch. 3.4.). On our way there, we have to review a number of concepts and theories. We use them on the one hand in order to achieve this synthesis, and on the other to better understand innovations and innovative performance. They are divided into three main parts: economics of innovation (ch. 3.1.), theory of the firm (ch. 3.2.) and financial theory (ch. 3.3.). Economics of innovations is a branch of economics that tries to explain innovations. In that chapter, we are interested in the determinants of innovations at the micro, meso and macro levels. The national system of innovation concept will be revisited. Because the innovations we study are produced by firms, we next review the theories of the firm. Finally, since the NSI overlaps with the financial system, in the following chapter we look for theories that can contribute to our understanding of financing innovations. This chapter on financial theory (3.3.) comprises three main sections: common ways of financing enterprises (3.3.1.), theories of capital structure (3.3.2.), and finance and the real sector: a complex relationship (3.3.3.). The theories of capital structure section is divided into two sections: neoclassical theory of the capital structure (3.3.2.1.), and new institutional theory of the capital structure (3.3.2.2.).

3.1. Economics of Innovation

Economics of innovation is a field of economics dealing with innovations. Presenting its main theories will help us on our way towards building a concept of financing innovations. The question of how innovations are financed needs to be raised both at the level of the firm and at the societal level. In order to draw a conclusion about financing innovations at the societal level, one needs to know how different innovations are financed at the firm level. As has already been pointed out, there are both firm-specific influences, as well as specificities at the macro level. Below, we shall review three large groups of economics of innovation concepts and theories: 1) theories in connection with long-run economic cycles, 2) neoclassical theories, and 3) institutional theories. These are generally recognised as relevant in the literature and may help us in further developing the approach pursued here.

The concepts and theories from the economics of innovation offer explanations of innovations at the micro level (level of the firm), meso level (level of industries or that of interaction between firms), and macro level (level of the NSI), as answers to relevant questions. The questions that can be raised at the firm level are the following: who innovates (e.g. the individual entrepreneur, who has his own firm; the entrepreneur who operates in a larger company – the so-called intrapreneur; the innovator who (still) operates with no affiliation to a company; a group of people from one company or from different companies/NGOs/government bodies/other institutions)? Can this innovation be properly managed (in terms of marketing, financing, and other business functions)? What are the effects of innovation on the firm that marketed it (e.g. increase in market share)? What are the most important obstacles to innovation activities? Is there any government support for innovation activities in the firm (not only financial support, but also advisory and marketing support?) Have the innovations been patented or not? Which other – formal or strategic – kinds of protection does the firm use in order to protect its innovations? Are innovations incremental or radical, and what is the (in)tangibility of innovation activities? At the meso level, one primarily needs to see whether the firm has cooperated on its innovation activities (with other firms/institutions/individuals not affiliated with this firm). Furthermore, the firm interacts with its environment, absorbing information which may be crucial as an input into innovations. Finally, there is the macro level, which raises the following topics that should capture our attention: do the innovations predominantly belong to a wider stream of innovations (e.g. to a Kondratieff wave – see below)? Is there a conception and strategy at country level that stimulates innovations? Is the macroeconomic, social and cultural environment generally conducive to innovations? What does the innovation policy look like?

The determinants of firms' innovativeness found in the literature speak in favour of the holistic view of the innovative process – that is, the one that encompasses the capabilities of firms (not necessarily depending on their size), as well as connections with their meso and macro environments. According to Teece (2000), there are two basic determinants of innovativeness of firms, and these are business strategy and business environment. In addition, four other factors could be discerned (cf. Teece, 2000): 1) human resources, capabilities and quality of decision making; 2) organisational structure and incentives; 3) organisational culture and values; 4) sources of financing and links (strategic alliances, networking, clusters). While drawing conclusions from the analyses of firms with successful technological innovations[57], Freeman remarks on the characteristics that such firms should have (Freeman, 1982: 112):

> 1. "Strong in-house professional R & D. 2. Performance of basic research or close connections with those conducting basic research. 3. The use of patents to gain protection and to bargain with competitors. 4. Large enough size to finance fairly heavy R&D expenditures over long periods. 5. Shorter lead-times than competitors. 6. Readiness to take high risks. 7. Early and imaginative identification of a potential market. 8. Careful attention to the potential market and substantial efforts to involve, educate and assist users. 9. Entrepreneurship strong

57 Or "technical", as Freeman calls them.

enough effectively to coordinate R&D, production and marketing. 10. Good communications with the outside scientific world as well as with customers".

We do not know which level of patent protection is optimal, since this level varies internationally, and may be more appropriate for some industries (e.g. pharmaceutical industry) than for others (e.g. in which knowledge is tacit) (Crespi, 2004). Furthermore, there are no clear findings as to the link between market concentration and innovativeness of firms (Cohen and Levin, 1989; Crespi, 2004). The size of the firm may not necessarily have a bearing on its innovativeness. Large firms' importance for some industries, as well as the importance of SMEs in general, is indisputable (Freeman, 1982: 131). It is a well-known fact that, because of the economies of scale, there is a predominance of large firms in certain industries (e.g. aircraft manufacturing industry, car manufacturing industry). However, this does not preclude SMEs from entering even these industries, for example as suppliers of various parts.

Innovations can be determined by cyclical fluctuations that occur over a longer period of time. According to a thesis[58] presented by van Gelderen, and later by its much better known proponent Kondratieff, capitalist economy (GDP) has cyclical fluctuations (phases) which are regular. These phases are: 1) prosperity, 2) recession, 3) depression and 4) recovery. Furthermore, each Kondratieff cycle lasts about 50 years (Sundbo, 1998). However, instead of being confined to economic phenomena (as researchers of the long waves usually assert), the long waves are also a social/institutional and political phenomenon – which follows from the fact that the economic system is only embedded in society. Moreover, it is questionable whether this phenomenon occurs simultaneously around the world or only in (some) capitalist economies of the western hemisphere. Although many factors have been identified by researchers as causes of the long (Kondratieff) waves[59], it seems more reasonable to presume that the causes could be many. However, some or all of the proposed causes overlap. Needless to say, it always boils down to supply and demand. The question should really be what stands behind them. As for the innovations, they are, according to Sundbo (1998), the best known cause of the Kondratieff cycles, as first introduced by Schumpeter[60].

58 Some call it a theory (cf. Sundbo, 1998).

59 Sundbo (1998) lists several possible causes identified in the literature: innovations, variations in the money or gold supply, investments, demand variation and social and psychological factors, food and raw materials, the development of population, demography and employment, and wars.

60 Schumpeter is definitely one of the most important economists in history, and he is especially important for the theory of innovations. His view of economics' problem was that of structural change, with the entrepreneur (innovator) being at the centre of his research (cf. Drucker, 1983).

As the Kondratieff waves are a macroeconomic phenomenon, one should also be able to explain the emergence of (certain types of) innovations at the level of particular sectors or firms. In order to do this, one can make use of Dosi's concept of technological paradigms. They "define the technological opportunities for further innovations and some basic procedures on how to exploit them. Thus they also channel the efforts in certain directions rather than others" (Dosi, 1988: 225). These directions can also be termed "technological trajectories", as Dosi (1988) suggests. Specific technological trajectories can be followed both at the level of industries and at the level of firms. As Dosi (1988) hypothesises, differences in innovations between industries depend upon "technological opportunities, appropriability regimes and demand patterns (…) together with the sector-specific nature of the knowledge on which innovations are based" (Dosi, 1988: 231). In addition to technological paradigms and technological trajectories, one can also distinguish technological regimes – in Geels' explanation, it is a phenomenon that emerges "[w]hen different firms share particular routines" (Geels, 2005: 58).

Five Kondratieff waves have been identified in the literature so far (see Sundbo, 1998; Santarelli, 1995), starting from the first wave in the second half of the 18th century. Santarelli (1995) identifies five technological styles, each coupled with a certain financial style during a particular long wave (see table 3).

Table 3: Technological styles and financial styles in historical perspective

Long waves	Technological styles	Financial styles
I	Specialised machinery for the textile industry ('water style')	Use of merchant capital in industrial activities
II	Railroading and transportation (exploiting *motive* steam power)	Stock exchange market
III	Chemicals, electricity, steel	International gold standard, Universal Bank
IV	Internal combustion engine, petroleum refining ('Fordist' style)	Central Bank control function
V	Information technology, biotechnology	Non-banking financial institutions (venture capital, pension funds, etc.)

Source: taken from Santarelli (1995: 81).

Each of the long waves can be viewed as characterised by a match between technological, economic and institutional subsystems. In the periods spanning the end of the current and the beginning of next long wave, the three mentioned subsystems do not act in concert with each other[61]. Similarly to the long wave theory, Perez (2002) identifies five technological revolutions, which overlap with the five long waves. These technological revolutions gradually diffuse and change not only the predominant system of production, but also the systems of governance, culture and ideology. Perez suggests that each of the technological revolutions starts with a particular "big bang" event that initiates the technological revolution. She associates the event with a year in which a major invention/innovation is either introduced to the market or tested/announced. Logically, each technological revolution had a core country or countries, in which the revolution and the process of diffusion of innovations started. A common characteristic of the approaches by Santarelli (1995) and Perez (2002) is that they claim that innovations are usually clustered around a central technology. Diffusion of technological revolutions goes through phases which are marked by various degrees of connectedness (coupling) or disconnectedness (decoupling) of financial capital and production capital. A functional separation of real and financial sectors, as well as the higher mobility of the latter, contribute to the instability of the capitalist system of production. As technological revolutions unfold and diffuse through the socio-economic system, financial capital at first supports (finances) new technologies, before decoupling from the real sector and creating an asset bubble. Namely, profitable opportunities in the financial sector leave financial capital to lead a life of its own, before the value of these investments in the financial markets comes down to the more fundamental values of the real sector. A recoupling of financial capital with production capital follows, before the inevitable decoupling – but by now the technological revolution has already spread, and a new one begins (Perez, 2002).

Each of the Kondratieff long waves, Sundbo (1998) argues, is characterised by a particular innovation system, meaning a particular way of organising the productions of innovations. Furthermore, this new innovation system paves the way for a new paradigm in innovation theory. Sundbo (1998) distinguishes three different basic theories, each in its own Kondratieff wave, starting from the recovery phase of the third one. These theories are: 1) the basic entrepreneur theory, 2) the technology-economics innovation theory, and 3) the basic strategic theory of innovation. He identifies the first and the second theory as paradigms, and defines the third one (present in the fifth Kondratieff wave) as a potential paradigm[62], since it is not yet fully developed. The basic entrepreneur theory over-

61 Dosi and Orsenigo (1988: 29) assert the former is a case of "smooth configuration", while the latter situation can be described as exhibiting "structural mismatchings" between the three aforementioned subsystems.

62 Economic paradigms are usually formed as parts of schools of economic thought (cf. Cvijanović, 1999) and should therefore be defined as such. Sundbo's (1998) classifica-

laps with the third Kondratieff long wave (1883-1937), as Sundbo (1998) claims. It is based on entrepreneurship as an innovation determinant, the founder of the company being the agent of innovation. However, this theory later recognises that entrepreneurship could also be embedded in a larger company (so called intrapreneurship). An entrepreneur is an individual who acts on his own, disregarding his social surrounding. However, with the development of the intrapreneurship concept, the literature indeed recognises that that individual utilises social contacts. The technology-economics innovation theory was developed during the fourth Kondratief wave (1937-1983), as Sundbo (1998) asserts. Technology development figures as an innovation determinant, and the technician is the agent of innovation. This theory is more focused on invention than on innovation. "The modern version of this theoretical tradition sees no significant distinction between incremental and radical innovations. It mainly deals with incremental innovations, in the belief that the radical innovations are rare and impossible to predict" (Sundbo, 1998: 175). The theory is focused on technological innovations, that is, on product and process innovations. Hence, it disregards social innovations (organisational and market innovations), and only marginally considers service innovations. "The present period – starting around 1983 and lasting until around 1994 or slightly later – is a recovery period, characterized by extensive innovative activity which can form the basis for a later prosperity period. A new innovation system is now being formed on the basis of a new paradigm", Sundbo (1998: 44) concludes. That is, the period in question is likely to be the fifth Kondratieff wave. The basic strategic theory of innovation, according to this view, has a big chance of becoming a predominant paradigm in the aforementioned period. Market-oriented strategy is the main determinant of innovation. "The innovator in this case is the whole organization. In this theoretical framework the interaction with the outside world must be extensive. The whole idea of the theory is precisely that the outside world, both the market in the narrow sense and society in the wider sense, is involved in interaction with the companies in the innovation process" (Sundbo, 1998: 181). The latter theory acknowledges different types of innovations and is thus the broadest of the three mentioned theories, as well as complementary to the NSI approach.

According to Grupp (1998: 51), we can basically distinguish three contending areas, in the explanation of which the innovation theories diverge (Grupp, 1998: 51): 1) technical change – is it endogenous or exogenous?; 2) economic system – is it in a state of equilibrium or in a developing process?; 3) interaction of economic actors – is it primarily based on a rational activity or could it be a self-

---

tion disregards this because it is primarily based on the question of who produces innovation, and not on the question of which methodology a particular theory uses.

regulating empirical learning process? We can identify two basic groups of innovation theories: neoclassical and institutional (cf. Grupp, 1998).

Neoclassical economics has never really gone into any depth regarding innovations, treating them as a sort of "black box" (cf. Rosenberg, 1982, as quoted in Tremblay, 2004). It is typical of the neoclassical microeconomic approaches that they normally address innovations from three different angles or theories (Grupp, 1998): 1) competition (e.g. Arrow and Von Hayek), 2) decision theories (e.g. Kamien and Schwartz) and 3) game theories of innovation (e.g. Scherer, Dasgupta and Stiglitz). Ad 1) the approach that puts variables like the size of a company and market structure into the forefront as determinants of innovativeness (Grupp, 1998: 56-57). Ad 2) these innovation models introduce new variables, like "uncertainty, dynamics, externality plus other aspects as typical for the innovation event." (Grupp, 1998: 59). Ad 3) "Game theory approaches make it possible to model certain interactions between the R&D decisions of the company, and thus can be regarded as an extension of the decision theory approach." (Grupp, 1998: 60). As for the neoclassical macroeconomic approaches, one can distinguish two theories: 1) neoclassical growth theory, and 2) new growth theory. While the former treats technical advance as exogenous, the latter approach endogenizes technical advance, while incorporating rational expectations (Grupp, 1998: 62). Although neoclassical innovation theories can be elucidating, they generally "cannot plausibly construct all of the aspects important for an understanding of innovation" (Grupp, 1998: 63). Rosenberg (1976, as quoted by Grupp, 1998: 64) was critical of neoclassical innovation theories, mentioning that the neoclassicists neglected important connections between science and technology, and that they disregarded the findings of technology historians and various technical disciplines. By definition, to study innovation means to study change, and the neoclassical school of thought cannot offer an adequate explanation because it is influenced by an ergodic[63] axiom (cf. Medić and Cvijanović, 2001). Besides, the neoclassical approach is a mechanical one[64], a closed-system approach[65]. In the words of Dunn (2000: 345), "[n]eoclassical economics is in crisis and dying, and the reason for this is the application of closed systems reasoning to open systems".

63 "[O]f or pertaining to the condition that, in an interval of sufficient duration, a system will return to states which are closely similar to previous ones: the assumption of such a condition underlies statistical methods used in modern dynamics and atomic theory" (Webster's, 1996: 483).

64 In the Newtonian sense of the word – the economic models of this school of thought function as a machine, since an event of type A always produces an event of type B. These models can also be called deterministic.

65 A closed-system approach is an approach that regards any phenomenon as determined by a limited number of variables, that is, without any influences from the environment.

Before we tackle the institutional approaches of our interest, let us consider concepts that throw light on some aspects of the innovation systems. From that standpoint, there exist either (two) linear models of innovation development or an interactive model (cf. Račić et al., 2004 and Fischer, 2001). We can identify two linear models in the literature: 1) (science) push model, and 2) (market or demand) pull model[66]. In the former model, innovations (or inventions) are based on research and development activity[67]. Finally, they end up on the marketplace, which, in this model, figures only as a kind of absorber of innovations. The linear model is characterised by an absence of any other linkages or feedback between phases of innovation activities (Edquist and Hommen, 1999). Some industries (and their innovations) have been more based on scientific knowledge than other industries (cf. Freeman, 1982: 107). However,

"this does not imply the acceptance of a 'linear' model of R & D with a simple one-way flow of ideas from basic science through applied research to development and commercial innovation. On the contrary, there has always been and there remains in the modern science-related industries a strong reciprocal interaction between all these activities, and in particular a powerful influence of technology upon science" (Freeman, 1982: 107)[68].

The latter model recognises the important role of the market in the sense that the tastes, preferences, wishes and, finally, demand of the "consumers" determine which kinds of innovations will be produced. Demand can also be spurred by the government, as Freeman observes:

"The strength of the German demand for 'ersatz' materials to substitute for natural materials in two world wars spurred on the intense R&D efforts of IG Farben and other chemical firms. The strength of the military/space demand in the American post-war economy stimulated the flow of innovations based on Bell's scientific breakthrough in semiconductors and the early

66 Schmookler was the first economist to pose the hypothesis that innovation is dependent on demand (cf. Brouwer and Kleinknecht, 1997).

67 There are two basic types of research activity: fundamental and applied research.

68 The (science) push model seems to have been on the minds of policymakers who voted for the Lisbon Strategy of the EU in March 2000. It set for the EU the goal of becoming "the most competitive and dynamic knowledge-based economy in the world, capable of sustainable economic growth with more and better jobs and greater social cohesion" by the year 2010 (World Economic Forum, 2006: 1). One of the goals set by the Lisbon Strategy is that expenditures for research and development in each country of the EU should rise up to 3% of its GDP. Although this specific task has been accompanied by a range of other tasks or measures that were meant to support it, it is clear by now that this target has not yet been met (cf. World Economic Forum, 2006). First of all, such an approach emphasizes R&D expenditures, and not the final result (innovations). After all, R&D expenditures do not necessarily have to "materialize" as innovations. Secondly, not all innovations require R&D (e.g. service innovations), and not all EU countries have industries that would absorb such a level of R&D expenditures.

generations of computers. The urgency of British wartime needs spurred the successful development of radar of all kinds while the German Government sponsored the development of FM networks, as well as radar" (Freeman, 1982: 109).

However, as Freeman (1982) asserts, a theory that strives to be close to reality should take into account both directions of influence upon innovation development. Also, an innovative firm with good management can obtain a competitive advantage only if it utilises both R&D from one side, and market information from the other. At this point it is worth noting the compatibility of Keynes's[69] ideas with the demand pull model. Although he never studied innovations, the link between his theory and the theory of innovations can be established through his assertion that demand should be stimulated in order for the capitalist economy to be kept at the full employment level. Taking into account the aforementioned demand pull model, one can come to the conclusion that aggregate demand movements have consequences that affect not only output and employment, but also innovation[70] (cf. Brouwer and Kleinknecht, 1997)[71].

As opposed to the linear models of innovation development, there is also an interactive model in which a systemic position is taken – innovation development is a result of many interacting factors (and institutions). According to this view, not only is the innovation process not unidirectional (cf. Račić et al., 2004), as both of the aforementioned linear models basically claim, but it is also determined by a societal context in which innovation activities are taking place. Although the interactive model could be represented on the level of the innovation activities of the firm, with some connections with the environment (cf. Fischer, 2001), we will be focusing on the approach called the national systems of innovation (which is broader in scale and scope).

---

69 His great contribution to economic theory was the negation of Say's law and the assertion that economy can be in a state of equilibrium even below full employment level. Keynes claimed that savings do not necessarily equal investment, so unemployment is a consequence of the capitalist system. In order to decrease unemployment and avoid severe depressions, government should stimulate aggregate demand (consumption plus investments). As opposed to Schumpeter (who was devoted to studying economic change), Keynes's concept of an economy is that of a closed system in equilibrium (cf. Drucker, 1983).

70 This has important consequences for economic (or rather innovation) policy, which should take into account the effects of aggregate demand on enterprises' innovativeness.

71 In contrast to Schumpeter, who believed that the individual (entrepreneur) was a force of change, in Keynesian economics "[t]he macroeconomy... is everything, with individuals and firms having neither power to influence, let alone to direct, the economy nor the ability to make effective decisions counter to the forces of the macroeconomy" (Drucker, 1983: 2). Besides, Keynes considered innovation as an exogenous variable, which is again contrary to Schumpeter (cf. Drucker, 1983: 3).

We shall call alternative approaches in innovation theories institutional. In broad terms, their characteristics are that they recognise and study institutions as determinants of socio-economic activities, along with other socially embedded structures, such as conventions, rules, and habits. As opposed to the neoclassical approaches, institutional innovation theories recognise their objects of analysis to be open systems.

Each country has a specific innovation performance, differing not only in the quantity of inventions/innovations, but also in their quality (technological and other characteristics) (cf. Casper and van Waarden, 2005; Nelson, 1993). Institutional approaches to innovation, as seen here, focus on the interdependence between institutions, organisations and innovations. Economists of the institutional tradition have come up with the concept of the national system of innovation[72] (NSI). However, their attention was primarily concentrated on institutions as societal resources (like finance, human resources, knowledge), and not on institutions in the sense of socially rooted structures. Even when recognised as such, the normative and cognitive aspects of institutions were missing from the analysis (Geels, 2005: 35). Since the essentials of this approach have already been described[73], we shall briefly revisit them at this point, in order to stress some of its features. Freeman (2006) notes a distinction between the narrow and broader conceptualisations of the NSI. While the former incorporates only the institutions that conduct R&D and disseminate its results, the latter comprises a range of institutions and socio-economic factors that have a bearing on innovations. Considering only R&D means focusing only on a single segment of innovation activities. Švarc (2009) observes that the new understanding of the NSI is broader, ingrained in the socio-economic system. The new concept of the NSI incorporates the business sector, intermediary institutions, education and scientific system, political system and socio-economic environment (Technopolis, 2000, modified and extended by Kuhlmann – as in Švarc, 2009). Although a systemic perspective on innovations, such as the NSI, might disregard the role of agency as a driver of change because of seeing institutional structure as overly powerful (cf. Meeus and Oerlemans, 2005: 61), it does not exclude the possibility of change from the analysis. Moreover, an analysis of relevant institutions and agents can provide answers as to possible sources of change.

The national systems of innovation perspective is complemented by the network alignment approach, as advanced by von Tunzelmann (2004). The network

72 It is otherwise called the national innovation system.

73 See section 2.4.2.

alignment approach[74] stems from the evolutionary/institutionalist recognition of complementarities between elements of the innovation "ecosystem". These are the necessary institutional and resource prerequisites for an innovation "ecosystem" observed at the level of national economies, but its output is visible at the firm level. Global trade and financial flows (foreign direct investment and portfolio investment) have strongly influenced the national economies, in particular the small and open economies in transition – whose national systems of innovation have not yet been fully set up (see below). Von Tunzelmann (2004) distinguishes between three main types of networks – functional networks, resource networks, and geographical networks. Functional networks comprise various elements of the national system of innovation – at the micro, meso, and macro levels – that serve a certain function, like marketing, finance, higher education institutions, etc. Resource networks comprise linkages of the firm with different resources required for production. By studying geographical networks, one can obtain an answer to the question of where influences on (local) production stem from.

Similar to the idea of the NSI, there is a concept of the idea-innovation chain, which comprises "loosely to tightly connected – networks of (sub)organizations involved in the various functional domains (classes of required competences) of the innovation process: basic research, applied research, product development, manufacturing, quality control and marketing" (Casper and van Waarden, 2005: 9-10). The idea-innovation chain, according to Casper and van Waarden (2005), consists of five different layers. Taking a top-down approach, the first layer is represented by meta-institutions (regulations, policies, cultural values); the second layer is called national institutional frameworks, and is comprised of finance, banking, industrial relations and company law, education and training, as well as inter-firm relations; the third layer consists of sectoral resources (capital markets, labour markets, and knowledge and information); the fourth layer is called the idea-innovation chain, comprising firms and universities in their various organisational forms, which contribute to firms' innovativeness; the fifth level of analysis contains performance indicators: ideas, patents, new products and trade balance.

Whatever theory or model is accepted as appropriate, one should not neglect the notion of path dependency in (technological) innovations, as Dosi (1988) defines it – implying that innovative activities show cumulative characteristics. They are cumulative because the innovators/firms, who are the actual actors of

74 Network is hereby seen as one of the forms of governance, together with firms and states (being designated as hierarchies), and markets. "In a strict sense networks are here seen as all kinds of interrelations (among individuals, organizations, etc.) without market exchange or the *direct* exercise of power" (von Tunzelmann, 2004: 24).

innovative activities (functioning in a certain innovation system and being locked in into certain technologies[75]), use specific information and knowledge easily accessible in their environment. Dosi (1988: 225) basically claims that history matters, which means that firms are inherently bounded in their innovative possibilities. "[t]echnological progress proceeds through the development and exploitation of both public elements of knowledge, shared by all actors involved in a certain activity, and private, local, partly tacit, firm-specific, cumulative forms of knowledge" (Dosi, 1988: 226).

Due to the various determinants of innovation activities (and therefore innovativeness), the application of the NSI seems to offer a solution to the problem of how to observe innovative performance on a macro level. But rather than proposing that there is a unidirectional and deterministic influence of the NSI on firms, it is well established that there is an interaction between them (cf. Hollingsworth, 2000). The NSI influences the capacity of firms to innovate, although the outcomes of some firms may be different from the "average". The overall performance of the NSI cannot be measured with precision, but only approximated by a series of indicators capturing the inputs and outputs of the NSI, as in the case of the European Innovation Scoreboard (European Commission, 2008; See also table in Appendix III).

In order for something new to be created, an individual or a firm must apply their cognitive capabilities – knowledge –, and in this process new knowledge is created, tacit and/or codified. By inventing, firms (or individuals) create new knowledge (Foray, 2004). Knowledge sources are crucial in explaining innovation performance (cf. European Commission, 2008). But the question is how firms get the knowledge needed to produce innovations, and what kind of knowledge is needed for innovations. The first question is a matter of the source of knowledge that firms utilise in order to incorporate it in their innovation (activities). There may be many sources of knowledge, some being internal to the firm, such as R&D (Foray, 2004) and other innovation activities that have already been mentioned. External sources of knowledge may come from knowledge spillovers and cooperation (Knell and Rojec, 2007). Cooperation is necessary, especially when firms' competencies do not suffice for the development of some products (van Waarden, 2005: 232-233). Cohen and Levinthal (1990) argue that outside knowledge is crucial in the development of firm's innovative

75 The concept of lock-in (as regards technology) is usually applied to firms. A firm could be said to be locked in into certain (old) technology because of the costs required to change it. Similarly, it can also happen because its employees have knowledge specific to the technology currently in use (either they would need more time or the firm would have to incur costs in order for them to learn how to use new technology).

capabilities. These authors claim that prior related knowledge is a critical determinant of innovative capabilities, and call this "absorptive capacity". Knowledge can spill over or leak (Afuah, 2003), since it is difficult to control or to keep only inside a firm (Foray, 2004). Therefore, by producing and using knowledge, there are both private and social benefits. Knowledge is widespread in a society, and not concentrated, as it used to be in Adam Smith's time (Vercellone, 2007). It follows that a firm may tap from this, essentially public, resource as well. In all this, the role of the NSI is crucial in facilitating knowledge creation and knowledge transfer. But knowledge may spill over from firms that are across the border. There could be formal and informal channels of international knowledge transfer – the first four of the following being formal, the last five of the following being informal channels of technology transfer (Knell and Rojec, 2007: 40): "1. Foreign-owned firms; 2. Joint ventures; 3. Technical assistance programmes and other forms of aid; 4. Technology licencing contracts; 5. Imports from upstream suppliers, (...); 6. Exports to downstream customers; 7. Research collaboration; 8. Subcontracting agreements; and 9. People, including hiring skilled labour, education, training and trade fairs". "For most countries, foreign sources of technology are of dominant importance (90 percent or more) for productivity growth... Foreign sources of technology are more important for small (and relatively poor) than for bigger countries" (Keller, 2004: 776). The second question is related to the heterogeneity of innovations. Innovations may be incremental or radical, and in both cases based on own and/or external knowledge. This may have implications on the financing of innovations, as we shall later discuss.

3.2. Theory of the Firm

In order to understand the financing of innovations, financing theory should be viewed in conjunction with the theory of the firm. Different views and perspectives on the theory of the firm are reflected in the theory of finance. We shall first consider the neoclassical theory of the firm followed by its intellectual opposite: the new institutional theory of the firm on the one hand, and the institutional (competence-based and resource-based) theory of the firm on the other. Market-based analyses are not enough and definitely not adequate for studying innovation. It is the competence-based theories of the firm that are cognitively valid and close to reality. Our aim is to outline the theories of the firm by distinguishing the common underlying ideas. Finally, we discuss the relevance of institutional theories of the firm for studying the financing of innovations.

The neoclassical theory of the firm could be said to be the oldest of the three mentioned in the previous passage. It concentrates more on the market and less on production issues, and is, to a large extent, reduced to the categories of cost, revenues and profit. According to this theory of the firm, the firm is no more than a "black box" (Demsetz, 1997), since it is characterized as a production function that transforms input into output. The one and only motive for a firm's existence ("the bottom line") is profit, and there are no transaction costs, agency problems, or informational problems that would impede this goal. Firms, as well as households and individuals as economic agents, all behave rationally. The neoclassical school of economic thought has three main assumptions for a firm in perfect competition[76] (Demsetz, 1997: 428): 1) free functioning of the markets, 2) perfectly informed economic agents on prices and technology, 3) effective control of the assets by their owners. In such a world the firm becomes no more than a homogenous entity subject to "market dictate", as well as a price taker. The market form of governance is the only "legitimate" form of governance, since interference of the state into the market is wrong and undesirable. Moreover, the whole neoclassical theory is focused more on the process of market allocation, rather than on production and distribution. Hence, neither the management nor the ownership of the business make any difference. The restrictive assumptions of this theory diminish its value when it comes to explaining innovation activities and their financing. To comment on some limitations – we ought to be able to explain what is going on inside a firm in imperfect competition, as well as in an environment in which the market forms of governance are not necessarily predominant.

The new institutional theory of the firm follows the old institutionalism in that it recognises the institutions, but in many ways it emanates from neoclassical economics. It basically subsumes both the contractual (or contractarian) and transaction-cost approaches (cf. Hodgson, 1998), since they share the views that firms consist of explicit or implicit contracts (or mutual arrangements – see Eisenberg, 1999 for discussion), and that the relationship between costs and benefits in a world of imperfect information (cf. Hodgson, 1998) is fundamental for explaining the existence and functioning of the firm. The founder of this school of thought is Ronald Coase[77], while its chief successors and developers are Armen Alchian and Harold Demsetz, Michael Jensen and William Meckling (1976), and Oliver Williamson. The most important innovation by the representatives of this theory is that they analyse the firm from the perspective of the human agency, but they believe it is limited or incentivized by institutions, where transaction is the basic unit of analysis. The firm is more than just a homogenous element in the economy intertwined with (and functioning under the

76 Perfect competition is the basic model of the market structure for the neoclassicists, while imperfect competition (monopoly, duopoly, oligopoly) is more of an exception. Even in the most common and famous book (in the tradition of the neoclassical synthesis) for teaching the Principles of Economics – Samuelson and Nordhaus (2001).

77 In 1937 he published the famous article "The Nature of the Firm" in *Economica*.

dictate of) markets, as represented in orthodox neoclassicism. Another methodological novelty is the introduction of bounded rationality (Medić and Markov, 2003), instead of perfect rationality. Furthermore, information is imperfect, and both risk and (fundamental) uncertainty come into play. This has brought a range of issues into the theoretician's focus, such as issues of corporate governance, capital structure, as well as organisational issues. However, some things have remained beyond the reach of the new institutional economics, as it is mostly based on studying legal institutions alone, and has often used comparative statics, as typified by its most famous proponent, Oliver Williamson: "(...) the study of economic organization in a regime of rapid innovation poses much more difficult issues than those addressed here ... Much more study of the relations between organization and innovation is needed" (Williamson, 1985: 143-144, as quoted in Hodgson, 1998: 34). Because of the lack of analysis of nonlegal institutions (but not by all new institutional economists – see e.g. North, 1990), social capital (e.g. in the form of trust) cannot be recognised as a major factor guiding economic agents' actions. Since dynamic analysis is not utilised by this theory and individuals' preferences are given (Hodgson, 1998), one cannot fully comprehend the evolutionary development of firms. In the same vein, it cannot explain the complexities of the innovation process, although it does contribute to an understanding of some of its aspects. As noted by Dunn (2002), due to the suppositions of the new institutional theory of the firm that time is ergodic[78] and that economic agents are, in fact, rational (although they may act opportunistically), while trying to minimize transaction costs and maximize their benefits, the agents make decisions on a different basis than their counterparts in a non-ergodic environment[79].

Finally, we turn to perhaps the youngest group of theories of the firm: the institutional theories[80, 81]. "A firm's strategy orients and rationalizes its actions. The way a firm is organized, its structure, defines how it is set up to follow its strategy and work out the details" (Nelson, 1994: 244). In the classic paper by Wernerfelt (1997), resources are very much equal to assets, be they tangible or intangible; it is through them that the firms build barriers against other firms (so

78 This opinion is certainly not shared by all of the new institutional economics theoreticians (like the Nobel Prize laureate Douglass North), but Oliver Williamson implicitly uses this view in his analyses.

79 Namely, as the future is inherently and fundamentally uncertain, economic agents' decisions of today may prove to be wrong tomorrow.

80 Hodgson (1998) believes that institutional (evolutionary) theories are a subset of the competence-based (resource-based) theories of the firm. However, we shall group the two latter categories under the common denominator heading of institutional theories of the firm.

81 In this instance, we use the plural because we attempt to describe their common characteristics.

called "resource position barriers"). A resource-based view can be used to explain competitive advantage in the long-run, as Peteraf (1997) does. In her model, a firm can enjoy competitive advantage if it has resources that are imperfectly mobile (i.e. they stay in the firm in the long run) and are heterogenous (i.e. each firm has different resources that may be more or less efficient), with the conditions of *ex ante* limits to competition (i.e. firms can attain a competitive position before intensification of competition) and *ex post* limits to competition (i.e., once a firm has gained a competitive position, it does not get depreciated as much as it normally would if competition was in full swing) fully met. This view has gained in importance with the speedier technological changes of the latter part of the 20th and beginning of the 21st Century, when a firm's competencies cannot be easily preserved by simply relying on its basic resources. The core competencies approach represents a continuation of the resource-based view[82], in that it has accepted its basic premises and moved on from there to view resources as "the physical embodiments of core competencies" (Prahalad and Hamel, 1997: 241). Prahalad and Hamel (1997: 242, emphasis original) explain that "[c]ultivating core competence does *not* mean outspending rivals on research and development" by providing examples of well-known companies, such as Cannon and Xerox;

> "Nor does core competence mean shared cost, as when two or more SBUs use a common facility – a plant, service facility, or sales force – or share a common component. The gains of sharing may be substantial, but the search for shared costs is typically a post hoc effort to rationalize production across existing businesses, not a premeditated effort to build the competencies out of which the businesses themselves grow" (Prahalad and Hamel, 1997: 242).

It is through these views that the core competencies approach reinstates the role of finance in the growth of firms as derivative, or even secondary, in the competitive market processes. These authors maintain that cost variables (finance) are complementary to creating a strong market position. Core competencies can be gained and lost. This also lends support to the conceptualisation of the financing of innovations.

The resource-based approach has promoted the view that firms' choices can be different, rather than being predetermined by factor endowments (initial conditions) and probability, as the neoclassical theory claims (cf. Nelson, 1997). The decision process is hard, since the future is fundamentally uncertain, and agents are constrained by their individual and group competencies. Competences may be represented by skills and tacit knowledge sustained by the organisation (Hodgson, 2001: 247-248). Hence competence-based theories are perfectly

82 Hodgson (2001) notes that the terms "resource-based" and "competence-based" are often used to denote the same things, although the use of the latter term has increased.

compatible with the concept of the NSI as previously elaborated, since competences at the firm level interact with the NSI.

The role of capital (money) in production is underrepresented in the literature on the theory of the firm, so Dunn (2002) proposes that the issues of uncertainty, capital, and power in production be reconsidered. Since the mainstream theory of finance continuously neglects the issues of production and innovation, there appears to exist a big gap regarding financing of innovation. Krahnen (1998) reminds us that one of the most famous German theorists of the firm, late Erich Gutenberg, saw the finance function of a firm only as subsumed under production. On the other hand,

> "[p]resent-day theory of finance takes a radically different approach. It has become a comprehensive theory of decision-making within enterprises, and its models seek to establish links between the claims and behaviour of all the parties involved in the enterprise in terms of Pareto-optimal agreements. Production issues, if considered at all, appear in a highly condensed form, e.g. in the sense of a postulated production function or postulated alternative investment projects, and thus play no more than a peripheral role" (Krahnen, 1998: 5).

This may be connected with the change of circumstances in the economy.

> "In Gutenberg's time, financial managers were perceived as petrol pump attendants who filled up the corporate vehicle with the liquidity it needed to run on. From a present-day viewpoint, financial managers are seen as having traded in their pump attendant's uniform for a chauffeur's cap. Their task is now to direct and control the company's creation of value added and therefore – to retain the vehicle image – they now have their hands on the steering wheel." (Krahnen, 1998: 5-6).

3.3. Financial Theory

3.3.1. Common Ways of Financing Enterprises

Because of the uncertainty a new enterprise faces on the market, internal financing[83] would be ideal to solve that problem. The reason lies in small direct and indirect capital costs. However, an enterprise cannot make profit (from which it could internally finance itself) before it has placed its products on the market. Internal financing will have positive effects on financing from other sources, because it signals commitment on the part of the innovator/entrepreneur[84] and

83 Internal financing means financing from internal sources (retained earnings and depreciation). There are some advantages to using internally generated funds, like avoiding the cost of securities issuance. By using tax policy, the government can influence the capacity of firms to finance their innovations from internal sources.

84 The innovator and the entrepreneur could be the same person or two different persons. When the former is the case, the innovator is also in charge of placing his innovation

reduces informational asymmetry between the innovator and the other owners of the enterprise (cf. Niederöcker, 2002: 249). However, in this overview, we are interested in external financing.

Innovation can be financed by means of debt, equity, or financial instruments that share the characteristics of both debt and equity. The characteristics of debt and equity are listed below in Table 4.

Table 4: Debt and equity characteristics

Criterion	**Characteristics of equity (e.g. amount invested into joint-stock companies)**	**Characteristics of credit (e.g. banking credit)**
Availability of money	Not limited in time	Limited in time
Share of profits	Participation depending on amount of profit received (high potential return)	Fixed interest rate (limited return)
Coaction and control rights	Right to manage the business (complex relationship)	No coaction and control rights (relatively simple relationship)
Position in the case of bankruptcy	Residual right	Creditor with primary rights to return
Cost of capital	High	Low

Source: Bigus (2003: 19), modified and customised by the author.

As can be seen from the above table, credit has different implications for both investor and investee from those that apply in the case of equity. In essence, they are different as governance relations (see below): credit has a limited duration and a limited return on the upside, equity does not. Furthermore, equity enables an investor to control (monitor, manage, and supervise) an enterprise, while the typical relationship of a bank with an enterprise is much shallower. Legally and governance-wise, a more complex instrument would be a convertible security, resembling a corporate bond that has the characteristics of debt but, under specific conditions, can be converted into equity. In that case, the return for the investor may be unlimited on the upside.

The financing of growing technology-based SMEs from their beginnings on – especially the financing of innovation activities – can best be interpreted in the framework of phases of a company's growth. The typologies are different, but all of them include the phase of research and development (or a phase when a concept of the enterprise and its products are developed), the start-up phase (or introduction of products to the market), followed by the phase of growth of the

on the market, that is, he also acts as an entrepreneur. In the latter case, innovating and the entrepreneurial function are separated.

enterprise (usually called expansion phase) (cf. Hielscher, Dorn and Lample, 1982; Sabadinowitsch, 2000; Kulicke, 1997). A possible exposition of the growth phases of an enterprise is given in Table 5 – along with the main activities performed during a given period and some financial implications.

Table 5: Activities and financing of a young technology-based enterprise

Growth phase	**Main activities**	**Incoming payments and disbursements**
Generation of a concept	Conceptual development of products, devising of a strategy, market research and acquisition of capital.	Small disbursements and no incoming payments.
R&D	R&D and activities in connection with R&D – setting up a team, cooperation with partners.	High disbursements (e.g., several million Euro); small incoming payments if some R&D results could be marketed; high net needs for capital.
Introduction of product to the market	New product gets presented, promoted and distributed; at the same time, there is a need for expansion of production facilities.	High disbursements (like in the previous phase or even several times higher); incoming payments are coming from sales of the product; high net needs for capital.
Diffusion and growth phase	Expansion of sales, development of product line, etc.	Disbursements depending on expansion of production; incoming payments could be high if commercialisation of products has been successful; net capital needs are negative as company makes profit and there is no need for external capital.

Source: Kulicke (1997: 130-131), modified and adapted by the author.

Enterprises in general, and SMEs in particular, are troubled by informational problems which may be manifested as credit rationing (Stiglitz and Weiss, 1981) and equity rationing (Greenwald, Stiglitz and Weiss, 1984; Hellmann and Stiglitz, 2000). Stiglitz and Weiss (1981) show that credit markets are (or could be) as imperfect as their (much discussed in the literature) counterpart – the labour market. Namely, they show that the system ("the law") of supply and demand is not always sufficiently flexible to clean the market. Credit could be rationed, meaning that not all borrowers will get it, even if they are willing to accept a higher interest rate. Hence, an equilibrium can be achieved at an interest rate where there is an excess of demand for loans over supply of loans. Both increase in the price (that is, the interest rate) and increase in the collateral cause an adverse selection effect, since the borrowers who are willing to pay higher interest on a loan are more likely to engage in riskier activities. Consequently, the price and the collateral will have certain incentive effects on the behaviour of the borrowers. Regarding equity rationing, as in Greenwald, Stiglitz and Weiss (1984), an enterprise can have problems raising equity because of either incentive problems that the equity and/or debt have caused or the signalling effects. The former is a manifestation of the principal-agent problem, since managers are differently incentivised by debt than by equity. Imposition of bankruptcy costs, as well as debt, makes them more disciplined, as opposed to equity (more on that below). The latter means that because firms that accept debt may be perceived as good, firms that issue equity may not be perceived as good – the market may react negatively to new equity issuance. "Moreover, the announcement of a new equity issue is usually bad news for investors, who worry that the decision signals lower future profits or higher risk" (Brealey and Myers, 2003: 579).

Financing innovation is especially problematic in the early phases of company development. There are several reasons for that: 1) the position of an early stage firm in negotiations with a financier is weak because its innovation is still unprotected by patent; 2) once the innovation is protected by patent, potential investees approach the investor; 3) early stage financing is also relatively undeveloped because of the high fixed screening costs of an investee, who usually demands a relatively small amount of capital (Bigus, 2003: 38-39). With regard to specific investments[85], there is the underinvestment problem. Namely, if a party to a contract makes a specific investment, then the other party will have an incentive to try and achieve a higher value for itself in future negotiations. Since the former party anticipates this, it is likely either to underinvest or not to invest at all (Bigus, 2003: 40). Berger and Udell (1998) assert that not all firms are financially structured to be heading for an initial public offering (IPO), giving an example of family enterprises. The borderline between internal and external finance in SMEs cannot be clearly drawn, since the owner often pledges his or her own collateral in order to gain financial means[86].

85 Investments into equipment, human capital or dedicated assets that are not easily redeployable, and whose value is relatively low outside the current business relationship (cf. Bigus, 2003: 39-40).

86 A general problem concerning research on small business finance is posed by the intertwinement of a firm's finance and its owner's finance (Ang, 1992 in Berger and Udell, 1998).

Different financing sources are applicable for different phases in the growth of firms, as their age and information availability increase (Berger and Udell, 1998[87]). In Table 6, we specify financial sources established as the most frequent ones.

Table 6: Financial sources use according to phases of growth of the enterprise

Financial sources	Phases of growth of the enterprise		
	R&D	**start-up**	**growth**
Informal sources	+	+	-
Business angels	-	+	-
Venture capital	-	+	+
Government credits/subsidies	-	+	+
Strategic alliances and partnerships	+	+	+
Banking credit	-	+	+
Trade credit	-	+	+
Private placements	-	-	+
Mezzanine capital	-	-	+
Public equity and debt instruments	-	-	+

Source: depicted by the author, drawing on Berger and Udell (1998) and Timmons et al. (1990: 408), as in Bygrave and Timmons (1992: 16).

In the R&D phase, entrepreneurs normally use informal sources, but also strategic alliances and partnerships. The former are taken here to mean a variety of sources ranging from the personal sources of the owner, his family and relatives, as well as friends who are willing to put some money into the (future) company. The latter refer to all forms of cooperation, which do not include integration in the legal sense (cf. Noteboom, 2004). The amount of money raised from informal sources is usually relatively small, while in the case of strategic alliances and partnerships it does not have clear upper limits (cf. Timmons et al. (1990: 408), as in Bygrave and Timmons (1992: 16))[88].

---

87 Berger and Udell (1998) call this concept "financial growth cycle". "Small business may be thought of as having a financial growth cycle in which financial needs and options change as the business grows, gains further experience, and becomes less informationally opaque" (Berger and Udell, 1998: 622).

88 Affiliated companies (not listed in Table 6) are likely to get financing from free cash flow from other companies of the group. In that case, information asymmetries are

Venture capital financing can be divided into informal venture capital (business angels) and formal venture capital (venture capital firms[89]). As a subset of the formal venture capital sector, there could also be public sector funds. The technology of financing of venture capital firms (VCFs) differs from that of a bank. Namely, while a typical (European – universal) bank (when analysing credit application) is mainly concerned with the probability that the credited firm will repay the credit, a venture capital firm (when considering investment in an investee) will be primarily concerned with the potential profitability of the investee that can be realized in the market (cf. Ratz, 1988: 57). Another mode of venture capital financing comes from corporate sources and can be either direct or indirect (through specialised funds). Its primary interest lies in the control of new technologies, unlike the VCFs, whose motive is primarily financial (Tidd, Bessant and Pavitt, 2001: 361).

The informal venture capital market is represented by business angels (BA), who are normally defined as high net worth private individuals who invest their own money into companies, and actively participate in a company's management. They are usually, especially in the USA, involved in financing earlier phases of a company's development and individually invest smaller sums of money than the VCFs. BAs follow not only financial, but also non-financial goals (personal satisfaction, development of personal skills, philanthropy, etc.). Besides, they are normally flexible regarding their investment horizon and do not necessarily expect regular payouts (cf. Niederöcker, 2002: 224-225). Therefore, they are more likely than the VCFs to finance risky and uncertain investments in the earlier phases of a company's development. In addition, business angels' investments have positive signalling effects on other financial sources because of the management support and equity capital that BAs provide. The information asymmetry that BAs face is small relative to the information asymmetry that other providers of capital face (with the exception of investments from family and friends). BAs are normally familiar with the industry they want

smaller than in the case of outside financial institutions. Therefore, this form of financing has some advantages over financial sources from outside institutions.

89 One of the names used for such forms of venture capital organisations is "fund management company" (cf. Gate 2 Growth, 2002). However, we will use a more common name for these formal, (mostly) privately owned venture capital organisations: "venture capital firms" (VCFs). Although there are many forms of venture capital funds, the usual practice is that a VCF acts as a management company for the venture capital fund. However, private equity is sometimes understood in a broader sense than venture capital, in that it includes more financing options. Hence, if it is not applicable to refer strictly to the VCF we may also use the abbreviation VC/PE (venture capital/private equity). At several points in this work we will also refer to VC investments, the VC market, VC financing, and similar. When we do so, we mean venture capital as a financial source intended for propulsive enterprises.

to invest in and, therefore, have specific knowledge which can be very useful in the everyday running of a company. BAs can also be grouped together in so-called business angel networks (BANs), which can have multiple functions, such as the creation of awareness of this form of financing, provision of training (not only to entrepreneurs, but also to BAs), business matching, syndication, etc. (cf. Gullander and Napier, 2003). In such a way, BANs contribute to the mobilisation of both the demand and the supply sides of the informal VC market.

VCFs comprise the formal venture capital market[90]. They usually do not invest before the start-up phase, because of high uncertainty and inability to estimate the rate of return. A high rate of return is expected, especially when they exit an investment over the stock exchange. It is not yet clear in the literature whether the firms financed by venture capital are more innovative than others (Callahan and Muegge, 2003). Nevertheless, venture capital influences the growth rates of firms in which it is invested (Audretsch and Lehmann, 2004). Bygrave and Timmons (1992) made a survey of academic journals and found that compound annual rates of return were generally above 10%, with some investments reaching returns between 20% and 30%, or even breaking the 30% mark. Information asymmetry in VCF's investments is not so high because of VCFs' active participation in the management and the supervisory board of an investee. The costs of venture capital financing are much higher than financing with banking credits, because venture capital companies have higher monitoring costs. But people who work for VCFs need to understand the technology they want to invest in, and they have to possess (financial) management skills, as well as knowledge of the local (financial) market. In addition to excellent qualifications, another prerequisite for such experts is learning-by-doing (cf. Schertler and Stolpe, 2000: 13). The financial markets of the analysed post-transition countries are not as developed as the financial markets of Western European countries or of the US, and venture capital has been present there for a very short period of time. This may imply that venture capitalists in those countries are less experienced. The case of introduction of VC to Israel and the relevant literature show the venture capital needs to be built upon a strong base of enterprises, especially high-tech ones, in which VC can be invested (Avnimelech and Teubal, 2004[91]).

We can distinguish three main forms of credits important for innovation financing: banking credits, credits from affiliated companies, and trade credits (credits from suppliers and customers). It is usually hard to get credits from outside (financial) institutions in the early stages of a company's development. Banks offer

90 Public sector funds may stimulate the formal venture capital market as well – by (co)investing this money.

91 See also Teubal and Andersen (2000), as referenced in Avnimelech and Teubal (2004).

a variety of different credit arrangements, but (usually) not credits for innovations[92]. Therefore, in order to finance innovation activities, companies usually take classic banking credits against collateral. Because of a bank's lack of specialization and low levels of monitoring, information asymmetries between the bank and the lender are especially high. Companies are not very likely to get credit from a bank in the early phases of their development, because of their lack of reputation and absence of turnover (cf. Niederöcker, 2002: 229), (possible) lack of collateral, inability of the bank to estimate the probability the enterprise will pay back its credit, low or non-existent turnover and/or profit. A high debt-to-equity ratio could also be a reason for a bank to deny credit to a company[93]. Besides, a bank is normally reluctant when it comes to providing credit for financing intangible assets (which one has to invest in when pursuing innovation activities). Government programmes providing credits and/or credit guarantees and subsidies are especially important in the early phases of a company's development when uncertainty and risk are high. But in comparison with equity, credit may burden the cash flow of enterprises in their early phases[94].

In the growth phase, companies may make private placements, issue their securities publicly, or get mezzanine financing. Private placements include selling securities to one or several investors directly (cf. Francis, 1991), while public issuance means selling them to a wider public. Small firms prefer (or are limited to) private markets, not public ones – because of the costs of issuing securities and informational opacity (Berger and Udell, 1998: 628). Same authors believe that, small business owners will be more likely to issue debt than equity, because they wish to retain control over the enterprise. Mezzanine financing comprises elements of both debt (since interest is paid) and equity (e.g. via warrants). It allows for greater flexibility than a banking credit, since e.g. interest and principal could be designed to be paid on exit from the investment (Temple, 1999: 50). These characteristics make it a hybrid financing instrument (cf. Sabadinowitsch, 2000: 22-23).

92 Although banks are not typical providers of credits for innovations, they can finance innovations indirectly, since it is normal for banks in developed economies to invest in a venture capital fund.

93 However, high debt-to-equity ratios are more likely in the later stages of a company's development.

94 OECD (1995: 110) asserts that credit as a financial instrument is not ideal for financing innovations because of the following reasons: 1) need to secure the credit with a collateral, which newly founded firms usually do not have; 2) predetermined purpose for using the credit, which should not be too risky; 3) relatively short repayment period; 4) need to be liquid in order to pay back the credit, which could burden the business of a newly founded firm; 5) relatively inactive role of the credit institution, which does not participate in the strategic management of the firm and, therefore, does not help its business.

In the transition countries, there is generally a lack of long-term finance (Klapper, Sarria-Allende and Sulla, 2002; cf. Caprio, Jr. and Aslï Demirgüç-Kunt, 1997[95]). There is especially a lack of finance suitable for financing innovation (cf. Inzelt, 2003; cf. Račić et al., 2004), in particular for the very early stages of (future) companies' growth, as before and during the start-up stages. However, the problems of financing are by no means limited to transition economies alone (cf. The Gallup Organization, 2007), but are just more acute there.

3.3.2. Theories of Capital Structure

In this overview, we will consider only some theories or concepts that may have repercussions on the capital structure of firms. This section covers two of them: the neoclassical theory of the capital structure (3.3.2.1.) and the new institutional theory of the capital structure (3.3.2.2.).

In the case of high-tech SMEs, many of which are, presumably, high-growth companies, talking about optimal capital structure would be unrealistic. Even if not in the high-growth segment, SMEs do not have access to all of the financing options available to large enterprises. This is especially the case with companies in their early stage, which may not be able to even get a credit from the bank (because they do not have a business history), not to speak of issuing equity in the capital market (since a discount on shares would be too big). Moreover, the owners/managers of SMEs may not be willing to issue debt or equity because of reservations or lack of knowledge about this way of financing (this point may apply for large companies alike). This type of behaviour is more likely to occur in bank-based financial systems like the ones we study here (Hungary, Slovenia and Croatia as well as all other (post-)transition economies). The question is which of the concepts or theories can be applied to the financing of innovations in SMEs functioning in imperfect product and capital markets under the aforementioned constraints. Furthermore, how can we incorporate (elements of) the following concepts and theories in the innovation financing framework applicable here?

There are some theories we will not be considering here. For example, we will not be considering theories based on contests for corporate control, since their explanatory power is limited to situations of hostile takeover. We also skip models of capital structure based on industrial organisation characteristics, since

95 This particular study helps to understand the financing process, especially in the "developing countries". However, the authors do not specify which countries they mean under the term "developing countries". We believe these findings can also be applied to transition economies, that is, countries that their paper does not class as belonging to the "developed" category.

they only consider debt and are, therefore, partial. The problem with models dealing with optimal capital structure is that they often presuppose that the firms in question are active participants in public capital markets, that is, issuers of debt and equity securities (see Harris and Raviv, 1991). We do not use signalling models, since the link between capital structure and signalling has not been sufficiently elaborated in the literature (Frielinghaus, Mostert and Firer, 2005). Furthermore, calculating with financing, having in mind their growth opportunities, seems to be a plausible course of action for companies (Cornell and Shapiro, 1993: 205). Myers (1993) *de facto* acknowledges that one cannot identify the optimal capital structure, and that investment is more important than financing. "Financing decisions should be shaped to support the firm's investment strategy, not vice versa" (Myers, 1993: 142). The same author also quotes Brealey and Myers's Fourth Law: "You can make a lot more money by smart investment decisions than by smart financing decisions" (Myers, 1993 quoting Brealey and Myers, Principles of Corporate Finance, 2nd edition). The two latter "principles" are in line with our understanding of the innovation financing process.

3.3.2.1. Neoclassical Theory of the Capital Structure

The financial theory of the first generation is represented by the 1958 Modigliani and Miller theorem. According to their theorem, "*the average cost of capital to any firm is completely independent of its capital structure and is equal to the capitalization rate of a pure equity stream of its class*" (quoted in Williamson, 1988: 577; Emphasis original). This implies that there is no difference with regards to the cost of capital between credit and equity. However, this is not surprising, given the rather unrealistic assumptions of the model. First of all, the authors model the firm as a production function. Furthermore, in the model there are no taxes, no transaction costs, and information is a public good. Modigliani and Miller (1963) corrected their initial view and noted the importance of the tax advantages of debt, remarking that its advantage was not absolute. Retained earnings could be cheaper than debt because of the existence of personal income tax. One of the avenues that this model later advanced is the one called the "static tradeoff hypothesis". According to this hypothesis, a firm should maximise its value by balancing between equity and debt. The firm should use debt in order to utilise the tax advantages associated with it, but not to an extent that might seriously jeopardise the financial viability of the firm by involving the probability of default (Myers, 1984). Bradley, Jarrell and Kim (1984) found relative debt levels to be dependent on the industry (with their analysis focusing on specific industries). They also found volatility of firm earnings, relative expenditures on R&D, as well as advertising expenditures, to be inversely related to leverage. Hence, their findings lend some validity to the theory of optimal capital structure. On the other hand, Fama and French (1998) do not find taxes in relation to financing to have any influence on the value of the firm.

3.3.2.2. New Institutional Theory of the Capital Structure

The financial models of the new institutional economics are represented by the positive agency theory (AT) and transaction cost economics (TCE), or transaction cost theory (cf. Iliev, 2000). The AT and TCE models are considered to be complementary (Williamson, 1988: 567-568), since they share many similarities, but also show certain differences. The similarities between the AT and the TCE are that they recognise the role of institutions (e.g. debt and equity contracts), informational asymmetries, and incomplete contracting. We shall first describe the AT approach, followed by the TCE approach, before judging their respective suitability.

In the AT, agency costs are caused by a divergence of interests between principal(s) and agent(s), since their alignment requires incurring three kinds of expenditures: 1) monitoring, 2) bonding, and 3) the residual loss (Jensen and Meckling, 1976: 308). Various interests could also result from financial claims on the income of the firm by either shareholders or debtholders. In the former case, the shareholders face the problem of how to induce managers to conduct business in owners' interests. In the latter case, it is the debtholders who have to cope with the (potential) costs of owners' irresponsible behaviour. Namely, when a high debt-to-equity ratio is involved, the owners might want to invest in highly risky activities which may bring them considerable gain on the upside, whereby the debtholders' gain is limited, but the loss is potentially huge (Jensen and Meckling, 1976). Debt can influence the capital structure in two major ways: 1) because interest on debt is tax-deductible, and 2) because debt can discipline managers, since it obliges them to deal with interest payments, instead of leaving them with a free cash flow which might be invested at rates below the cost of capital. However, the control property of debt does not hold for all firms at all times – it is not valid for high-growth firms with very profitable investment opportunities without free cash flow (Jensen, 1986: 323-324). The AT literature focuses more on the influence of debt on managerial behaviour than on equity, while sticking to an *ex ante* approach (Kochhar, 1996).

The TCE is a more developed theory than the AT, in that it deals with both debt and equity as exerting an influence on governance, while considering both *ex ante* and *ex post* contracting (Kochhar, 1996). According to the TCE, debt should be used when asset specificity is low, while equity should be used when it is high. "[D]ebt is a comparatively simple governance structure" (Williamson, 1988: 580) and is rule-governed (Williamson, 1988: 580). "[E]quity finance (...)

is a much more complex governance relation" (Williamson, 1988: 580) allowing for discretion (Williamson, 1988: 581). Equity enables a governance relation that is more durable than debt, and the power of shareholders can be used at practically all times. On the other hand, the power of debtholders does not come to the fore if the company services its debt. Hence, the TCE recognises both debt and equity as governance structures, but perceives equity as the more important (Kochhar, 1996).

The TCE is a broader theory than the AT, as it leaves us better equipped to study the financing of innovations. The TCE tackles both debt and equity as governance structures and captures the *ex-post* behaviour of the firm. It is hard to believe that contracts could be perfectly written *ex ante,* and it seems impossible to describe financing of the firms only in terms of debt. Kochhar (1996) mentions studies corroborating the TCE, in which R&D intensity and uniqueness of a product (both as indicators of asset specificity) were inversely related to leverage. A summary of the respective strengths and weaknesses of the AT and the TCE is given in Table 7 below.

Table 7: Agency theory and transaction cost theory compared

AGENCY THEORY		TRANSACTION COST THEORY	
strengths	**weaknesses**	**strengths**	**weaknesses**
• financial intermediaries are acknowledged as important • gathering information and monitoring are perceived as costly activities • recognition of incomplete contracting and informational asymmetries • acknowledgement that market failure could be caused by credit rationing • firm can be seen as a nexus of contracts	• *ex ante* approach to contracts; failure to analyse post-contractual behaviour, inadequate for understanding financing innovation • debt financing and its disciplining qualities are emphasized; equity financing and its repercussions are relatively disregarded • observes financing mostly as a monitoring activity • capabilities and knowledge generation out of focus • it captures only risk, rather than uncertainty of innovation processes • does not focus on knowledge generation • a static and not a dynamic approach	• finance is observed through the lense of corporate governance • acknowledgement of incomplete contracting, observation of post-contractual behaviour and recognition of asset-specificity makes it more suitable to observe innovation activities • firm is seen as a governance structure • can be used to analyse disputes, in particular how to avoid and process them • it observes informational asymmetries	• it does not capture capabilities and generation of knowledge • does not explain innovation capabilities

Source: Iliev (2000: 11), Williamson (1988), Kochhar (1996) and Christensen (1992: 76-77); Modified by the author.

On the other hand, there is the pecking-order theory of capital structure, which does not presuppose that there is an optimal capital structure. Rather, according to this theory a firm prefers one financial instrument to the other. In short, there is a preference for internal finance. In case a firm needs more finance for its projects, it will first issue debt, then perhaps a hybrid, convertible security, and only then equity (Myers, 1984: 581). Myers (1984) admits this theory cannot explain everything, but claims that it is consistent with some historical evidence[96]. A

96 A summary of the literature speaking in favour or against pecking-order theory is given by Harris and Raviv (1991).

justification for firms to conform with a pecking-order when choosing financing sources lies in information problems – investors cannot see everything that goes on inside a company. In other words, there is a problem of credibility that the management has to overcome. Since investors will require a premium for risk, which makes financing more expensive, there is a reason to use retained earnings first, and before proceeding with more expensive sources – as Myers claims (Cornell and Shapiro, 1993: 198-199). But Myers (1984) and Myers and Majluf (1984) (both referenced in Watson and Wilson, 2002: 561) suppose that the interests of managers and shareholders are aligned, so the only "troubling" informational asymmetry is the one between insiders (managers and shareholders) and outsiders (potential investors in the firm). This, on the other hand, makes it more applicable to SMEs, since the ownership and management functions are frequently incorporated in one person (cf. Lopez-Gracia and Aybar-Arias, 2000).

A common denominator to the AT, TCE and pecking-order theories is that they belong to the literature called the economics of information. According to Christensen (1997: 8-9), it suffers from several shortcomings: financial models are usually one-period models, therefore concentrating on *a priori* relationships between the lender and the borrower. They usually neglect the dimension of trust formed in longer-term relationships. Hence, these models misrepresent reality, since more complex lender-borrower relationships take place, and trust mitigates the problem of asymmetric information. Also, institutions which determine the environment in which innovation is financed are neglected, and the financial models are usually stylized and not tested empirically. Finally, uncertainty is replaced by risk, and there is no consideration of opportunistic behaviour by the lender, but only by the borrower.

3.3.3. Finance and the Real Sector: A Complex Relationship

Bartlett (2002) notices that there are two main approaches to development in the literature: 1) the money first approach, and 2) the "institutional" approach. The representatives of the former (Bartlett here refers to Pissarides, 1999) basically claim that a larger variety of financial instruments, as well as lower costs of capital, can overcome barriers to the development of SMEs. The representatives of the latter approach maintain that the financial side is just not enough in the absence of supportive institutions and favourable social capital. While the former approach implies a strict causality, the latter implies complexity. We believe that the second approach is the more realistic one. Furthermore, this dilemma illustrates what we shall explain below – that the financial and the real sectors are involved in a complex relationship. In order to explain it, we have to consider a number of factors beyond functional ones, i.e. beyond what mainstream finance theories would suggest (cf. Ardalan, 2004). As neither markets nor firms are perfect, financing does not happen automatically.

The neoclassical finance theory follows a positive approach to economics (cf. Frankfurter and McGoun, 1999). There are other worldviews a scientist can adhere to when studying finance, and not only from the point of view of the functionalist paradigm (see Ardalan, 2003). In the orthodox (neoclassical) approaches, money is not viewed as a social relation (as in the view of the institutionalists, Marxists and post-Keynesians), but as a means simplifying transactions. Consequently, neoclassical economics does not elaborate on the evolution of money, but rather considers it, more or less, as "natural" in market economies (cf. Wray, 2006). Brennan (1995) asserts that financial economics from the 1970s to 1995 has gone through changes, as finance theories have shifted from encompassing a largely comparative, static analysis to a more dynamic one, and have recognised bounded rationality and the changing preferences of economic agents. Furthermore, finance theories have recognised legal institutions, although the nonlegal ones have largely remained outside their scope.

The perfect market hypothesis (of neoclassical economics) has been seriously contested since the rise of the stream of literature called economics of information in the 1950s (Stiglitz, 2000). The main focus of the economics of information has been the problems of selection and incentives. It claims that information is asymmetrically distributed, conveyed not only by prices, but also by other characteristics of an economic agent (Stiglitz, 2000).

One needs to stress that the economics of information has recognised institutions, as well as deepened our understanding of the role information plays in economic (trans)actions. It is essentially an approach of the new institutional economics. It was the post-Keynesians who exercised its most vociferous critique from a heterodox perspective. They observe economic phenomena, especially financial markets and institutions, through the lens of fundamental uncertainty. While fundamental uncertainty presupposes a non-ergodic environment, the notion of asymmetric information, on the contrary, presupposes an ergodic environment and a certain amount of information (Van Ees and Garretsen, 1993). Another post-Keynesian, Wolfson (1996), argues that it does not necessarily have to be asymmetric information that causes credit rationing, but that it could also be fundamental uncertainty and asymmetric expectations about the future (that affect both borrower and lender). Hence, he suggests a broader definition of credit rationing – where it refers to a situation in which a bank refuses to provide credit to a particular borrower, no matter how high an interest rate he is willing to pay.

It is hidden action[97] and hidden information problems that render the perfect market hypothesis ineffective, and cause the agency problems (Stulz, 2000). The hidden information and hidden action problems (and, hence, the cost of capital) depend on the financial structure of a country, which Stulz (2000: 14) claims to consist "of the institutions, financial technology and rules of the game that define how financial activity is organized at a point in time". Based on the existence of hidden action and hidden information problems, Stulz (2000) makes an important distinction between the cost of external funding (for the entrepreneur) and the cost of capital (for the investors). Namely, assuming that investors do not completely trust the management, they could try to impose restrictive covenants on it. These implicit costs that the management has to face constitute the difference between the two types of costs of capital. In conclusion, Stulz (2000) sees staged financing (by both banks and venture capitalists) as the optimal solution for financing an entrepreneur's projects.

Lately (starting from 1970s), there has been a profusion of literature on financial development and growth. Wachtel (2001) describes it as a change in developmental economics, which has given way to market-based approaches in analysing the link between financial development and economic growth. Instead of the supposed interventionist policies, one should focus more on the development of financial markets and institutions. Although Wachtel (2001) warns of possible financial crises, a proper sequencing of financial development would, according to him, minimize their incidence. He lists the functions of the modern financial sector which: 1) improves the screening and monitoring of enterprises, 2) encourages savings, 3) reduces the costs of financial institutions through the economies of scale, and 4) enhances possibilities for risk and liquidity management.

There is a positive correlation between the quality of legal institutions and the development of the financial system – this is generally posited by the body of literature called "law and finance", written predominantly in the new institutional economics tradition. La Porta et al. (1997) find that better protection of investors – observed through the quality of legal institutions – means deeper equity and debt markets. In a study going a step further, the same authors (La Porta et al., 1998) find that civil-law countries[98] have weaker protection of creditors and shareholders than countries of the common-law tradition[99]. Pistor, Raiser and Gelfer (2000) apply a law and finance framework to transition economies,

97 E.g. when the actions of the management remain hidden from the shareholders (Grant, King and Polak, 1996: 4).

98 For example, those of continental Europe, such as Germany, France and the Scandinavian countries.

99 For example, the USA and UK.

and find that it is not only legal institutions in the written form which are important, but also their implementation. Namely, transition economies very quickly adopted new and improved standards of investors' protection, but their implementation remains weak, which hinders the development of their financial markets and institutions. We believe that this body of literature should be seen together with Ha-Joon Chang (2002), who claims that imposing the institutions of the developed countries on the developing ones is plainly wrong. Namely, the developed countries of today used different interventionist and protectionist policies and institutions during the period of their development, whereas it is only now that they are more fully liberalized. Imposing modern standards of the developed economies of today on developing countries has, in actual fact, only slowed down their economic growth. However, the law and finance literature does offer a multidisciplinary insight into how wealth is created, albeit by its prevailing use of functional analysis.

As to the importance of financial development for economic growth, King and Levine (1993) and Khan (2000) find some empirical evidence to support this link. Levine and Zervos (1998) confirm the important correlation between the two groups of factors that determine financial development and economic growth. However, they contend that the relationship between financial development and long-run economic growth has still not been properly understood. Berglöf and Bolton (2002) argue that the financial sector in transition economies has contributed relatively little to the restructuring of the manufacturing firms, and that government policy and institutions are of primary importance for both economic and financial development. Wachtel (2001) underlines that it is not just the amount of finance provided that counts, but also the effects the financial sector has on enhancing the efficiency of the real sector. On the macroeconomic level, the correlation between the financial system and economic growth and development is positive, as the two variables are mutually reinforcing (Berthélemy and Varoudakis, 1996). A more developed economic system is more likely to have a bigger financial sector in terms of percentage of the GDP (Goldsmith, 1969; Demirguc-Kunt and Levine, 2001). However, the recent literature covering the link between economic growth and financial growth/structure does not answer the question of how to proceed in order to develop a financial system and achieve economic growth.

Following one of the main institutionalist propositions, "it is the whole organizational structure of the economy which effectively allocates resources and distributes income, not just the market mechanism" (Foster, 1991: 211). Coordination and regulation of economic activities by the market is just one of the forms of coordination in an economy[100]. As previously elaborated, culture and social capital, as well as institutions, all influence economic activities as well (we will be revisiting this later). Macroeconomic, mesoeconomic, and microeconomic conditions can make a difference, too. By macroeconomic conditions, we mean the overall conditions of a national economy, usually depicted by a time series of macroeconomic indicators and variables, like GDP per capita, unemployment rate, inflation rate, exchange rate, etc. Moreover, it makes a great difference whether an economy is in a recessionary state or in an expansion phase. It can also be in a stable state, whereby the rate of growth of the economy is below the potential, but some variables (such as high unemployment rate) indicate a latent crisis. On the mesoeconomic level, one observes the industrial and market levels, while on the microeconomic level it is firm-specific conditions that one has to focus on. We shall try to explain the importance of the mesoeconomic level of the analysis below – especially regarding financing conditions.

When considering financing, there is an important distinction to be made between the supply-side finance constraint and the demand-side finance constraint (Cressy and Olofsson, 1997). They are both imperfections of the capital market, the only difference being whether the constraint is outside the firm (the case of the supply-side finance constraint) or inside the firm (the case of the demand-side finance constraint). While the former is much discussed in the literature, the latter form of constraint has not been analysed that well (see Hutchinson, 1995 as well). Cressy and Olofsson (1997) provide an example of the demand-side finance constraint – when the owners of firms face a trade-off between intention to grow and unwillingness to give up a certain share in ownership (e.g. when issuing equity). Discussing the same issue, Berger and Udell (1998) mention that some businesses are not financially structured to go through a financial growth cycle. These latter authors make a distinction between "mom and pop" enterprises and "life-style ventures", both being founded with the intention of keeping the business within the family, or for oneself. This issue is well-known in the literature (Tidd, Bessant and Pavitt, 2001; Young and Cvijanović, 2006). Although these issues seem to be recognised as inherent to SMEs, Cressy and Olofsson (1997) rightly emphasize that the issue of the demand-side finance constraint is underresearched concerning the impact it has on business performance and economy in general. Consequently, neither of the related policy issues were familiar. It could be the lack of information or knowledge of financing sources causing it. Van Auken (2001) shows that the technology-based SMEs, from research in the USA, face various constraints in raising external finance. These are, in particular, the limited knowledge of some capital sources (especially those normally used to finance growth), and a lower capability to negoti-

100 Other forms are state planning and direct state intervention, coordination as in networks, and similar.

ate and price equity investment than to negotiate and price debt investment (at any stage of a firm's development). Freel (1999) analyses a sample of SMEs from England and finds that, regardless of their innovativeness level, they are the least familiar with equity financing. Consistent with that finding, the SMEs from that study used short-term debt as their principal way of financing. Long-term financing instruments were less frequently used among the firms in the sample. Furthermore, innovative SMEs were more likely to fail to obtain funding from a bank than non-innovative SMEs.

The previously mentioned issue of financing is inextricably connected with the growth of the firm. One can distinguish three types of SMEs regarding growth – low-growth SMEs, moderate-growth SMEs and high-growth SMEs (Hanks et al., 1993, McMahon et al., 1993, McMahon, 2001 – all referenced in Račić, Aralica and Dadić, 2005). Growth is not only associated with any intentions that a firm founder has in his or her mind when founding the firm[101], but also with other factors, such as: the sector the firm is active in, the market structure of the respective industry, whether its business is based on corruption or not, the industrial structure of an economy, the structure of the demand in a particular sector of an economy, as well as other factors from Porter's diamond, management style, etc. We will provide a few examples to illustrate the above-stated. The growth prospects of a firm in a non-tradable sector are rather limited, compared to a firm active in a tradable sector of an economy. The difference in the growth prospects of firms is also determined by the respective market structure. A monopolistic firm has a better potential to grow than a firm in an oligopolistic market, not to mention firms in a "perfectly competitive" market. If one defines corruption as gaining advantage through illegal non-market transactions, a firm equipped with political connections to high-ranked officials is "better off" than a firm without them. As for the industrial structure of an economy, it could be based on low-tech or medium-tech firms active in importing, retail, and the service sector in general (which is the case with Croatia – cf. Račić, Aralica, Dadić, 2005), or on high-tech sectors driving the growth of the GDP (which is the case with the USA). In the former case, the ability of a high-tech firm to grow might be very much constrained due to, among other things, a lack of appropriate suppliers, a lack of big companies driving the demand for their products, etc. As previously mentioned, even the structure of the demand can be crucial for enterprises willing to grow. For example, Japanese teenagers probably demand much more sophisticated products than teenagers from a less developed country. Hence, the financial sector may find fewer opportunities for financing when firms have a limited growth potential.

101 Basically the issue of ownership, as mentioned above.

As elaborated above, a company's turnover can be based on factors other than innovations, such as monopolistic rent. But when it comes to the financing of innovations, the problem is often that there is nothing to finance: that is, no innovations. Moreover, the quality of innovations can differ (incremental vs. radical), and so can the financial instruments for their financing. Even when we presume that the firm in question is acting in a perfectly rational way and is looking for the optimal means of financing. Moreover, the financing of innovations should be viewed *vis-à-vis* other profitable opportunities in a (still) national economy. If there are profitable opportunities in the market, involving relatively little risk and uncertainty, it is reasonable to assume capital will be channelled into these, rather than into the financing of (uncertain and risky) innovations. There are two things that lead us to that conclusion: possible high opportunity costs and the myopia of modern capitalism (market myopia). As investing into innovation is inherently uncertain, and can potentially bring only long-term benefits for the investor, a risk-averse investor, faced with profitable short-term opportunities, is more likely to invest into the latter. An additional reason for such a decision could be found in market myopia: that is, a focus on short-run profitability neglecting possible gains in the long-run.

These findings have important theoretical consequences and policy implications. First of all, it can be said that "imperfections" can also be on the part of producers, and not only in the market. Secondly, policies addressing problems only present on one side cannot solve them. In the neoclassical usage, imperfection usually refers to the way the market works, as when it allocates resources in a non-perfect way. If capitalist economy is inherently stable, and producers and consumers know exactly what they want (maximize utility or profits) and how to do it (since they are perfectly rational), the state should not interfere with the market in such a major way as to cause distortions (so claims the neoclassical economics school). Neoclassical economics wants to reduce the state's involvement in the economy as much as possible, although it recognises that the state needs to take care of public goods, as well as to provide and uphold "the rules of the market game". Market imperfections are, according to the neoclassical economics, just what the word says – cases when markets do not function *perfectly,* and, therefore, exceptions (e.g. rigidity of prices, asymmetric information)[102]. However, in line with the aforementioned findings, policy makers should address both the supply and the demand sides (e.g., both the supply of finance and the demand for finance). While stimulating the supply side may be more straightforward (e.g., by introducing pension funds and removing barriers for stock market development), it remains unclear how to deal with the demand side

102 Besides, the neoclassical theory does not doubt the utility maximization goal of producers and consumers, whose motivation is solely instrumental.

if entrepreneurs are generally not willing to relinquish a share in their companies' equity, or are not sufficiently informed about financing sources.

Hence, the relationship between finance and the real sector is complex, since their interaction results in multiple outcomes, depending on the relevant socio-economic variables. It may be said that finance is not primary or fundamental for the growth of firms. To put it in the language of mathematics, it is necessary, but not a sufficient condition. Likewise, our explanation of the financing of innovations cannot only include the financial side of the market processes (supply of funds), but must also cover the demand side (behaviour of entrepreneurs). Production is primary vis-à-vis exchange, but it is inextricably linked to monetary economy (Dunn, 2002: 75, referring to Davidson, 1972). In post-Keynesian economics, entrepreneurs must plan their production in advance: that is, they must predict effective demand before formulating their demand for credit (cf. Arestis, 1992), although it is questionable to which extent it will be accommodated (Dow, 1991). A focus on strategic decision-making by entrepreneurs in an uncertain and risky environment, in which production takes time, calls for the integration of money into the theory of the firm. Since entrepreneurs are active in a monetary economy, in which money is needed throughout the production process (and appears at the end of this process, once products have been sold in the market), a firm should be seen in conjunction with the monetary sector (cf. Dunn, 2002). Post-Keynesian economists believe that "[m]oney and production are regarded as being so fundamentally interrelated in the functioning of capitalist economies that it is rarely helpful to separate the two for analytical purposes" (Dow, 1991: 189). Hence, we do not need, and it would even be misleading, to discuss which of the two spheres was the first, and which the second[103].

Therefore, money (finance) cannot be considered as just an external variable, like in neoclassical economics. While neoclassical (monetary) economics presupposes causality running from the money supply to the real economy, the post-Keynesian approaches, such as the circuitist approach, assert that there has to exist a demand for money coming from the real sector, based on entrepreneurs' assumptions as to the level of aggregate demand (cf. Gnos, 2006; Realfonzo, 2006). We believe the post-Keynesian economics view to be realistic, as it claims that money and finance (the notion that includes all forms of money and financial instruments) are created endogenously (Palley, 1996). Therefore, not only does the central bank have the power and authority to "pump" money into the economy, but money (finance) is created in the economy, during and through economic processes. To that end, entrepreneurs may prolong paying their due obligations, engage in barter (both of these used extensively in Croatia

103 This problem is colloquially called "the chicken and the egg problem".

during the 1990s), move their production to countries where wages are lower, or engage in financial innovations (the latter was especially true for the USA from the 1980s onwards). The latter view has developed into the endogenous money theory. Banks and other financial intermediaries do not have to respond to the request of entrepreneurs, so production is necessarily uncertain (cf. Realfonzo, 2006). Palley (1996) proposes an even broader hypothesis – that of endogenous finance. His usage of the term subsumes the whole financial sector, not just banks.

3.4. Principles of Financing Innovations

Literature on the economics of innovation is quite abundant, as is literature on financial economics. However, we have noticed a chronic lack of literature on the interrelationships of finance and innovation, and this opinion is shared by some other authors (e.g. Guinet, 1995, Janeway, 2007). The theoretical aspects of innovation financing are hardly present in the literature, and there is not yet any theory of innovation financing, as Stratmann (1998: 11, 15-16) claims. However, Schumpeter is the first scientist that started to work (intensively) on problems of innovation financing. The characteristics of Schumpeter's work on financing innovations could be summarized as follows (cf. Stratmann, 1998: 16-21): 1) there is a time-lag between payments that have to be made during innovation activity and receipt of payments from already existing innovations; 2) there is a difference between a normal credit (issued against a collateral) and credit whose issuance depends on trust that future value will be created; 3) the financier takes over the whole risk of an innovation, and not the entrepreneur; 4) Schumpeter believes that internal financing is a very important way of financing innovations. Since large enterprises ought to have bigger internal financing possibilities, he concludes that they also have a higher innovation potential than smaller ones; 5) the main tasks of a money market and a capital market are to secure means for financing innovations. Schumpeter believes that entrepreneurs are crucial for innovations, but they have to be financed by banking credit (Christensen, 1992: 30).

The financial and real sectors are functionally separated, but structurally intertwined. The former means financial institutions and non-financial firms are separate legal entities liable to different regulations. The latter means they are interrelated, as, in the long run, the financial sector fundamentally depends on the real sector. Because of this relationship, one can encounter problems with the financing of innovations – both at the firm level and at the level of a national economy – which manifest themselves as lack of finance for innovations in general, and for particular phases of innovation activities in particular, and/or financial sources being too expensive. In the short run, the financial sector does not have to support the real sector, especially not its innovation activities. As we have already elaborated, that could be the case when "financial bubbles" emerge at the stock exchange, that is, the financial sector decouples from the real sector. The separation of the two sectors causes information problems between financial institutions and firms from the real sector. But it would be too easy to say overcoming information problems could solve the problem at the macro level, as entrepreneurs themselves may not be willing to receive an investment and/or may not have knowledge about certain financial sources and/or may not be skilled enough to be able to produce innovations of a certain quality. In other words, there could be demand-side financing constraints. Furthermore, the financial sector has liquid sources, as opposed to innovative companies which are in need of liquidity now, in order to be able to make innovations and produce liquidity at some future point in time. This generally makes the financial sector act conservatively when it comes to financing innovations, since there is considerable uncertainty and risk involved. In our approach, the firm, with its innovation activities and innovations, is considered as a primary player of the innovation game. Its capabilities and strategy are the necessary preconditions for the production of innovations, with the financial sector playing a complementary role.

Moreover, these two sectors of an economy have different capabilities to act – finance is more mobile than the real sector, because capital can easily move from one place to the other, while the real sector may be tied to a certain location. An overly strong role of the financial sector can disrupt innovation processes based on learning and social capital, although it does speed up technological change. Building knowledge capable of creating a competitive position of an enterprise is a long-term endeavour rooted in the social fabric of a society. This stands in sharp contrast to the quick reactions of the financial sector (Lundvall et al., 2002: 225, referring to Jessop, 1999). The predominance of short-term financial logic (not only in the financial sector, but also in the manufacturing sector) and the incentivisation structure it has created have been one of the primary factors that caused the demise of the manufacturing sector in Great Britain. These processes have not only gone hand in hand with the expansion of the financial sector – especially the capital market – and the arm's length relationships it has reproduced, but also with the actions of the political elite and stimulation of high real interest rates by monetary policy (Walker, 1993: 163). Other reasons for the diminishing role of manufacturing in Britain lie in the neglect of education and training in the civil sector, as well as social (overly accentuated individualism and social conflicts as obstacles to coordination) and cultural (lack of a culture capable of sustaining technology) factors (Walker, 1993: 188).

Our approach, one that we shall describe here and apply in the empirical part below, follows Lundvall's (2007) procedure of analysing the core of an innovation system (see above). As we see it, it should analyse the innovation activities of firms in order to determine how new knowledge for innovations is created (e.g., what is the relative importance of R&D as opposed to other innovation activities, such as design, training related to innovation activities, etc.), what is the importance of the knowledge sources that firms utilise in their innovation activities, and what is the quality of the end product (is it incremental or radical innovation?). Hence, it counts whether an innovation is incremental or radical, and whether it predominantly has endogenous or exogenous knowledge content[104]. In other words, it is important to understand whether innovations predominantly result from own knowledge (and innovation activities) or external knowledge (and innovation activities). In that respect, R&D is of primary importance, especially in high-tech manufacturing. The specific weight that R&D activities have stems from their role in the creation of new knowledge and expansion of firms' potential innovation capabilities[105]. This distinction has serious implications for financing. Namely, if innovation activities are predominantly done outside the firm, that is, if a firm only contributes little to the innovation, then there is no need for abundant financing. If improvements in a product are only incremental and relatively simple, then this is not a project that needs specialised financing sources. For example, if a firm imports PCs, and the only improvement it makes is mounting keyboards containing letters peculiar to the firm's country (making a slight adjustment in the software as well), then this can be considered an incremental innovation, but it is in itself not financially demanding. Furthermore, although financiers like venture capitalists also contribute to value added through their managerial, marketing and technological services to the investee, they are not going to be the ones actively managing innovation activities. In other words, firms have to be capable of managing innovations. Although their task is complemented by specialised financial institutions like VCFs, it is primarily the capabilities of firms that count. Furthermore, the innovation capabilities of firms that are built through innovation activities are no more than a subset of their capabilities or competences, although a very important one. Firms make market transactions in markets where they are faced with competition. But they are involved in a more complex set of relationships by engaging in cooperation and networks. Cooperation is a more general term than network, the latter being one of the three main organisational forms, among hierarchies (like firms) and markets[106]. Their examples in technology-intensive industries can be found among strategic alliances and partnerships, joint ventures, collaborative research pacts, etc. (Powell, 1990). Firms do not just cooper-

104 Zoran Aralica brought this point to my attention.

105 This is elaborated in Chapters 2.1. and 2.2.

106 "In network modes of resource allocation, transactions occur neither through discrete exchanges nor by administrative fiat, but through networks of individuals engaged in reciprocal, preferential, mutually supportive actions" (Powell, 1990: 75).

ate with other firms, but also with higher education institutions, scientific research institutions, and other institutions that make up the knowledge infrastructure (see e.g. Meeus and Oerlemans, 2005) in an NSI. Cooperation and non-market-based linkages between innovating firms and other organisations enhance learning, increase the competencies of firms, and create opportunities for financing innovations. These are accessed not just through co-financing of collaborative projects, but also through other forms of cooperation: joint applications for research projects, clustering of firms that cover the whole production process and, hence, enable cost sharing, sharing of some administrative functions in a technological park[107], etc.

The relationships between the players in the financial sector on the one hand and representatives of the real sector on the other evolves over time (cf. Lerner, 2007 and Lamoreaux and Sokoloff, 2007). However, this relationship does not occur in a vacuum, but rather interacts with the corresponding institutional environment (like the system for protecting intellectual property and markets for inventions and innovations), inventors and entrepreneurs, as well as the state. Furthermore, the government, through its funding, seems to play a very important role in the development of some innovations, in addition to setting up a stimulative institutional environment (cf. Lamoreaux and Sokoloff, 2007).

We have already discussed the properties of bank-based and capital market-based financial systems, and noted that the former are generally more stimulative of incremental, and the latter of radical innovations. In our approach, it is important to note that the opposite holds true as well: incremental innovations will have a positive feedback on the bank-based financial system, while radical innovations will have a positive feedback on the capital market-based financial system. Namely, incremental innovations in general have a smaller knowledge content, and are not as financially demanding, uncertain, and risky as radical innovations. Likewise, radical innovations should be more profitable as they are new to the market. Although we cannot consider this an absolute truth, but no more than a tendency, VCFs are more likely to finance firms with radical as opposed to incremental innovations, since they want to reap above-average re-

107 A variety of terms is used for such organisations (e.g. technological centre, technological park, business incubator). Pleschak (1997: 229-230) names several goals of technological centres and business incubators (*Technologie- und Gründerzentren*): to advance the founding of enterprises; to establish future technologies; to facilitate synergies between science and industry, as well as of networking on regional, national and international level; consulting; regional economic development.

turns[108]. Since capital markets are complementary with VCFs (VCFs can exit an investment by IPO), we conclude that radical innovations go hand in hand with capital market-based systems. This is in line with evidence drawn from the literature, as presented above. Incremental innovations are more likely to be supportive of financing by banking credits, since they are more suited to banks' limited monitoring capabilities and financial demands for relatively safe investments.

108 The profit motive for VCFs is not always predominant, since VCFs can have other objectives, such as regional development. See, for example, the website of the Community Development Venture Capital Alliance (www.cdvca.org).

4. National Systems of Innovation in Hungary, Slovenia and Croatia

As recognised by the reviewed theory, the national system of innovation is a crucial determinant of national innovative performance. It is, at the same time, the object of our research in the empirical parts of this study. The capacity of a firm to innovate is not only influenced by its own capabilities, but also by the totality of institutions and organisations – including innovative firms – that we call the national system of innovation. The financial system interacts with the NSI, in that innovative performance is conditioned on financing, and financing depends on innovations (as discussed in the previous chapter). The NSI and financial system themselves are fundamentally influenced by culture and social capital, as they require strong linkages between different players that reach beyond market-based relationships.

The national system of innovation (NSI) is deeply rooted in the socio-economic system in which it functions, and can, therefore, only change in the long run. Consequently, the NSIs of post-transition economies still exhibit some features inherited from the previous socialist system. This is why it is important to outline the historical legacies of these economies. Path-dependence is a useful concept for examining transition (Vincensini, 2001), although Estrin (2002) finds that historical legacies have not had such a great impact on the economic performance of the Central and Eastern European Countries. Slovenia, Hungary and Croatia have shared some characteristics throughout their modern history, but have taken quite distinct paths to economic growth and development. They had also shared some features of the socialist system before the transition started. Even before World War II, they had capitalist economies and western legal systems (Estrin, 2002: 105). The former Yugoslav countries (since 1952) and Hungary (since 1968) broke with the strict, Soviet-type economic planning. The countries of the former Yugoslavia were based on self-management. In all of the aforementioned countries demand influenced production decisions at least to some extent. That was paralleled by the relative autonomy of the enterprises in the area of employment, production, investment and sales (cf. Estrin, 2002). However, the system coexisted with soft budget constraints[109]. The recent socialist past has had an impact on these countries, although it is difficult to measure its magnitude with any precision. During the socialist times, Hungary, Yugoslavia, and Poland launched their market-based reforms long before the transition started[110]. In Hungary, it was the development of the "second econ-

109 This term was originally coined by Kornai, denoting careless loss-making behaviour by state-owned firms stemming from their bailing out by the state. For a survey of the uses of the concept in the literature see Kornai, Maskin and Roland (2003).

110 In 1989 in Hungary and in 1991 in Slovenia and Croatia.

omy", meaning privately-owned businesses in which people were employed either part-time or full time, proliferation of the black market ("moonlighting activities"), and non-market activities pursued by people for their own livelihood – since the late 1960s. In 1970s Poland, change was made apparent by the growth of the second economy and the rise of the Solidarity movement. The former Yugoslavia started to decentralize and liberalize from the 1970s on. Therefore, entrepreneurship in both Hungary and the countries of the former Yugoslavia had had time to develop two or more decades before the transition started (Szelenyi and Szelenyi, 1994).

The historical legacies of Hungary (ch. 4.1.), and of Slovenia and Croatia (ch. 4.2.) can provide us with a better understanding of the socio-economic systems of these countries in general, and their NSIs in particular. Therefore, these introductory notes are followed by chapters on their respective legacies. In the chapter on the general features of the national systems of innovation in Hungary, Slovenia and Croatia (4.3.), we try to assess the kind of macroeconomic and socio-cultural environment in which the national systems of innovation in Hungary, Slovenia and Croatia have been developing. We shall review the innovative performance of these countries and the shared features of their NSIs. Having dealt with that, we shall analyse their financial systems in an effort to learn about the impact on their respective NSIs, or rather, on their innovative performance. In the subsequent chapters, we would like to elaborate on the innovative performance of each of these countries. We explain the distinguishing features of their NSIs with regard to high-tech industries and financial systems. The national system of innovation in Hungary is explained in ch. 4.4., the national system of innovation in Slovenia in ch. 4.5., and that of Croatia in ch. 4.6.

4.1. Historical Legacy of Hungary

Hungary's transition to economy based on markets can be characterized as a gradual one, since it started in 1968. Market reforms in Hungary, dating from 1968, exerted a significant influence on the market way of thinking of Hungarian business people, as well as on consumers (Fogel and Zapalska, 2001). The reforms were especially visible in the 1980s, once the market institutions had been built (e.g. the bond market, commercial banking, liberalization of prices, etc.), and private businesses could be founded since 1981 (Hare et al., 1992). Futo, Hoggett and Kallay (1997) claim that the rapid expansion of small and medium-sized firms in Hungary after the start of transition had roots in the partially liberal economy during the socialist times. They recognise that the culture of being informal, training of people's entrepreneurial skills in the so called "second economy", possibility of self-employment in agriculture and artisanship, as well as decentralized decision-making in large state-owned enterprises, all made the economy relatively well adapted to the market way of regulation and coordination of economic activities. Sik (1994) defines the second economy in terms of "all activities which fell outside the direct control of the socialist state" (Sik, 1994: 47). One of Sik's points (1994) is that the second economy was a double-edged sword. On the one hand, it helped sustain families and build social capital important for a market-based economy, while on the other it influenced the formal economy in a negative way. The latter characteristic was present because the actors had learned to plan only for the short term, and because their practices were distortive for the formal economy. In Sik's (1994) words,

> "[t]he transformation can do a great deal, but it cannot erase ingrained habits and cannot influence people to sacrifice their previous investment into human and network capital on the altar of change. The former practices will continue to survive for a considerable time, so the spirit and the practice of the second economy haunts the new Hungarian market economy and shapes the institutional conditions and the behaviour of those who want to enter the formal economy" (Sik, 1994: 65)[111].

Sik (1992, as referenced in Futo, Hoggett and Kallay, 1997) noted that too heavy a reliance on informality had both advantages and disadvantages. On the one hand, being informal facilitates economic transactions because of how easy it is to strike a deal. On the other hand, informality may discourage entrepreneurs from being growth-oriented, since the smaller the firm, the easier it is for it to evade taxes.

Although the liberalization of Hungary's economic system during the communists' rule represented a welcome contribution to the development of entrepreneurship, the liberalization policy was inconsistent and, therefore, created an uncertain environment for sustaining entrepreneurship. The entrepreneurs never knew what was coming next, which made them biased in favour of the short-term. Changes in the economic system during the transition were very intensive and also contributed to uncertainty in doing business (Danis and Shipilov, 2002). However, the SME sector had made a giant leap forward by the end of the 1980s, when it contributed 1/3 to the country's GNP (Fogel and Zapalska, 2001: 36). Hisrich and Fulop (1995) contend about a cultural misfit in Hungary regarding the predominantly negative perception of entrepreneurship and entrepreneurs, however less than before 1989. Cultural change enabling a positive entrepreneurship climate is likely to take a considerable amount of time. Hisrich

111 Path-dependency in Hungary's economic development is also visible as regards the uneven development of the country. Hungary was highly centralized before World War I and it is quite unevenly developed today, with Budapest being not just the country's capital, but also the centre of its economic activities. Although Hungary achieved a more even development during socialism, this period of time obviously had no influence on the long-term situation regarding economic development (Lorentzen, 1999).

and Fulop (1995) assert that the 1980s were particularly important for the development of entrepreneurship in Hungary, with flourishing private or semi-private companies becoming serious competitors to the state-owned enterprises.

4.2. Historical Legacies of Slovenia and Croatia

As opposed to other socialist countries, socialism in the former Yugoslavia relied to a great extent on markets, as well as on social ownership and the famous "self-management" principle (Minniti and Polutnik, 1999). However, that did not happen before 1955, because up until that time central planning, along with state-owned enterprises, prevailed (Ostojić, 2001, as referenced in Cvijanović, 2007: 63). From 1976 on, foreign investment was tolerated in some cases (Bojnec, 1999). In spite of the prevalence of large enterprises and chronic lack of SMEs, there was room for private initiative in the form of trades (Cvijanović, 2007). Indeed, both Slovenia (Minniti and Polutnik, 1999; Bojnec, 1999) and Croatia have benefited from entrepreneurial activities during the socialist period in terms of easier adjustment to a capitalist economy. Due to the lowering of legal restrictions from 1988 onwards, and because of the introduction of market-based laws, the number of SMEs in Slovenia has grown rapidly since 1990 (cf. Minniti and Polutnik, 1999).

Slovenia and Croatia had much better initial conditions than Hungary with respect to many fields of the socio-economic life: enterprises and banks were more independent in Slovenia and Croatia, and they had less state control. Some Slovenian enterprises in particular were oriented towards exports to the OECD markets, which made them more competitive (cf. Whitley, 2000). Slovenia was the most developed of all the former Yugoslav republics, and also the one most oriented towards the West in terms of both international trade and joint ventures with foreign companies. Croatia was in second place regarding the GDP per capita. Its international trade was not as oriented towards the West, but rather towards the inner market of the former state and other east European markets (Kušić and König, 1999).

We could, therefore, say that these very favourable initial conditions, including the *institutional* level, were something that Slovenia and Croatia shared before the transition started. But they took different paths during the transition process. After the proclamation of independence on 25 June 1991, Slovenia[112] entered into a military conflict with the Yugoslav Federal Army, but it only lasted for a week (cf. Mencinger, 2004). Croatia was engaged in a prolonged war (1991-

112 Croatia declared independence the same day.

1995) with disastrous human and socio-economic consequences[113], in which not only the Yugoslav Federal Army was involved, but also paramilitary forces. If Slovenia's transition can be characterised as gradual (Mencinger, 2004), with reforms pursued from the very beginning of the transition[114], then we could call Croatian transition a protracted one. Croatia's occupied territories were only re-integrated as late as January 1998, and Croatia is still struggling with reforms that are required for EU accession and important for a functioning economy[115].

4.3. General Features of the National Systems of Innovation in Hungary, Slovenia and Croatia

After reviewing the macroeconomic characteristics of the three countries, we shall tackle the issue of privatisation. It has been one of the key elements in the process of creating a strong and viable private sector. Its results varied not only according to the macroeconomic conditions, but also with regard to the socio-cultural conditions. The latter shall be outlined in relation to social capital and conditions for entrepreneurship, along with resultant entrepreneurial activity. Following the aforementioned preconditions of the NSIs, we next review the innovative performance and competitiveness of these countries, focusing on high-tech industries. After briefly summarising the general features of their NSIs, we describe how the financial system interacts with these.

Restructuring an economy and moving towards greater innovative performance is a time consuming process that requires systemic changes. In terms of overall economic – and especially innovative – performance, these countries do not belong to core economies. To put them into the European context, we could divide European countries into core countries, semi-peripheral countries, or peripheral countries, as Rončević (2002) did[116]. Societies need to be civilisationally competent and to possess a certain quality and levels of social capital in order to be (or move toward) the core countries group. In this regard, some influence is exerted

---

113 In terms of the economic consequences, Kušić and König (1999) mention that about a third of Croatia's territory was occupied and approximately a third of Croatia's economic potential was ruined.

114 Mencinger (2002) argues that the initial conditions and path taken during the transition were especially important for Slovenia's transition.

115 Croatia, just as any other country in the process of accession to the EU, has to harmonise its legal system with the *acquis communautaire* of the EU. By 23 December 2008 it had not yet opened a number of chapters that are very important for a functioning market-based economy, such as the free movement of capital and market competition (for this and other information on the current status of negotiations between Croatia and the EU, see Ministry of Foreign Affairs and European Integration, 2009).

116 The framework that Rončević (2002) applies is the already mentioned heuristic model of socio-cultural factors of developmental performance as in Adam et al. (2005).

by historical legacies. However, the classification of countries into core, semi-peripheral, and peripheral by Rončević (2002) is, by his own confession, quite arbitrary. The key criteria for such a classification are the threshold of 1 US utility patents per million inhabitants[117] and a shared border (on land or at sea) with the "key innovators", that is, one of the "European core" countries. Also observed was the level of the GDP. In his classification[118], Slovenia (with 8 US utility patents per million inhabitants in 2000, a shared border with key innovators, and the GDP per capita in PPP in 2000 17.367 USD) and Hungary (with 3,6 US utility patents per million inhabitants in 2000, a shared border with key innovators, and the GDP per capita in PPP in 2000 12.416 USD) were among the semi-peripheral countries. Rončević (2002) did not place Croatia into any of the groups, but following his criteria we could place Croatia either together with the other countries of the "European semi-periphery" (possibly at the bottom of the list) or perhaps even with the countries of the European periphery (far ahead of the rest of the group). Namely, in 2000 Croatia had 0,85 USPTO utility patents per million inhabitants, the GDP in PPP per capita of 8,091 USD, and a shared border with a core innovator (Italy) at sea (UNDP, 2002; USPTO, 2008 and own calculations).

Slovenia is the most successful of the three compared countries if we consider just the GDP and its growth over the period from 1990 to 2007 (see Figure 2). Transition depression could be noticed in all three countries, although with different levels of intensity. While Slovenia's economy shrank by 14% and Hungary's by 15%, Croatia's GDP level sank by 36%. Slovenia was the first to reach its 1990 GDP level, which it achieved in 1996. It took three more years for Hungary (that is until 1999) to get to its 1990 level of GDP, and 8 more years (until 2004) for Croatia to do the same (United Nations Statistics Division, 2009 and own calculation). The steep fall of the GDP and its protracted recovery in Croatia can to a great extent be explained as a result of the 1991-1995 war, and of the politically and economically unstable situation that continued into the late 1990s (cf. Cvijanović and Kušić, 2002: 886). Economic growth in Hungary has been fuelled by a huge inflow of foreign direct investments (FDIs), which was made possible by a liberal privatisation strategy open to foreign investors (Cvijanović and Kušić, 2002: 888). Slovenia has managed to fully utilise excel-

117 Rončević (2002) was inconsistent in this paper, as he "softened" the criteria for Slovakia. It only had 0,7 US utility patents per million inhabitants, but also a shared border with one of the core innovators, as well as a GDP level comparable with the other semi-peripheral countries.

118 Rončević (2002) takes the data for US utility patents from The Global Competitiveness Report 2001-2002 (World Economic Forum, New York, Oxford: Oxford University Press) and the GDP per capita (PPP in USD in 2000) data from the Human Development Report 2002 (New York, Oxford: Oxford University Press).

lent conditions at the start of the transition process. Its reform process has been gradual in socially stable conditions (Mencinger, 2002). However, all three countries were hit by the economic crisis that originated in the USA in the summer of 2007[119]. While their real GDP could still grow in 2008, it is expected to experience a sharp drop in 2009. The real GDP growth in Hungary in 2008 was only 0,4%[120], and in 2009 it is expected to fall by 6,1% (OECD, 2009a). The real GDP growth rate of Slovenia in 2008 was 3,5%[121] (Bank of Slovenia, 2009: II.-93), while a fall by 5,8% in 2009 is forecasted (OECD, 2009b). Croatia's real GDP grew by 2,4% in 2008 (CNB, 2009a). The economic crisis has started to have an impact on Croatia towards the end of 2008[122] (CNB, 2009b: 33). Croatia's real GDP is predicted to fall by 3,5% in 2009 (IMF, 2009: 15).

Figure 2: GDP in Hungary, Slovenia and Croatia in constant 1990 prices in million US dollars

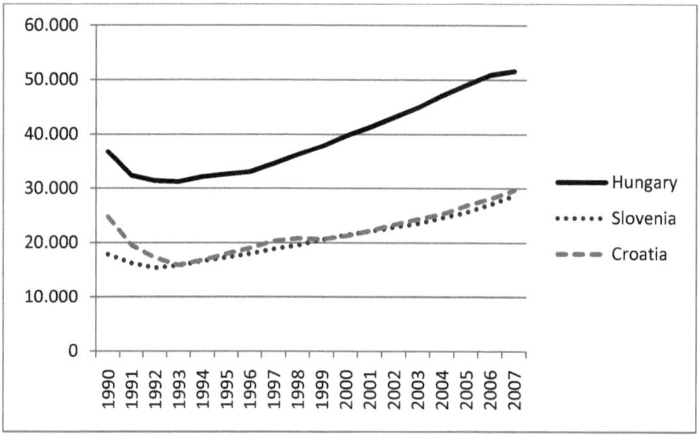

Source: United Nations Statistics Division (2009).

119 This economic (financial) crisis, usually called the sub-prime crisis, had two main channels through which it spread: one was increasing risk premiums and the other was harder access to liquid funding sources (Király, Nagy and Szabó, 2008: 38).

120 It fell in the last quarter of 2008 by 2,5% year-on-year (HCSO, 2009). Output of manufacturing industry (value added) fell in the first quarter of 2008 (Eurostat Database, 2009a), reflecting the economic crisis.

121 It only fell in the fourth quarter of 2008 by 0,8% year-on-year (SORS, 2009). Output of manufacturing industry (value added) fell in the third quarter of 2008 (Eurostat Database, 2009a) because of the impact of the economic crisis.

122 The growth of the GDP slowed down in the fourth quarter of 2008, as the growth rate of that quarter in year-on-year terms was only 0,2% (CROSTAT, 2009). Output of manufacturing industry (value added) fell in the third quarter of 2008, just like in Slovenia (Eurostat Database, 2009a).

Enterprise restructuring in transition economies is positively correlated with privatisation in all transition economies, except in those of the Commonwealth of Independent States (Djankov and Murrell, 2002). Privatisation has been one of the main pillars of transition and has been important for a functioning economy largely based on markets[123]. It could be seen as a bridge between the past structure of the economy and the future structure of the economy. Croatia and Slovenia employed the same primary and secondary privatisation methods of management and/or employee buy-outs, and voucher privatisation respectively (probably due to path-dependency – cf. Kušić, 2005; See also Kušić, 2001). Hungary used direct sales as a primary privatisation method, and management and/or employee buy-outs as a secondary method (Estrin, 2002: 109). Estrin (2002) notes that privatisation effects have been slow to manifest themselves. In the earlier transition literature, negative results were dominant. In the recent literature on transition economies, positive effects were prevalent. The results of Hungary's small-scale and large-scale privatisation have been assessed as better than either Slovenia's or Croatia's in 2008[124]. In connection with that, Hungary has also managed to better reform its enterprises than either Slovenia or Croatia[125]. Kušić (2001) believes that Hungary's privatisation was more efficient because it was open to strategic investors who had to pay for their shares in newly privatised enterprises. In his opinion, it was especially important that privatisation was open to foreign strategic investors, which resulted in higher competitiveness for Hungarian economy. Although the methods of privatisation were similar in Croatia and Slovenia, Račić and Cvijanović (2005) describe Croatia's privatisation as an especially sluggish process that has not contributed to the restructuring of enterprises. They claim that it was mismanaged and coupled with ample illegal practices. Due to war and a prolonged transition depression, its negative effects were amplified.

Newly privatised firms and newly founded firms have been confronted with hard conditions for doing business due to institutional insufficiency and a rather restraining cultural environment. Croatia seems to have worse conditions for entrepreneurship than either Slovenia or Hungary. Namely, although the Doing Business 2008 report (WB and IFC, 2007) listed Croatia among the top ten reformers 2006/07 according to the ease of doing business indicator[126], Croatia

123 A mixed economy, as it is called in the literature.

124 EBRD's index of small-scale privatisation and large scale privatisation, respectively, for each of the countries under analysis, has been as follows: Hungary – 4,3 and 4,0; Slovenia – 4,3 and 3,0; Croatia – 4,3 and 3,3 (EBRD, 2009).

125 EBRD index of enterprise reform in 2008 for Hungary is 3,7, and for Slovenia and Croatia 3,0 (EBRD, 2009).

126 The ease of doing business indicator is calculated as an average of country's rankings in ten different indicators: starting a business, dealing with licenses, employing work-

was ranked 97, Slovenia 55, and Hungary 45 among 178 countries from the 2008 survey. Furthermore, Croatia lags behind the other two countries in two indicators that measure the quality of institutional framework[127] of World Economic Forum (2008): the legal definition and enforcement of property rights and intellectual property protection[128]. Cultural factors, mostly due to socialism, have further restrained the development of entrepreneurship. Public opinion on entrepreneurs in Croatia, according to Karajić (1998)[129] is twofold: some believe domestic entrepreneurs are agile and that their activities contribute to the development of the economy. However, Croatian entrepreneurs have unequivocally been perceived in a rather negative way. First and foremost, they have been considered as closely connected with the political elite and as not being moral (that is, as exploiting others and being unfair)[130]. Being an entrepreneur is not valued

---

ers, registering property, getting credit, protecting investors, paying taxes, trading across borders, enforcing contracts, and closing a business.

127 Moreover, institutions for SMEs in Croatia have started to develop only in the mid 1990s, with a sector within the Ministry of economy in charge of SMEs. The first major official event covering SMEs was established in 1998 – a national conference on SMEs. It was not before 2000 that the first Ministry for Crafts and SMEs was founded and economic policy towards SMEs was institutionalised. Another important governmental agency was founded in 2002 – the Croatian Agency for Small Economy (HAMAG), which is very important for implementation of policy towards SMEs. With a new government elected in 2003, the Ministry of Economy, Labour and Entrepreneurship emerged, and it took over the role of the main institution for SMEs, which it remains. Another important governmental institution supporting SMEs is the Croatian Bank for Reconstruction and Development. Other quasi-governmental associations are the Croatian Chamber of Economy and the Croatian Chamber of Trades and Crafts. The former was explicitly established by Law in 1991, the latter only indirectly prescribed by another Law (Singer and Lauc, 2004).

128 The indicators were measured in 134 countries. In the former indicator, Croatia is ranked 84, Slovenia 57, and Hungary 46; in the latter indicator, Croatia is ranked 58, Hungary 45, Slovenia 41 (World Economic Forum, 2008).

129 Karajić's (1998) research was conducted in November 1996 on the perceptions of the general public regarding the transition in Croatia. The empirical research encompassed questions on democracy, political participation, and economy. As regards the latter category, it was further broken down into segments concerning the desirable relationship between the state and markets, privatisation and, finally, characteristics and activities of Croatian entrepreneurs. Citizens who answered questions were found to be in favour of the so called "state and political paternalism" and social state concept on the one hand and "market as regulatory mechanism" on the other. Karajić explains state and political paternalism to denote total control of the market by state institutions as well as political decisions. He considers the paternalistic preferences of the population to be rooted in habits developed during the socialist times.

130 However, the attitude towards entrepreneurship is not the same in all parts of Croatia. Čengić (2002) focused on the Croatian county of Međimurje (bordering on Slovenia and Hungary) because of some extraordinary characteristics of the local entrepreneurs. Čengić notices an above-average share of manufacturing SMEs in comparison with

in the Slovenian society (Močnik, 1999, as referenced in Ženko, Mulej and Marn, 2004: 184). A culture not as conducive to innovation is not just a result of religion (Roman Catholic), but also of the socialist heritage, in which solidarity has been valued more than competition (Ženko, Mulej and Marn, 2004: 184). In Hungary, cultural factors (Kuczi, 2000 and Szerb-Ulbert, 2002 – referenced in Szalavetz, 2007: 64) influenced by the socialist past were found to be limiting firms' growth (Szalavetz, 2007: 64).

Social capital (Paldam and Svendsen, 2000) and trust (Whitley, 2000: 210-212), as one of its components, were being continuously destroyed in the "old communist" countries of Central and Eastern Europe. In terms of entrepreneurial culture and social capital, research shows that Hungary, Slovenia, and Croatia are relatively similar countries. Thomas and Mueller (2000) measured four different characteristics of entrepreneurs and found that there were differences in entrepreneurial culture across these societies. These four characteristics were as follows: 1) innovativeness (in terms of being creative), 2) risk-propensity, 3) energy level (in terms of commitment), and 4) internal locus of control. The USA served as a benchmark, and eight other countries were analysed: Singapore, Canada, Ireland, Belgium, Germany, China, Croatia and Slovenia. Entrepreneurs in the USA scored highest in all of the characteristics except in risk-taking propensity, where they came in third. All of the aforementioned measures of entrepreneurial culture varied across the countries, except innovativeness. Croatia and Slovenia scored lowest at risk-taking propensity – Croatia coming in last, and Slovenia at second-to-last. Slovenian and Croatian entrepreneurs' energy level is at the bottom end of the scale among the countries from the sample. The situation is similar with regard to the internal locus of control. Raiser, Rousso and Steves (2003) use the 2002 Business Environment and Enterprise Performance Survey to find that Croatia, Slovenia and Hungary had similar levels of trust in 2002. The authors use prepayment demanded by suppliers as a measure of (dis)trust – the lower the level of prepayment demanded the higher the trust in a country. Moreover, trust is found to be strongly correlated (at the country level) with higher income per capita and good governance (Raiser, Rousso and Steves, 2003: 17). Rus (2002) analyses the results of the survey on social capital in three countries of former Yugoslavia: Slovenia, Bosnia and

service SMEs, as well as a share of successful SMEs, and a lower rate of unemployment. The proximity of two EU-members did not prove a significant predictor of success of SMEs. His results point to a prevailing influence of the recent past, before the transition started, especially of the following: a positive political attitude in relation to craftsmanship in the 1980s and a relatively long tradition of crafts in that region. Both Čengić (2002) and Poljanec-Borić (2002) notice a large share of Croats who returned to the county from abroad. The latter factor might have contributed to the success of SMEs in the county of Međimurje.

Herzegovina and Macedonia. He finds that social capital in Slovenia is on average higher than in the latter two countries. In a nod to Putnam, Štulhofer (1998) denotes social capital as socio-cultural capital and stresses its two main components: the associational dimension (measured by the number of members of various voluntary non-governmental organisations) and the trust dimension (as the number of people in the population who believe that people can be trusted). Although Štulhofer did not have strictly comparable data (since he used data for 1995/96 for Croatia and for 1990/91 for other countries), he found the level of Croatian socio-cultural capital to be approximately at the level of the other countries he compared (Slovenia, Hungary, Poland, Romania and Bulgaria). Furthermore, he found the level of socio-cultural capital in Croatia to be much lower than in developed European countries. Svendsen (2005) also finds some measures of social capital (general level of trust and voluntary organisations' density) to be higher on average in Western European countries than in Eastern European countries. Data from the World Values Survey (2009) confirm the general level of trust to be similar in all of the three countries that we study (see Figure 3). It is interesting to see that the general level of trust declined in Slovenia in 2005 in comparison with 1999. However, it did remain at a level similar to the other two countries. Such low levels of trust in the studied countries could negatively influence innovation activities. Namely, it may hinder linkages of the firm with its environment that could be crucial because of knowledge sources and options for (co)financing innovation activities. As high-tech industries require intensive cooperation and financing, they may be particularly affected.

Figure 3: General level of trust in Hungary, Slovenia and Croatia

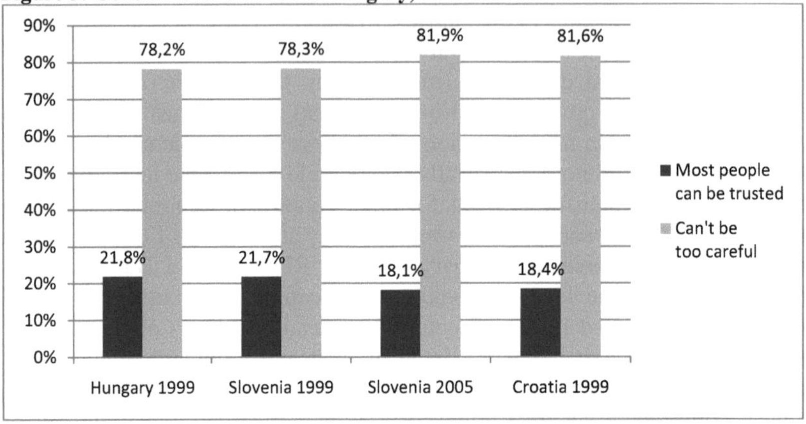

Note: Data refer to 1999 for all countries, and for Slovenia there are additional data for 2005. The question the respondents were asked was the following: "Generally speaking, would you say that most people can be trusted or that you need to be very careful in dealing with people?"
Source: World Values Survey (2009).

The aforementioned cultural, social and institutional conditions have had an impact on the viability of entrepreneurial activities. It is a factor that can contribute to our understanding of the overall capacity of the NSIs of Slovenia, Croatia, and Hungary to produce innovations[131]. This capacity will be higher the more growth-oriented enterprises there are. Indicators presented below bear witness to Croatia's position being relatively worse in comparison with Slovenia and Hungary[132]. An especially important indicator to consider is the one measuring total

131 To judge the extent and quality of entrepreneurial activities, as well as to approximate the success of transition from the start-up to the mature stage of the enterprise, we use the Global Entrepreneurship Monitor (GEM). This report was a part of a large international project which analyses differences in entrepreneurial activities across countries.

132 Some other indicators measured in the GEM study reflect the relatively similar position of all three countries. In 2006, all three compared countries performed poorly regarding cultural and social norms. Croatia ranked 30th (out of 37), Hungary 35th, and Slovenia 36th. Another indicator covering cultural and social norms is represented by the attitude towards women in entrepreneurship. Here we find Hungary at the bottom of the list (37). Croatia is 32nd, and Slovenia 20th. Hungary was ranked 13th regarding primary and secondary education, and was followed by Croatia (15th) and Slovenia (25th place out of 37). Education after secondary school is again best rated for Hungary (11th place), again followed by Slovenia (19th) and Croatia (20th). Government policies for regulations in 2006 were best ranked for Croatia (22nd), followed by Hungary (30th), and Slovenia (35th). The same can be said for government programmes, with Croatia ranked 16th, Hungary 27th, Slovenia 28th. Croatia made a sig-

entrepreneurial activity (TEA index[133]), which can be split into two: the TEA opportunity index (indicating the number of people who start a business because they have perceived an opportunity) and the TEA necessity index (indicating the number of people who were forced to start a business because of their poor socio-economic situation). There was at one point an upturn in entrepreneurial activities in Croatia, reflecting itself above all in growing entrepreneurial activities[134]. A negative side to it was that this was mainly achieved through the growth of the TEA necessity index. In fact, Croatia was the only country in 2005 with a ratio of TEA necessity index to TEA opportunity index higher than 1 (Singer et al., 2006)[135]. This is considered bad for viability of economic structure, since entrepreneurs driven by necessity are normally not so centred on business growth, unlike entrepreneurs driven by opportunity (Singer, 2007). Such negative results for Croatia are presumably due to its rather late recovery from the transition depression, coupled with particularly slow and inefficient privatisation, and weak institutions that delayed the restructuring of enterprises (cf. Račić and Cvijanović, 2005). Another noteworthy trend is the growth of small enterprises with a growth potential. The methodology of the GEM research enables us to compare Croatia with both Slovenia and Hungary. As regards the already mentioned ratio between TEA opportunity and TEA necessity indexes (which represent entrepreneurial motivation), Croatia was placed last among 35 countries in 2005. Slovenia occupied 7th place, Hungary 30th. As to the success of transition from the start-up to the mature stage of the enterprise, Croatia was placed 27th, while Slovenia (7th) and Hungary (11th) fared much better (Singer et al., 2006). According to the GEM research for 2006 (Singer, 2007), entrepreneurial activity in Croatia (measured by the TEA index) advanced. This remains a step forward even after the decomposition of the TEA

---

nificant leap forward regarding financial support in 2006, as it was ranked 19th, closely followed by Hungary (21st) and Slovenia (26th) (Singer, 2007).

133 "**Entrepreneurial activity of the nascent entrepreneurs and new business owners**, measured by the ratio of the number of such entrepreneurs (start-up) in the sample of the adult population, aged 18 to 64, is expressed through Total Entrepreneurial Activity (TEA) index" (Singer, 2007: 13, emphasis original).

134 In the 2002-2005 period, a decrease in regional disparities in entrepreneurial activities in Croatia is noticeable as well. The entrepreneurial environment in Croatia has generally been developing in the fields of entrepreneurial education, business, professional and physical infrastructure. The downside has been a pronounced one regarding the underdevelopment of the financial markets, as well as weak government policies (Singer et al., 2006). In 2006 in Croatia, entrepreneurial activity became more equable across the regions, with general improvement in the entrepreneurial environment (Singer, 2007).

135 Singer et al. (2006) have surveyed 2000 people from Croatia's total population each year (inquiring about their entrepreneurial activity), as well as 36 experts (on specific questions regarding the entrepreneurial environment).

index into TEA opportunity and TEA necessity indexes, because the ratio between the former and the latter is finally positive. However, the only country behind Croatia, according to this motivation index (TEA opportunity/TEA necessity), is Brazil. That means Croatia occupied 41st place in 2006. Slovenia managed to rank 9th, and Hungary was ranked 21st. In the transition from start-up to maturity, enterprises in Croatia (place 35 out of 42 countries) did worse than both those in Slovenia (place 13) and those in Hungary[136] (place 7).

Innovative and competitive performance of Hungary, Slovenia and Croatia lags behind more developed (or "core" countries), but there are differences between them. Slovenia's innovative performance appears to be ahead of both Hungary and Croatia. As competitiveness and innovation are interdependent, it is not surprising to observe Slovenia is more competitive than both Hungary and Croatia. This can be seen in the overall score in the Global Competitiveness Index 2008-2009: out of 134 countries, Slovenia is ranked 42, Croatia 61, and Hungary 62. We have graphically depicted indicators for each of the areas of Porter's diamond in Appendix III, Figure 19. It is particularly noticeable that Slovenia has no negative outliers in any of the dimensions shown. In comparison with Hungary and Croatia, Slovenia is particularly strong in the "strategy and operational effectiveness" of its firms, as well as in supporting industries. It is also noticeable from Figure 19 that the three countries have relatively similar position in the "efficiency enhancers" area (which partially corresponds with Porter's factor conditions). Better innovative performance of Slovenia is reflected in the European Innovation Scoreboard 2007 (European Commission 2008), which – on the basis of the Summary Innovation Index (SII)[137] (see figure 4) – distinguishes between 4 groups of countries with different qualities of innovation performance. These are: 1) innovation leaders (Sweden, Switzerland, Finland, Israel, Denmark, Japan, Germany, the UK and the US), 2) innovation followers (Luxembourg, Iceland, Ireland, Austria, the Netherlands, France, Belgium and Canada), 3) moderate innovators (Estonia, Australia, Norway, the Czech Republic,

---

136 Entrepreneurship in Hungary has differed in scope and scale since the beginning of transition. Acs et al. (2006) claimed it had gone through distinctive stages. The first (1990-1996) was marked by large-scale creation of new firms. However, most of these firms had to close down due to, at least partially, "Hungarian entrepreneurs' limited experience and inability to meet the requirements of the market economy and competition" (Acs et al., 2006: 23). From 1996 through 2000, the rate of growth in the number of new companies declined, as a consequence of inward foreign direct investment. Nowadays, there is a rise in the biotechnology sector. However, some claim (Inzelt and Szerb, 2004 – as mentioned by Acs et al., 2006) that public policy lacks an emphasis on entrepreneurship.

137 Summary Innovation Index is a number that indicates relative performance of a respective country's national innovation system. In its calculation the most recent statistics are used, as displayed in Appendix III.

Slovenia, Italy, Cyprus and Spain), and 4) catching-up countries (Malta, Lithuania, Hungary, Greece, Portugal, Slovakia, Poland, Croatia, Bulgaria, Latvia and Romania)[138]. The countries in the first two groups are above or around the EU27 average (actually, only Canada is below that average), and the countries in the last two groups are below the EU27 average. However, it should be noted that the SII cannot provide the exact measure of national innovation performance due to some data problems (it is calculated on the basis of the latest *available* data – i.e. those available in the table in Appendix III[139]) and the weight it gives to specific indicators of innovation performance. Moreover, the classification of EU27 and other countries can be criticised on the grounds of subjectivity. However, the SII can indeed be used to compare various countries, particularly in over a longer time period.

An indicator that links innovativeness and competitiveness is certainly the number of patents registered[140]. The more patents there are, the more revenues can be created on the basis of them. Slovenia is the best performer among the three countries regarding the number of EPO, USPTO and Triad patents per million population. However it is below the EU average (see table in Appendix III). Such patenting activity is reflected in terms of revenues from innovative products. In a recent survey (The Gallup Organization, 2007) conducted among the SMEs of EU27, as well as those of Norway, Iceland, and Turkey, Hungary's SMEs were found to be in third place on the list ranking the SMEs of countries that reported no revenue from innovation. On that same list, Slovenia comes in second-to-last (The Gallup Organization, 2007: 75). Slovenia tops the list of countries, where SMEs are concerned, that did report revenue from innovative products and/or services, with 24% of turnover coming from innovative services/products. Hungary's innovative SMEs reported only 13% of turnover from innovative services/products, so Hungary is closer to the EU27 average, but still above it (The Gallup Organization, 2007: 76).

138 This group also includes Turkey, although its SII score is less than half of Romania's, which is the second worst on this list.

139 Indicators that measure output of the NSIs we compare here are the following indicators: exports of high technology products, (no. 17), sales of new-to-market products (no. 18), sales of new-to-firm products (no. 19), European Patent Office (EPO) patents per million population (no. 21), United States Patent and Trademark Office (USPTO) patents per million population (no. 22), triad patents per million population (no. 23), community trademarks per million population (no. 24), community industrial designs per million population (no. 25). Other indicators could be viewed as inputs into the NSIs. Unfortunately some data are missing, especially for Croatia (10 out of 25 indicators). In the remaining indicators, Croatia is ahead of Slovenia and Hungary in only 2 indicators (youth education attainment level and public R&D expenditures) (see table in Appendix III, as well as explanations of each indicator).

140 I thank Zoran Aralica for bringing this point to my attention.

Figure 4: Summary Innovation Index (SII) from the European Innovation Scoreboard

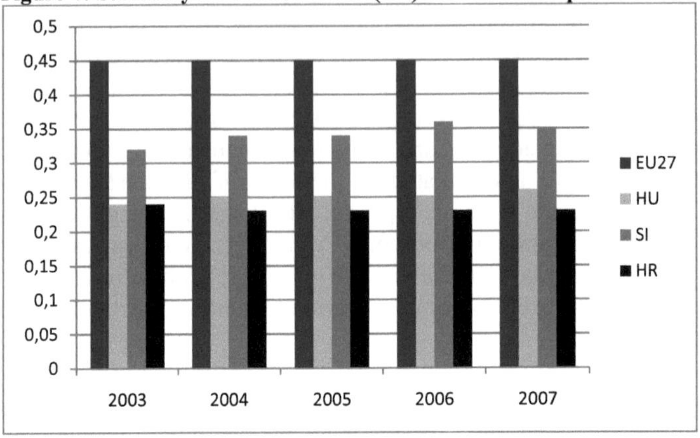

Note: 27 members of the European Union are denoted as EU27; Hungary is abbreviated as HU, Slovenia as SI, and Croatia as HR.
Source: European Commission (2008: 51).

A developed NSI is especially important for high-tech industries with high investments into R&D. We can estimate the relative importance of these industries (i.e., their products) by the share of exports of high-tech products in total exports, as in Figure 5. At the same time, the value of this indicator demonstrates their competitiveness in foreign markets. Hungary has had a fairly stable value for this indicator, above 20% - higher than the EU average (see table in Appendix III). However, as we shall explain below, the knowledge intensity of Hungarian high-tech exports is questionable. Slovenian high-tech exports have been quite stable around the 5% level, and Croatian high-tech exports have been a little bit above 5%. However, they fell in the period 2004-2006, as shown in Figure 5.

Figure 5: Exports of high-tech products as percentage of total exports

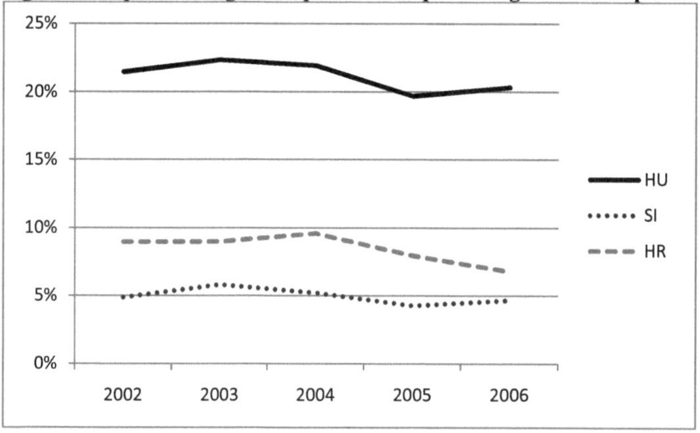

Note: Similarly to the classification of manufacturing industries according to the level of technology, there is an alternative classification of high-technology products that is used in this figure. The main criterion for such classification of products is the value of the R&D intensity indicator, calculated as R&D expenditure/total sales. The classification that Eurostat uses for this indicator is based on the Standard International Trade Classification (SITC, revision 3) (Eurostat Database, 2009b – see there Reference Metadata in Euro SDMX Metadata Structure (ESMS)); See also Aralica (2007) for detailed explanations on classification of products).
Source: Eurostat Database (2009b); Author's calculation.

The weaker innovative performance of the countries in focus as compared to the EU27 average is primarily due to their underdeveloped NSIs. The national systems of innovation of Central and Eastern European Countries (CEECs) have many similarities, and still suffer from some problems. Radosevic (2006: 47) points out that these problems have a dual nature, since enterprises have not been strong generators of innovations, and links between the main innovation agents have been weak. As for the former, there is a small share of innovators among SMEs (except in Estonia). The latter regards linkages with all other agents in the national systems of innovations. Policies to address these problems have generally been insufficient in CEECs, and have even concentrated too much on R&D and high-technology sectors. In that sense, they have disregarded firms producing at other technological levels, whose share in these economies is the largest. They have also disregarded empirical evidence suggesting that firms in CEECs make little use of R&D (Radosevic, 2006: 48). From the network alignment perspective, we could say that national systems of innovation in post-transition economies have not managed to develop in a satisfactory way. They function below their optimum and, therefore, do not manage to fully support the needs of enterprises. In the socialist times, firms in formerly socialist economies

– especially those behind the "Iron Curtain" – were generally involved in quite linear and hierarchical linkages with other actors in the national systems of innovation (von Tunzelmann, 2004). However, nowadays, firms need to be independent and competitive business organisations with multiple linkages in the NSI. Because of the fact that firms in formerly socialist economies used to be essentially organised around production activities, they now have weak capabilities to link with other actors in the supply chain (Radosevic and Yoruk, 2004). Freeman (2006) does not explain the former socialist economies' lagging behind developed countries only in terms of R&D, although he emphasizes that the R&D sector remained strongly concentrated in public institutes. He believes that the main reasons for the weak performance of those countries in the period after the Second World War and before 1990 lay in their totalitarian regimes, too high a share of military investments (like in the USSR), and bureaucracies that stiffened the socio-economic systems, whereby R&D at the enterprise level was low or nonexistent. At the time, a rather weak role was played by universities in terms of fundamental research, with some exceptions in Poland, Czechoslovakia and Hungary. Freeman (2006) did not provide an account of what was going on in this regard in the former Yugoslavia, nor particularly in Hungary, but he certainly did identify factors which should be considered in the analysis of the national systems of innovation of today. We claim that the weaker innovative performance of post-transition countries in general, and Hungary, Slovenia and Croatia in particular, is amplified through their financial systems. This happens because of the already identified interdependency between (their) bank-oriented financial systems and (their) innovative output of lesser quality (or incremental innovations). In a mutually dependent relation, banking credit is more likely to support incremental innovations, and because of incremental innovations enterprises will not (be able to) demand more sophisticated financial sources, such as VC. This makes the structure of the economy rigid.

Namely, the financial systems in Croatia, Slovenia and Hungary resemble that of the European Union[141]. The latter is predominantly bank-based and is much more developed in the old member states of the EU than in the post-transition countries that joined the EU in 2004 and 2007 – in terms of assets of financial institutions to the GDP (Allen, Bartiloro and Kowalewski, 2007). For example, in 2002, financial sector assets in the 15 old member states of the EU were 5,58 times larger than the GDP, and in the 8 post-transition countries that joined EU in 2004 that figure was 1,17 in the same year (2002) (Bakker and Gross, 2004: 36-37). Croatia's financial system has shown a similar performance both in terms of being bank-oriented and in terms of its size (cf. Dalić, 2002 and CNB, 2007). However, the financial markets of the analysed countries are less sophis-

141 Financial systems in transition economies are bank-based (Egert, Backe and Zumer, 2007).

ticated than the EU27 median (see World Economic Forum, 2008). In comparison with Slovenian and Hungarian firms, Croatian ones are much more dependent on banks, which can be seen from Table 8. The percentage of firms using banks to finance investments is a lot higher than the OECD average, while Hungary and Slovenia hover around that average. The highest percentage of firms that identify access to finance as a major constraint are registered in Hungary, while the percentage of both Slovenian and Croatian firms that have such problems is lower than in the OECD countries. Among the EU countries, Hungarian firms are the second most limited in access to finance, behind Malta. Slovenian firms are slightly more limited in access to finance than the EU average (The Gallup Organization, 2007). Hungary and Croatia lag behind Slovenia in terms of ease of access to loans (see World Economic Forum, 2008)[142]. In terms of credit financing of enterprises, this does seem to give Slovenia an edge in comparison with Hungary and Croatia[143].

Table 8: Some financial statistics for Hungary, Slovenia and Croatia

Finance	Croatia	Hungary	Slovenia	Region - Eastern Europe and Central Asia	OECD	All countries
% of Firms with Line of Credit or Loans from Financial Institutions	67,29	-	-	51,19	-	33,34
% of Firms Using Banks to Finance Investments	60,00	22,30	29,60	16,75	22,40	16,13
% of Firms Using Banks to Finance Expenses	63,22	33,61	38,57	25,63	34,65	27,56

142 In Hungary and Croatia, it is much harder to obtain a bank loan than in Slovenia, with a sound business plan and without collateral. Namely, in the World Economic Forum (2008) survey Slovenia occupies place 27, Croatia 72, and Hungary 74, out of 134 countries.

143 In terms of strength of investor protection, Slovenia is ahead of the other two countries: out of 128 countries, Slovenia is ranked 19, Hungary 86, and Croatia 98. But Slovenia is behind as regards protection by law and enforcement of minority shareholders' interests: out of 134 countries, it occupies place no. 100, just below Croatia (99), and far below Hungary (66) (World Economic Forum, 2008).

Value of Collateral Needed for a Loan (% of the Loan Amount)	120,87	168,99	145,55	154,23	130,22	137,67
% of Firms Identifying Access to Finance as a Major Constraint	3,62	24,96	9,63	16,61	13,01	29,24

Note: the data for Croatia are from 2007, for Hungary and Slovenia from 2005.
Source: Enterprise Surveys (2008).

As these countries have bank-based financial systems, it is important to look at some quantitative data regarding credits by banks. Although banks have a limited role as financiers of innovations – both in terms of funds available for such purpose and in terms of governance capabilities –, banking credit provides a basis for the normal functioning of an economy. Banking credit to the private sector was on a steady rise from 2000 to 2007 (see figure 6), supporting the aforementioned positive developments in manufacturing production and the GDP. Banks continued with expansion of credits to enterprises even in 2008 – as will be described in the fifth part (see below).

Figure 6: Total outstanding credit by banks to households and enterprises, as percentage of GDP at year end

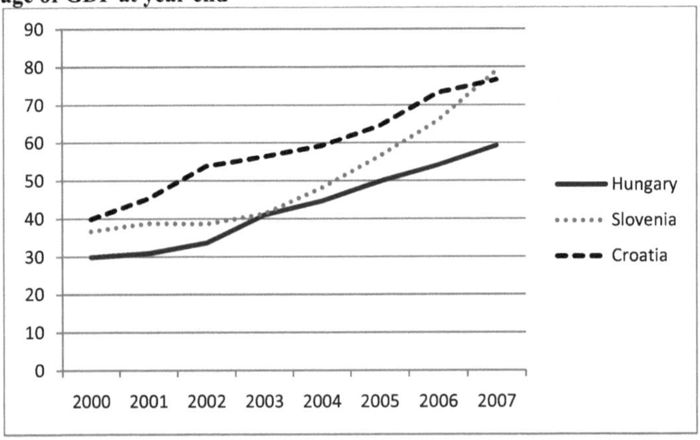

Source: EBRD (2009).

As regards capital markets in these countries, they are primarily characterised by the small numbers of newly listed companies, as well as by low liquidity. In the period from 2003 to 2008, the difference between newly listed and delisted companies at Ljubljana Stock Exchange and Budapest Stock Exchange is mainly negative, as can be seen from Table 9. Although we cannot strictly compare the data with the Zagreb Stock Exchange, because listing some shares at this Stock Exchange was obligatory (see below), we can mention data on public offerings. Namely, until the second half of 2008 there were only 9 public offerings (Stojić and Barać, 2008: 180). Liquidity[144] is highest in Hungary and lowest in Croatia, with Slovenia in the middle – but quite low in comparative terms[145]. The aforementioned characteristics suggest that the capital market plays a marginal role for companies that may want to issue shares in order to finance innovations.

Table 9: Difference between newly listed and delisted shares at stock exchanges in Ljubljana and Budapest

Stock exchange /year	2003	2004	2005	2006	2007	2008
Ljubljana	-1	7	-24	-20	-13	-3
Budapest	1	-3	-2	-3	-1	2

Source: WFE, 2009; Author's calculation.

In terms of the NSI, a crucial role in these countries could be played by both formal and informal venture capital, since it targets SMEs with a growth potential. Venture capital is more available in Slovenia than in Hungary or Croatia (World Economic Forum, 2008)[146]. However, venture capital firms' role in Central and Eastern Europe (CEE) is still small. Iliev (2006) mentions the biggest problems – high transaction costs for VCFs, deal flow of low quality, lack of other financial instruments and/or institutions. Because of these problems, minimum investment amounts are even higher than 500.000 USD, which is more than in the case of the UK's early-stage VC investments. VCFs are investors that strongly and proactively influence investees' strategies, their goal being to sell the company after holding onto it for a few years. There are several exit (divestment) routes, the most common ones in CEE and EU being trade sale[147] and initial public offerings. Due to underdeveloped capital markets in CEE,

144 Calculated as turnover divided by market capitalization.

145 As for 2007 and 2008, the figures are as follows, respectively: Hungary (110,3% and 158,4%), Slovenia (16,6% and 18,9%) (Eurostat Database, 2009 and own calculation), and Croatia (6,2% and 11,9%) (ZSE, 2009 and own calculation). That is much lower than in the United Kingdom, with its capital market-based financial system. Comparable figures for the United Kingdom are 286% and 306,3%, respectively (Eurostat Database, 2009 and own calculation).

146 In terms of venture capital availability for entrepreneurs with innovative projects (that involve risk), Slovenia is ranked 42, Hungary 82 and Croatia 83 – from a total of 134 countries ranked in the World Economic Forum (2008).

147 Trade sale is the sale of a firm to some other firm.

trade sale is even more pronounced in CEE than in developed countries. This makes CEE susceptible to demand from companies interested in buying investees (Iliev, 2006). Szerb and Varga (2002, as referenced in Iliev, 2006) noted that the strategies of VCFs are aligned with the interests of multinational corporations. A study by Szerb et al. (2007) analyses informal investment (investment by business angels, also known as informal venture capital) in Croatia, Hungary, and Slovenia on the basis of data from the Global Entrepreneurship Monitor (from the years 2001-2004). They found informal investment in these countries to be both low in amount and rare in occurrence. This was explained by "a lack of (1) start-up and business management skills and expertise, (2) opportunity perception and (3) personal acquaintance with an entrepreneur" (Szerb et al., 2007: 269). Another shared characteristic applying to the three countries is that "real" business angels' investing in third-party-firms is quite rare. However, the Slovenian business angel market differs from that of Croatia and Hungary, in that it is more developed. More importantly their estimates show that there is a finance (equity) gap for start-up enterprises. Szerb et al. (2007) believe that their findings are expected on account of the similar level of development of the economies they observe, in particular *vis-à-vis* financial imperfections (of markets and firms) from both the supply and the demand side. As for the supply side issue, the problem can be described in terms of inadequate supply of financial (re)sources. The demand-side issue is emphasized to lie in "the local entrepreneurs' limited knowledge and market economy experience" (Szerb et al., 2007: 259). In that respect, they believe that the countries of Central and Eastern Europe are still in transition, not only because of institutional factors, but also because of cultural and infrastructural ones (Szerb et al., 2007: 258).

Furthermore, the "equity capital culture" in the USA or the UK (Cumming and McIntosh, 2004, as referenced in Vadnjal, Letonja and Berginc, 2007: 2) stands in contrast to the predominant culture in Croatia (Young and Cvijanović, 2006), Slovenia (cf. Vadnjal, Letonja and Berginc, 2007; Andritzky, 2007), and most probably Hungary as well (cf. Karsai, Wright and Filatotchev, 1997). Namely, the predominant culture in those post-transition countries is bank-oriented. We expect this characteristic to have implications for entrepreneurs' rejection of such equity sources as venture capital. An equity-oriented culture is more stimulative of equity in general – from venture capital to equity invested through an IPO. In a bank-oriented or credit-oriented culture, entrepreneurs are likely to prefer debt to equity.

There are a few concluding remarks that we need to make before moving on. Financial systems (sectors) in these countries are: 1) bank-oriented, and 2) less developed and/or sophisticated than financial systems in developed countries. As for the former, the relatively larger role of banking credits, as opposed to financing via capital markets, makes these countries less susceptible to radical innovations and/or innovations in high-tech sectors. The latter means that there is little influence of banks and capital markets on the corporate governance of SMEs. This influence stands in sharp contrast to a situation in the USA, where capital markets act as a correction "mechanism" for decisions of the management, and in Germany, where banks exert great influence on corporate governance through their ownership and/or control rights over enterprises, as well as through their financing. The smaller size and lower sophistication level of financial systems (sectors) translates into fewer possibilities for innovation financing, because of fewer financial instruments being available and fewer funding opportunities. This has negative feedback on venture capital firms, which on the one hand use credit to finance their investments, and, on the other, need a strong capital market to make exits from their investments. The role of the government could therefore be important in supplementing the role of the financial sector in financing innovative ventures.

In addition, there is generally a lower level of competitiveness in the analysed countries in comparison with developed ones. This is apparent in all four elements of "Porter's diamond". As these countries are all small and open, and, as a result, face strong competition from abroad, there is a need to strengthen their competitiveness. More competitive (foreign) products alone can have negative influence on the domestic industry and facilitate its defensive restructuring. Therefore, market influence should be corrected by the "visible hand" of the government. As innovativeness is a crucial prerequisite for greater competitiveness, policy makers should pay attention to their NSIs, in particular to facilitation of linkages between its various elements.

4.4. National System of Innovation in Hungary

Hungary is positioned very high on the rankings of countries with a high share of high-tech industries in the GDP, even at the world level. The same can be said of the country's exports (Srholec, 2006). Nevertheless, Srholec (2006) finds that information and communication technologies (ICT) high-tech exports of emerging countries like Hungary and Estonia, or China and Mexico, do not have a high knowledge intensity. A brief look at the innovation indicators in the table in Appendix III provides some support for this finding, in that, e.g., Hungary's superior performance in terms of exports of high technology products is not adequately accompanied by output in terms of intellectual property rights (patents, community trademarks, and community industrial designs)[148]. Hence, we

148 Hungary is below Slovenia and the EU average as concerns patents, community trademarks, and community industrial designs, and behind Slovenia in employment in medium-high/high-tech manufacturing. At the same time, Hungary performs better

would say that the innovation activities producing such output are not so knowledge-intensive and are consequently more tangible. In the words of transaction cost economics, we have a case of low asset specificity. Such innovation activities are on average more suitable to be financed by banking credit than by VC.

The problem is aggravated by the "dual economy syndrome" mentioned in several studies (European Commission, 2007b; Acs et al., 2006; Havas and Nyiri, 2007). The duality is reflected in the higher efficiency of foreign-owned firms, as opposed to firms owned by domestic owners. The report by Hemmings (2005) also identifies a duality syndrome in the Hungarian economy, because high-tech production is normally organised by foreign subsidiaries whose R&D is done outside Hungary (Hemmings, 2005: 5). This problem is caused by a huge inflow of foreign direct investments that has not been accompanied by adequate restructuring of domestic firms – in spite of policy measures supporting technology transfer between foreign-owned and domestically owned firms (Vincentz, 2002: 17). It could otherwise be seen as a discrepancy in foreign and domestic sources of knowledge.

The report by Havas and Nyiri (2007) finds mixed evidence on the general framework conditions for innovation activities, in which it includes macroeconomic performance, entrepreneurial culture, and conditions for doing business. This conclusion was reached primarily because of uncertainties in connection with fiscal performance ("[g]overnment spending in 2001-2006 undermined the fiscal stability of the country – without major achievements in terms of underpinning long-term, sustainable socio-economic development." (Havas and Nyiri, 2007: 6)); Besides, frequent regulatory changes have created some instability for the companies, there are high costs of doing business (cost burden put on enterprises by the tax system and administration charges is higher than the OECD average), and negative trends in the domestic entrepreneurial culture (decrease in the birth/death ratio of enterprises and predominance of necessity entrepreneurs). The report by Havas and Nyiri (2007) also identified positive signs in the general framework conditions for innovation in the Hungarian economy, like the sound economic role of SMEs, good competition, and intellectual property rights regulations.

Public policies regarding innovations in Hungary have been improving, although they cannot yet address all the requirements of a modern economy. Science, technology and innovation policy-making (from the point of view of private sector involvement) in Hungary has a lot of room for improvement (Inzelt, 2008).

---

than the EU average in exports of high-technology products and employment in high-tech services, and even in share of medium-high/high-tech R&D (see indicators from the European innovation scoreboard, Appendix III).

Hungary and other transition economies did not have a participative tradition of policy making (cf. Inzelt 2008: 85), but rather a centralized, top-down, politicized model of policy making. However, Inzelt (2008: 92) believes that Hungary has gradually been developing forms of cooperation between private and public actors, although "STI policy-making continues to be characterised by the predominance of the public sector" (Inzelt, 2008: 92)[149]. Inzelt (2008) asserts that there is a model in the literature and in practice of EU policy making that is characterized by a network of agents (from both public and private sectors) where government is a catalyst of the policy making, not its main driver (Inzelt, 2008: 83). Government programmes in Hungary (since 1995) have been encouraging arm's length cooperation in the university-industry-government triangle (Inzelt, 2004). Indeed, she finds that the cooperation in the triangle has increased. However, not many enterprises are interested in closer relations with universities, which is also the problem in other transition economies (Inzelt 2004: 993). Hemmings (2005) finds that public policies have started to address specific innovation issues. Most notably, the issue of lagging behind the more advanced EU members as regards spending on R&D has been tackled. The Hungarian government has introduced a number of schemes to boost investment into R&D. A major problem, as identified in Hemmings (2005), is financing R&D. To that end, a few public policy schemes have been implemented. First of all, there is the innovation contribution (a levy that firms pay to the so called innovation fund; if the enterprise does R&D, it can reduce its levy; this scheme, therefore, functions as a stimulus for companies to do more R&D), and then there are additional governmental funds which enterprises that do R&D can obtain from both the government (state budget) and the Hungarian Development Bank. Besides, there are also EU funds. As regards the latter, innovation funding is normally connected with the Economic Competitiveness Development Operational Programme (ECOP) (Hemmings, 2005: 13-14). Funding programmes are not only focused on supporting R&D, but also on assisting SMEs in obtaining intellectual property rights and in marketing. Far from being concentrated only on funding, there are other innovation policy schemes that address issues in which Hungary is lagging behind: regulation that makes it easier for universities to establish spin-off companies, as well as reforms in tertiary education (Hemmings, 2005: 12). The problem with VC financing is that SMEs offer projects of very little value, and are consequently not financed by VC (Hemmings, 2005: 15).

Additional problems with R&D concern either too little spending on R&D or uneven distribution of R&D expenditures. Public R&D expenditures as percentage of GDP were 0,5% in 2005, whereas in the same year business R&D expen-

149 STI refers to science, technology and innovation.

ditures as percentage of GDP were 0,41%. Comparable figures for EU were 0,65% and 1,17%, respectively (from Table 16 in Appendix III). The Hungarian business sector has only 35,6% of the researchers (data for 2006), which is lower than in developed countries. In addition, business R&D expenditures are unevenly distributed: over 70% of business R&D expenditures is spent by large enterprises and multinational enterprises. In terms of sectors, the pharmaceutical sector has a share of more than 40% in these expenditures (ERAWATCH, 2009b). This could also be seen in Figure 7 (business expenditures on R&D by high-tech sectors). After manufacturing of pharmaceuticals, the next sector with the highest business expenditures on R&D (BERD) in total NACE[150] is manufacture of radio, television, and communication equipment and apparatus (DL32). It is followed by manufacture of medical, precision and optical instruments, watches and clocks (DL33), and manufacture of office machinery and computers (DL30). For manufacture of aircraft and spacecraft (DM353), data are not available for recent years.

150 See note below Figure 7.

Figure 7: Business expenditures on R&D (BERD) by high-tech sectors as percentage in all NACE sectors in Hungary

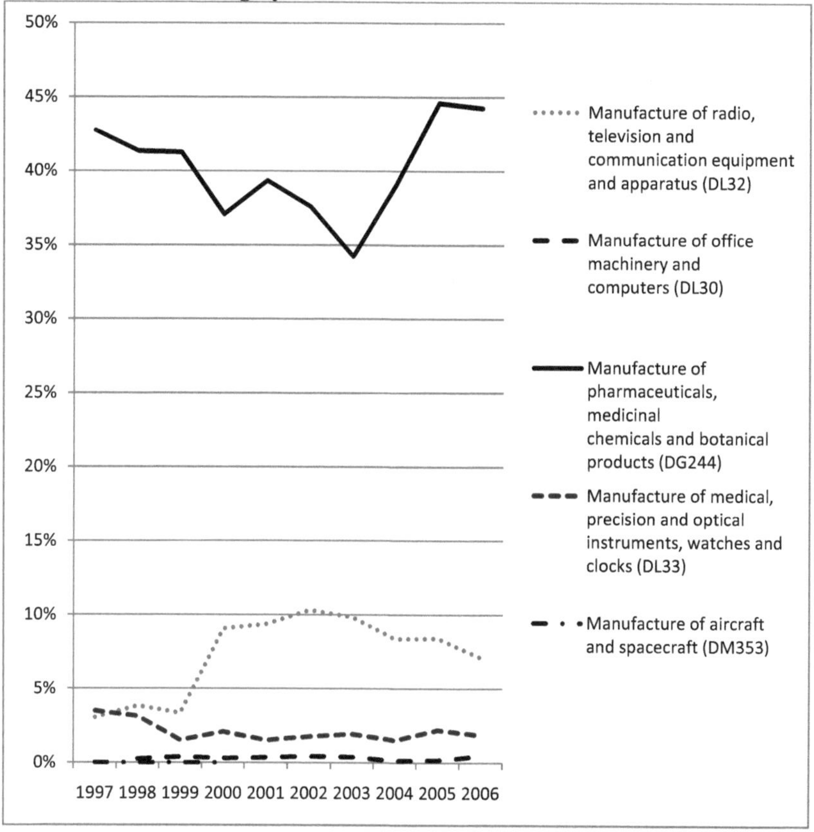

Note: NACE classification corresponds to ISIC classification. Pharmaceutical sector is incorporated in DG244 (manufacture of pharmaceuticals, medicinal chemicals and botanical products) as there were no detailed data.
Source: Eurostat Database (2009b) and author's calculation.

The innovative performance of Hungarian companies has improved a lot since the beginning of transition. Inzelt (2003) analysed the first (pilot) Community Innovation Survey in Hungary, done in 1994, for the 1990-1993 period. Although only 110 firms responded to the questionnaire, she found a low level of innovation activities. That was coupled with a low R&D to sales ratio (just 7,3% of the companies from the sample spent more than 8% of their sales on R&D), and the importance of the domestic as opposed to foreign markets (companies would innovate in order to open up new domestic markets and not new foreign ones; besides, the sources of information important for innovation activities would normally be domestic). The results are understandable for the first transition phase marked by the transition slump. Moreover, most of the companies surveyed were former state-owned enterprises or cooperatives that were not as sensitive to the market signals. However, the results for the financial factors as impediments to innovation activities are ambiguous. Unsurprisingly, the firms identified the lack of financial sources as the major obstacle to innovative activities. Inzelt (2003) traced the reason back to the high interest rates, little funding appropriate for innovation activities, as well as the nonexistence of venture capital. Nevertheless, Inzelt (2003) concludes that "policy-makers must understand that, even if a company has enough financial resources, it may remain non-innovative. *Supplying funds for research does not automatically increase the intensity of innovation activities*" (Inzelt, 2003; 868, emphasis original). Unfortunately, this apparent paradox was not further analysed, e.g., by comparing the aforementioned results with those of other countries. CIS number 4 in Hungary, examining the 2002-2004 period, identified only 20,9% of firms with 11 and more employees as innovative. That result left Hungary in 24th place out of the then 25 EU members (European Commission 2007b). The major challenges identified by the European Commission (2007b) are the following: 1) low share of innovative enterprises, especially SMEs, 2) weak cooperation in innovation activities, and 3) potential shortage of workers crucial for R&D and innovation. The last problem is indicated by the number of science and engineering graduates and the share of population aged 25-64 with tertiary education (see table in Appendix III). All three problems have been identified in the CIS and other studies. There have been a number of policy measures in Hungary to address the most important innovation issues, but it cannot be said whether they are working or not, since no *ex ante* and *ex post* evaluations of innovation policy responses exist (with one exception). The other problem shared by all post-socialist countries is that changes to the socio-economic system have been manifold, and have come from various directions (municipal, regional, national, EU, global). Hence, determinants of innovation activities have proved hard to analyse. The fact that problems with innovation activities have persisted has, therefore, been termed "a puzzle" (European Commission, 2007b: 40). Due to a multiplicity of policy schemes and measures, some of these have overlapped. Hence, the rational and systemic element of policy response has been lost (European Commission, 2007b: 42-43). The European Commission (2007b: 44) believes that the building of national innovation systems has not attracted enough attention or policy responses, in either qualitative or quantitative terms. Other issues have been given priority, as innovation issues have been perceived more from the cost angle than from the benefit angle. The Hungarian national innovation system has gradually developed over the years, and it cannot be said to be lacking in institutions or measures undertaken, but rather in their quality. The problems identified by the European Commission (2007b: 1) are twofold: firstly, the suboptimal way in which some of the institutions function and, secondly, their deficient networking and common policy responses. Furthermore, there are problems connected with lack of information (e.g., the first CIS study was conducted only in 2002, covering the 1999-2001 period), evaluation (it has been lacking for all but one policy measure), and monitoring (it has not developed). However, EU governance has been noticeable in the newest initiatives that could result in a culture of evaluation (European Commission, 2007b: 12).

In spite of the development of the financial sector, enterprises from the real sector still complain about financing sources. Hungary has managed to reform its banks and non-banking financial institutions almost to a degree comparable to that of Western Europe – the grades given by EBRD for both reforms for 2008 was 4 (4+ is the maximum grade). As already discussed, capital market has limited importance for domestic economy[151]. Venture capital market in Hungary has nevertheless been very developed if we compare it to other post-transition economies. It had started to develop even before 1990, and had reached a considerable size by 1995, becoming approximately twice as big as the VC market in the Czech Republic and Poland. Governmental VC funds played an important role in that. The entrepreneurs, however, are not well informed about VC, and prefer not to let VC firms buy an equity share in their company (Karsai, Wright and Filatotchev, 1997). However, the financial sector seems to have no importance for the financing of R&D in Hungary. It is mostly financed with internal sources (75,6%), followed by foreign sources (15,9%), and public funds (8,4%) (see European Commission, 2007b[152]). The data for the 2002-2004 period, as found in CIS4, show that financial obstacles ("financial constraints") represent the biggest limitation to the innovation processes. Among the latter, high innovation costs were the biggest impediment, and "lack of appropriate sources of finance" the second biggest (European Commission, 2007b). According to The Gallup Organization (2007), problems with financing innovation activities (high interest rates and problematic access to finance) in Hungary stand out from other problems connected with the process of innovation. However, that is no surprise as SMEs in EU27 accentuate this problem as the harshest one (The Gallup Organization, 2007: 77). Problem with human resources (lack of human resources and their cost) is ranked as no. 2 among the problems hindering innovation processes in Hungary's SMEs (The Gallup Organization, 2007: 79). Considering the aforementioned deficit areas in innovative performance and the NSI as iden-

151 Stock market capitalization in 2007 was 32,4% of GDP (EBRD, 2008). The financial crisis in 2008 manifested itself in the fall of the blue-chips index of the Budapest Stock Exchange by just over 50%, thus diminishing stock market capitalization (BSE, 2009).

152 The data refer to 2006.

tified in various reports (low knowledge content of innovation activities, low innovativeness level of enterprises coupled with their lack of cooperation and a deficit of skilled workers), the problem with innovation activities is a lot deeper than only its financial (supply) aspect. The deficits identified highlight the quality of innovation projects as the principal factor we should analyse.

4.5. National System of Innovation in Slovenia

Slovenia has managed to preserve its public institutions for R&D from the socialist times (cf. Sorčan, 2002, as referenced in Bučar and Stare, 2002: 427). The preservation of the knowledge base at public research institutions has proven important in our empirical field research as well, since some of the enterprises figure as spin-offs from the Jožef Stefan Institute[153]. It was possible to preserve the public research system because it was relatively strong, decentralized, and autonomous during socialism (Bučar and Stanovnik, 1999: 97, as referenced in Bučar and Stare, 2002: 427). However, it was hard to develop an NSI at a time when enterprises were undergoing defensive restructuring, as in early transition phases (cf. Bučar and Stare, 2002). But although it has survived the transition period until today, changes in the Slovenian NSI as a whole have not favoured the sustainability of innovation activities in the long run. Although it could be said that legally the Slovenian NSI is well-rounded, the linkages between industry and research institutions remain weak. The lack of an innovation culture (which is detrimental to innovative activity of not just firms, but also of all public organisations), weak implementation of public policies regarding the NSI, and a whole range of microeconomic and macroeconomic factors (Bučar and Stare, 2002) feature as the main obstacles to more intensive and better-quality innovation activities.

The main problems of the Slovenian national system of innovation[154], as identified by the European Commission (2007c), are the following: lack of formal coordination schemes between governmental organisations (that coordinate innovation policy), questionable impact or, rather, role of some institutions in the Slovenian NSI, lack of an impact assessment for certain regulations. These prob-

---

153 See cases of Slovenian enterprises no. 14 and 15 in the case study reports (Appendix II). The Jožef Stefan Institute is the leading Slovenian research institution in the fields of basic and applied research in natural and life sciences, as well as in engineering (JSI, 2009 – see also for more on the Jožef Stefan Institute).

154 Some of the so called "bridging" institutions within the Slovenian NSI, along with the years they were founded, are the following (European Commission, 2007c: 3): technology parks and centres (1994), incubators (2003), clusters (the formation of clusters started in 2001), technology networks (from 2003 on), technology platforms (2004), centres of excellence (2005), regional development agencies, Slovenian enterprise fund and others.

lems seem to be typical of less developed NSIs in post-transition countries. As regards the innovation policy, it has proven to be relatively effective, although not as efficient as the best practices. The EU conditionality has found fertile ground in Slovenia:

> "It can be argued that when Slovenia was a candidate country for EU membership, European innovation policy, the EU Action plan, and various monitoring and benchmarking exercises had had a positive impact on innovation policy in Slovenia, not only because of an abundant information inflow but also because awareness of innovation policy had increased substantially in government circles" (European Commission, 2007c: 13).

One of the main problems and challenges for policy-makers in Slovenia is how to increase the efficiency of R&D activities. Namely, in spite of the high share of R&D in the Slovenian GDP, high-tech exports in relative terms are behind some of the new member states of the EU (European Commission, 2007c: 35). Two major problems are identified in the European Commission (2007c: 38) regarding conduct of innovation policy: firstly, it could benefit from better information and, secondly, it is hard to measure its impact because of the multiple and simultaneous influences of other factors or policies. However, innovation policy in Slovenia has responded to some challenges quite well (see European Commission (2007c) for details). Moreover, Slovenia has an apparent mismatch between the structure of a desirable national system of innovation, in that R&D activities are conducted mainly in governmental and academic institutions, and an inadequate amount of industrial applications on the output side (Kos, 1998, as referenced in Bojnec, 2001: 71).

Public sector R&D expenditures in Slovenia are lower than business R&D expenditures, and they are altogether lower than the EU level – as can be seen from Table 16 in Appendix III. But the share of private sector researchers in the total number of researchers (39,1%) is lower than the share of public sector researchers (60,9%). R&D in the business sector, specifically in manufacturing, is particularly strong in the chemicals (most prominently in the pharmaceuticals sector) and machinery and equipment (electrical equipment) (ERAWATCH, 2009c). This can be seen in Figure 8, in which we analyse all high-tech sectors[155]. Just like in Hungary, the pharmaceutical sector dominates BERD, followed by manufacture of radio, television, and communication equipment and apparatus (DL32), and manufacture of medical, precision and optical instruments, watches and clocks (DL33). Sectors covering manufacture of office machinery and computers (DL30) and manufacture of aircraft and spacecraft (DM353) do not have BERD.

155 See note below Figure 8.

Figure 8: Business expenditures on R&D (BERD) by high-tech sectors as percentage in all NACE sectors in Slovenia

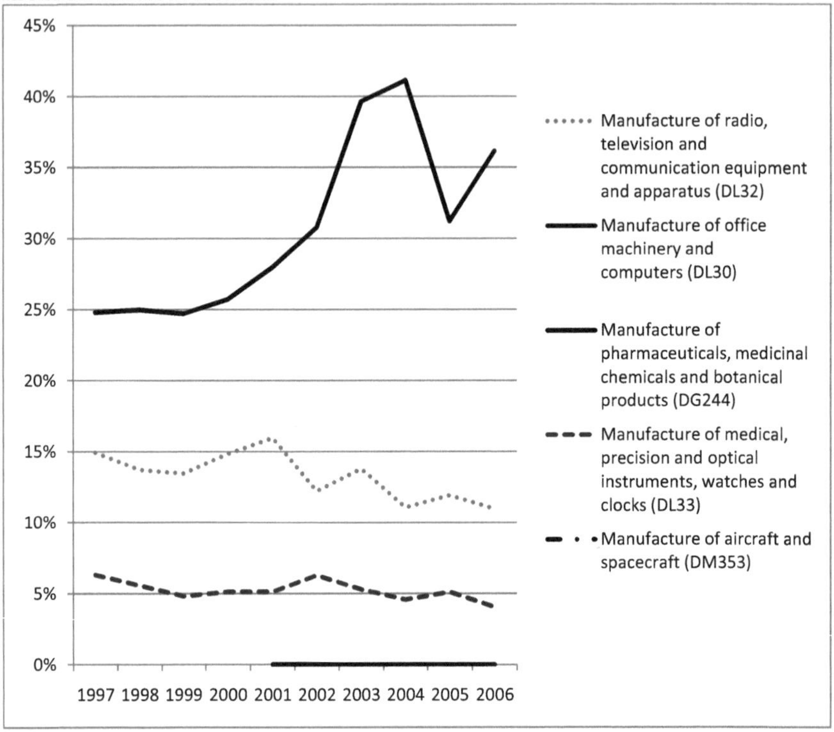

Note: NACE classification corresponds to ISIC classification. Pharmaceutical sector is incorporated in DG244 (manufacture of pharmaceuticals, medicinal chemicals and botanical products) as there were no detailed data.
Source: Eurostat Database (2009b) and author's calculation.

Unlike Hungary, Slovenia does not suffer from a dual economy problem. We may explain this by the fact that Slovenia had a much smaller inflow of FDIs than Hungary. On the other hand, its economy is more competitive than the Hungarian economy. However, foreign investment enterprises[156] in Slovenia have shown better performance than domestic manufacturing companies (Rojec, 1998, 2000b, as referenced in Rojec and Jaklič, 2004). However, Jaklič, Damijan and Rojec (2008), using a sample of data in the period from 2000 to 2002, found that foreign ownership of firms in Slovenia did not significantly determine innovation activity. The latter was significantly determined by innovation cooperation, as well as by spending on R&D.

156 Enterprises with a share in equity of 10% or more of foreign companies.

The financial sector in Slovenia is relatively developed for a post-transition economy[157], but it is not well-suited to financing innovations. Slovenia shares a problem with Croatia regarding the issuance of loans by banks. Namely, banks lend against collateral, and not on the basis of a firm's cash flow (Nowak and Murrow, 2006). The problem with lending against an asset is that enterprises need to have assets in order to get a loan, and not all enterprises have assets that can be used as collateral. Unlike in the USA, this does not have to be an income-producing asset. Such an arrangement makes it safer for the banks to lend, but less likely that some companies will obtain a loan (Nowak and Murrow, 2006). Capital market for trading both equity and debt in Slovenia is comparable to other post-transition countries that joined the EU in the last two enlargement waves. However it is underdeveloped in comparison with old members of the EU. In addition, there is practically no trading in securities other than shares and bonds. Equity market is relatively opaque due to off-market trading, and has low liquidity[158]. Connected with that is the low frequency of activities in the primary equity market. In a financial system (like Slovenia's) where banking credit dominates external financing, and banks predominantly credit existing clients, SMEs – especially innovative firms and start-ups – are disadvantaged (Andritzky, 2007)[159]. Moreover, the role of venture capital is only marginal in Slovenia. After estimating the demand for venture capital in Slovenian firms[160], Vadnjal, Letonja and Berginc (2007) conclude that the short-term supply of VC funds is larger than the respective demand[161]. This finding is similar to that of

157 Slovenia's index of banking sector reform in 2008 was lower than that of either Hungary or Croatia – it was 3,3, while the index of reform of non-bank financial institutions was comparable to Croatia's – 3,0. Stock market capitalization in 2007 was 57,2% of the Slovenian GDP (EBRD, 2008).

158 In 2008, the Slovenian Stock Exchange Index fell by about 2/3 (LJSE, 2009), further depreciating the relative importance of the capital market.

159 Bartlett (2002) conducts empirical research in Slovenia, Bosnia and Herzegovina, and Macedonia (in the autumn of 2000) and finds financial factors to be the most important driver of development of SMEs in the aforementioned countries. His finding that tax barriers do not have such a negative impact on development of SMEs does not seem to fit into this picture. Furthermore, he finds that institutional and other societal barriers do not effect this development. Hence he gives support to the money-first approach as a policy measure. However Bartlett (2002) did not investigate impacts of barriers on innovations and sectors according to the technology classification. Therefore he could not distinguish sources of growth of these enterprises – which could be based on some kind of rent (e.g. because of the monopoly position in the market). Hence, in this author's opinion, Bartlett (2002) could not develop his conclusions even further.

160 The firms were based in domestic university incubators, business incubators, and technology parks.

161 Vadnjal, Letonja and Berginc (2007) try to estimate the true extent of the market for venture capital in Slovenia. However, they find it hard to accomplish that task since

Young and Cvijanović (2006) regarding the venture capital market in Croatia. Slovenian entrepreneurs have little knowledge of various financing sources, in particular of VC[162]. Actually, they have an aversion towards VC (Vadnjal, Letonja and Berginc, 2007).

The problem regarding financing innovation activities is much less present than in Hungary (for Hungary see above). Namely, the availability and cost of human resources represent the biggest constraint to innovation activities in Slovenia. This problem is on average a lot more pronounced than in the EU27. The problem with financing innovation activities (regarding both access to financing sources and the problem of high interest rates) is at the EU27 average level (The Gallup Organization, 2007: 79). However, the financing of innovation activities is not entirely left to the market. R&D in Slovenia is up to 91% financed internally (European Commission, 2007c). Furthermore, a big role in the financing of innovations of SMEs in Slovenia since 2005 has been played by the Slovenian Enterprise Fund. It gives grants, credits and credit guarantees, and it also subsidises credit lines. The Slovenian Enterprise Fund has been financed by both the state and some European funds, such as the European Regional Development Fund and the European Investment Fund. In 2006, it invested EUR 34,19 million (European Commission, 2007c: 41). In view of the rather limited role of the Slovenian banking sector and capital market when it comes to financing innovations, the role of the public sphere seems necessary.

4.6. National System of Innovation in Croatia

Among high-growth enterprises in Croatia, not many are connected with innovations (cf. Mateljić, 2007). Račić, Aralica and Dadić (2005) analyse the performance of Croatia's exporting SMEs, focusing on their growth. They found that the SMEs that achieved above-average growth levels and had the highest potential to grow in the future mostly existed in the sectors of medium-high technology manufacturing and in services[163]. However, Račić, Aralica and Dadić (2005) be-

---

there have only been 20-30 private equity/venture capital investments in Slovenia, probably with no exits so far. Not all of these equity investments have been venture capital investments (meaning investments into propulsive and innovative SMEs), which means the number of cases one can observe are even fewer.

162 Vadnjal, Letonja and Berginc (2007: 7) quote one of their interviewees as saying: "it is impossible to integrate the venture capital concept into [the] exceptionally nonentrepreneurial culture".

163 However, this result was obtained with just under 10% of statistical significance. The actual growth of firms was calculated on the basis of a difference between average revenues (of the firms in the authors' sample) in 2004 and 1999. The threshold for splitting the sample into groups of low-growth and high-growth enterprises was set at 58% of nominal growth (between 2004 and 1999), which highlights the exploratory

lieve that "[i]n Croatia business undertakings with growth potential are relatively rare" (Račić, Aralica and Dadić, 2005: 11-12). As for other characteristics of enterprises, those with individual founders are more likely to have grown, as are enterprises owned by foreign owners (Račić, Aralica and Dadić, 2005: 12).

The first Community Innovation Survey (CIS) in Croatia, for the 2001-2003 period, reveals crucial findings regarding innovation activities in Croatia (Račić et al., 2004). On the basis of a response rate of 15%, the main conclusions are as follows: the innovativeness of an average firm grows with its size. The authors of the study find it logical that bigger firms (that operate with more resources and have more specialized human resources) are more innovative. SMEs showed a lack of innovative capabilities. Innovation management is neglected, and innovation cooperation is very poorly developed[164]. A problem with innovating firms in the manufacturing sector was that 33,81% of them (from this research) had no R&D activity whatsoever. Furthermore, out of all the innovating respondent companies from the manufacturing sector that did have R&D, its intensity[165] was meagre: 65,57% companies sacrificed less than 1% of total revenues for R&D, 24,94% companies invested between 1% and 4% of total revenues in R&D, and only 9,5% companies invested more than 4% of total revenues in R&D (Račić et al., 2004: 58 and own calculations). When asked about obstacles to innovation activities, the firms cited market-related reasons embodied by high costs and lack of appropriate financing sources. As for other factors, the respondents complained about too little support from the state. In the structure of innovation activities of the firms from the manufacturing sector in 2003, the biggest share (in monetary terms) was taken up by acquiring machinery and equipment, only then followed by marketing expenses, and expenses for intramural research and development. This structure of innovation activities was mirrored in the way innovation activities were financed: 79,24% by "own financing resources"[166], 9,57% by supplier credit, and 8,49% by banking credit. Leasing

---

nature of the authors' paper. This criterion for discerning growth enterprises was combined with the criterion of net increase in the number of employees in the observed period.

164 Probably the best example of a science-industry link is a subsidiary company founded by the Institute Ruđer Bošković, which is the largest public institute (therefore state-owned) in the field of natural and technical sciences in Croatia. The subsidiary company, founded in 2007, is intended to function not only as a sort of a technology transfer office, but also to commercialize inventions and innovations by the Institute. The company was founded with a credit from the World Bank, in cooperation with the Government of the Republic of Croatia (see Rudjer Innovations, 2008).

165 Indicator intensity of R&D measures R&D expenditures as a share in total revenues (Račić et al., 2004).

166 The phrase was coined for the purposes of this research study that the author participated in in order to make it more understandable to the respondents. The intended

participates in the total financing sources of innovation activities with just 0,91%, and additional investment of equity capital[167] with only slightly more than 1% (Račić et al., 2004: 58-59). This adds up to over 99%, with the obvious conclusion that credits of private individuals (which normally have high importance for start-up companies) and debt instruments (which are normally used by more mature medium-sized and/or large companies – at least in Croatia) are marginalized. Namely, they cumulatively account for less than 1% of total financing sources for innovation activities. Moreover, "additional investment of equity capital", which is a *sine qua non* of big investment projects which cannot be financed by leverage, is neglected in the business strategies of Croatian companies. Although equity could potentially be expanded in every phase of a company's lifetime, it certainly is crucial for micro, small and medium-sized enterprises, particularly for their start-up and expansionary phases.

There are three main innovation challenges according to the report by the European Commission (2007a): 1) level of R&D intensity should be increased as well as made more efficient, 2) innovations should be diffused more efficiently, and 3) innovation demand should be stimulated. Policy responses to the abovementioned challenges have not been entirely adequate. Croatia has the highest public R&D expenditures of the three countries. Croatia is also second among the three compared countries regarding the level of business R&D expenditures (see table in Appendix III). However, it lags behind Hungary and Slovenia in all 5 of the indicators measuring output in the European Innovation Scoreboard 2007 (see table in Appendix III). These five indicators are EPO, USPTO, triad patents, community trademarks, and community industrial designs. There are a number of reasons for this, and some of them lie in the fact that cooperation between science and industry is weak, and so is the level of protection of intellectual property rights. Furthermore, innovation is neglected in entrepreneurial strategies (cf. European Commission, 2007a). Croatia's NSI also shares some characteristics with that of Slovenia and Hungary, the most prominent ones being: 1) lack of data, studies, policy analyses, as well as a lack of evaluation of various instruments and/or measures, 2) lack of coordination between the two primary institutions in charge of the conduct of innovation policy (Ministry of Science, Education and Sports and Ministry of Economy, Labour and Entrepreneurship), and 3) frequent changes in the institutions comprising the NSI. Inno-

meaning was internal financing. The other possible answers comprised banking credits, supplier credits, credits of private individuals, additional equity investment, leasing, debt instruments (bonds, commercial paper) and other (under this option, the respondent had to specify the means).

167 Another phrase coined by the research team to incorporate all possible variants of equity investment, be it in a private sphere - e.g. venture capital - or in a public one - e.g. initial public offering.

vation policy exists only at the level of the central government, and not at the regional level – neither at the NUTS 3 level (level of the counties), nor at the NUTS 2 level (Croatia has been divided into 3 NUTS 2 regions since 2007) (European Commission, 2007a). However, as the European Commission (2007a) argues, Croatian policy makers have started to develop the national innovation system rather late, that is only from 2000. This is understandable, due to the fact that Croatia embarked quite late on a more or less stable transition path.

R&D is not evenly distributed. Only 15% of the researchers were in the business sector in 2005 (ERAWATCH, 2009a). Besides, "[a]bout 85% of all business R&D is performed by several large companies engaged in pharmaceuticals, ICT, agriculture and the food and beverages industries" (ERAWATCH; 2009a: 3). The leading high-tech sector according to BERD is pharmaceuticals – just like in Hungary and Slovenia (see figure 9). However, the relative level of BERD in pharmaceuticals, in comparison with total NACE, is several times smaller than in the other two countries. Manufacture of radio, television and communication equipment and apparatus (DL32) is second, followed by manufacture of office machinery and computers (DL30). Manufacture of medical, precision and optical instruments, watches and clocks (DL33) had no BERD in 2006, and there were no data for 2006 covering the manufacture of aircraft and spacecraft (DM353). The latter sector had very little activity in 2008, so we can conclude that its importance was negligible for the national economy[168].

168 In 2008, the manufacture of aircraft and spacecraft sector had 9 firms, with the total revenues of 7 of these amounting to nearly 1,2 million Euro. Data for two other firms are missing, but these are only smaller firms (Poslovna Croatia, 2009 and author's calculation).

Figure 9: Business expenditures on R&D (BERD) by high-tech sectors as percentage in all NACE sectors in Croatia

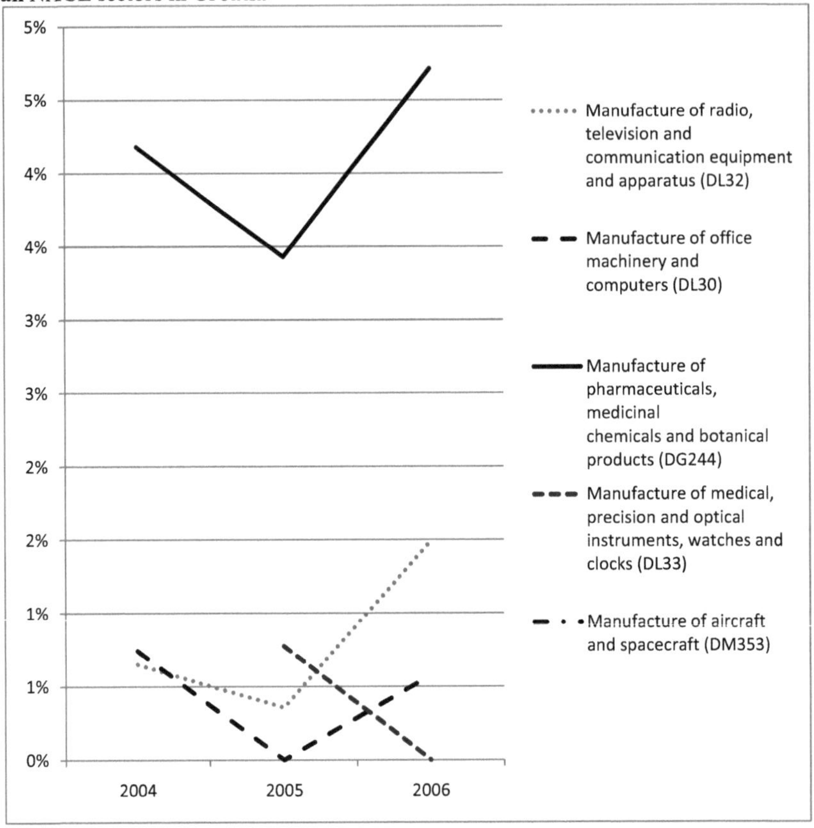

Note: NACE classification corresponds to ISIC classification. Pharmaceutical sector is incorporated in DG244 (manufacture of pharmaceuticals, medicinal chemicals and botanical products) as there were no detailed data.
Source: Eurostat Database (2009b) and author's calculation.

Croatia's financial sector is a typical post-transitional financial sector, with banks being its strongest part[169]. Progress in terms of both quantity and quality of the financial institutions' financing of SMEs in Croatia has been evident. Not

169 The European Bank for Reconstruction and Development graded the banking sector reform in Croatia a 4 (on a scale from 1 – lowest grade – to 4+ being the highest grade), and the reform of non-bank financial institutions a 3. Stock market capitalization has steadily grown to reach 119,6 per cent of the GDP in 2007 (EBRD, 2008), although it steeply fell in 2008 because of the fall in the stock market index of the Zagreb Stock Exchange by 2/3 (ZSE, 2009a).

only have banks increased the supply of their products and services, but they have also lowered interest rates, and made an enormous expansion in terms of credit volume directed towards enterprises (cf. Jambrović, 2007). Banking guarantees have been on the rise and have almost doubled from December 31 2004 (when they accounted for 1,475 billion EUR) to December 31 2007 (they were 2,771 billion EUR) (Gelenčer, 2008: 8). The small role of the capital market is becoming even smaller, since very few companies are expected to maintain their shares' listing on the capital market in Croatia. Namely, due to political pressure from the EU, Croatia will probably abolish the law stipulating that public joint-stock companies (JDD companies) have to list their shares on the stock exchange. According to some estimates, out of over 200 that have been listed, only 15 companies might actually maintain their shares as listed (Kiseljak, 2007). In addition to banks, some state institutions have also been involved in financing the SME sector of the economy, such as the Croatian Bank for Reconstruction and Development (HBOR), Croatian Agency for Small Economy (HAMAG), Regional Development Fund of the Republic of Croatia, Business Innovation Center of Croatia (BICRO), Croatian Employment Service, as well as several ministries[170]. Due to an absence of evaluation, it is unclear how these programmes work, and whether there is feedback from a project's result to the financing institution, i.e. policy learning.

Nonetheless, the problem of financing innovations is not necessarily solely in the financial sector, but rather in the interaction between the financial sector and the real sector. The current structure of financing sources seems to match the innovation activities that are conducted – as we have mentioned above. In Croatia's case, largely incremental innovations have positive feedback on credit financing, which in turn reinforces incremental innovations. We shall revisit this connection in our empirical field research, which follows below.

170 Ministry of Science, Education and Sports, Ministry of Economy, Labour and Entrepreneurship, and some other ministries as well.

5. Financing Innovations of SMEs in Hungary, Slovenia and Croatia

5.1. Introduction and Methodological Notes

The analysis that follows is a result of the empirical field research that we have conducted and presented below. It supplements the empirical desk research on national systems of innovation elaborated in the preceding chapter. After providing the methodological notes on the empirical field research, we shall elaborate our findings on cases from each particular country. At the end of this chapter, we shall present a conclusion.

Multiple case-studies of an explanatory and descriptive nature[171] have been undertaken, covering high-tech SMEs[172] with production activities in Slovenia, Hungary and Croatia. Qualitative and quantitative data were collected in two main ways: primarily through interviews, but also by drawing on secondary sources (databases, internet, brochures, and similar materials). Interviews were conducted, over the phone and in person, with qualified persons from each enterprise in the time period between November 2007 and March 2008. The case study database, which offers an overview of conducted interviews and most relevant findings, can be seen in Appendix I. For more detailed descriptions from each case study, readers should check the case study reports in Appendix II. We shall draw on the latter not just for more detailed observations, but also for anecdotal evidence that can be relevant to our conclusions.

We believe that the economic crisis, which started to affect Hungary, Slovenia, and Croatia in 2007 and/or 2008, did not influence our results in any significant way. Namely, the field research had been completed by March 2008. As we have already mentioned, real GDP growth rates in 2008 were positive, while only the last quarter of 2008 proved to be critical. Although the tightening of lending conditions in all three countries has been notable, especially in 2008, banks have managed to increase lending to enterprises from the real sector. As a result, the banking system, as the foundation of these countries' bank-based financial systems, remained viable. In Hungary, the tightening of lending conditions by banks towards the corporate sector was noticeable from the second half of 2007. The tightening resulted mainly from weak external economic conditions, such as an increase in risk for some categories of enterprises, and a decrease in liquidity due to the effects of the sub-prime crisis (MNB, 2008a). Although the tightening in lending conditions continued throughout 2008 (MNB,

171 The primary purpose was to have an insight into the subject matter and to explain phenomena and cases of financing innovations.

172 As units of analysis.

2008b; MNB, 2009a), banks could nevertheless expand their credits to the corporate sector in the same period (MNB, 2009b). In Slovenia, banks made their lending conditions more restrictive in the second half of 2007, particularly in the last quarter of the year. This was because on the one hand sources of financing were harder to obtain, and on the other because lower economic growth was predicted (Bank of Slovenia, 2008: 63). Nevertheless, in 2007 they were able to significantly increase their lending to non-banking institutions (Bank of Slovenia, 2008: 64), although the impact of the financial crisis from the US started to be felt. Loans by banks provided to firms strongly increased in 2007 (Bank of Slovenia, 2008). Bank lending conditions tightened in 2008 (Bank of Slovenia, 2009a), and lending to non-financial corporations grew further (Bank of Slovenia, 2009b). Restrictive monetary policy by the Croatian National Bank managed to slow down the expansion of assets of banks in 2007 and reduce it to 13,3% by the end of the year[173]. But in 2008, a slowdown was felt even more because restrictive monetary policy was coupled with a tightening of economic conditions in the country and abroad. Bank sector assets expanded by 7,4% at the end of 2008 (CNB, 2009b: 87). Banking credits at the end of 2008 expanded by 14%, and this was primarily because of expansion of credits to various units of the government and companies (CNB, 2009b).

At or around the time of our survey, banks started to implement the new Basel II[174] standards. Evidence of its influence on bank lending practices is quite mixed. Hungary started with the implementation of Basel II on 1 January 2008, and respondents to the Hungarian National Bank survey stated that it would not substantially alter their readiness to provide credit (only one bank in this survey indicated that tightening resulted from implementation of the Basel II standards) (MNB, 2008a). Banks in Slovenia started to fully implement the Basel II framework in 2008. In the survey, the banks assessed that Basel II would stimulate retail banking rather than other banking operations. Hence, they expected a certain negative impact on lending to enterprises (Bank of Slovenia, 2008: 124, 126-127). One of our interviewees from a Slovenian firm noticed difficulties in the availability of credit because of the implementation of Basel II. Croatia started implementing Basel II later, as the relevant law (Law on Credit Institutions) that introduced the framework for the Basel II standards, was only passed

173 Because of the huge increase of lending by the banks, the Croatian National Bank started to extensively use measures of restricting lending growth from 2004 on (Ivičić et al., 2008: 2-3).

174 Basel II is a short name for the European capital framework for credit institutions and investment companies. It is a three-pillar system that prescribes new capital requirements for banks, tighter risk controls and management, as well as disclosure of information by the banks to the public (Bank of Slovenia, 2008: 124).

in September 2008. Banks have time until 1 July 2009 to implement it (CNB, 2009b: 109)[175].

The interviews conducted were semi-structured, with the following open-ended questions:
1. Has your company produced innovations in the past 3-5 years? If not, why?
2. What kind of innovations do you normally come up with – incremental or radical?
3. If you have an idea about an innovation, but no money to finance it, who do you turn to for funding? Why?
4. Have you applied for funding by VC/PE firms? Why (not)?
5. What are the three greatest problems of financiers with whom you apply for funding?

Some firms were asked two additional questions:
1. How much does your final product cost, and how much did you invest in the innovation process?
2. Is financing really the greatest problem you face in the innovation process? Which other problems do you face in the innovation process?

As it was not possible to find a single adequate database for selecting the target group of companies for any of the countries covered here, several sources had to be combined. We put together a list of high-technology SMEs for each of the countries. We then randomly chose the SMEs that were active, in order to conduct an interview with their representative. In addition, we interviewed several enterprises from Hungarian, Slovenian and Croatian technology parks (or similar organisations – see below). Contact data on Hungarian high-technology SMEs were collected through a website[176]. The companies selected in that way were joined by two more from Pannon Mechatronics Cluster Zalaegerszeg, and one from the MEDIPOLIS Knowledge Centre in Pécs. Case studies from Hungary include a total of 13 companies. Slovenian high-tech companies were put together from various sources[177]. Out of the 16 firms from Slovenia, five are members of the Technology Park Ljubljana, one is from the Styrian Technology

175 Although banks started with preparations for implementing Basel II in the period from 2004 to 2007, their lending policy was, on average, made less strict (Ivičić et al., 2008).

176 This website was www.europages.co.uk (Europages, 2009).

177 The sources were two websites, a presentation, and a recommendation. The websites were: 1) Technology Network ICT (2006), and 2) Technology Network PCT (2006). The presentation was sent to the author by Marko Jurančič of the High Technology Products Manufacturers' Cluster. Moreover, several firms were recommended to the author by Matija Rojec of the Faculty of Social Sciences, University of Ljubljana. This is gratefully acknowledged.

Park in Pesnica pri Mariboru, and one from the Business Incubator Network Pomurje in Murska Sobota. Croatian firms were initially (and mainly) chosen from the database on Croatian exporting firms[178]. In addition, there were some randomly picked firms from the business weekly "Lider", and some were suggested to us. Furthermore, the author made case studies of a Croatian firm and a trade that belong to the Technology – Innovation Centre of Rijeka, and two firms that function in the premises of the Technology Park Zagreb. Hence, the total number of Croatian enterprises included in this research is 15.

Figure 10: Distribution of Hungarian enterprises belonging to high-technology industries encompassed by interviews

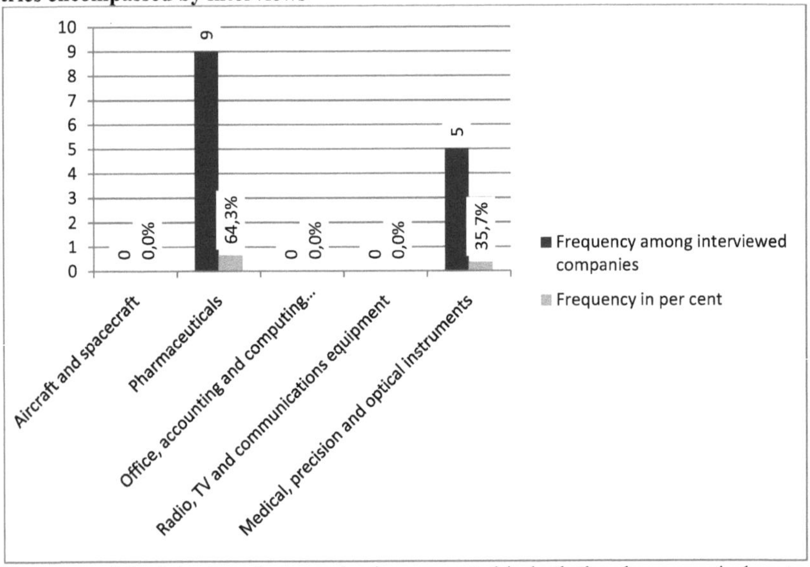

Note: one of the interviewed companies is represented in both the pharmaceutical sector and the sector of medical, precision and optical instruments. The number of companies is 13.
Source: own research.

As concerns the high-technology industries represented in the empirical research, we have managed to cover the most important industries – as previously analysed. In both Slovenia and Croatia, four out of five industries are repre-

178 The database used is called "Središnja baza hrvatskih izvoznika", BETA version from 2006 (Exporters Database, 2006). Firms chosen from this database were additionally filtered according to the following criteria: 1) they had to have a website, and 2) they had to have experienced a growth in revenues of at least 10% from 2005 to 2006. In order to make this selection, an additional database was used – "Business Croatia", version 2007 (Business Croatia, 2007).

sented. Only manufacturing of aircraft and spacecraft is missing. However, as this industry is negligible in Croatia, and has no R&D investments in Slovenia, we still have a good representation of high-tech sectors in these two countries. Two high-tech industries were covered by the research done in Hungary: the pharmaceutical industry (spending more on BERD than any other high-tech industry), and manufacture of medical, precision and optical instruments (this industry has the third largest share of BERD among the high-tech industries). The exact distribution of companies across the different industries is given in Figures 10-12.

Figure 11: Distribution of Slovenian enterprises belonging to high-technology industries encompassed by interviews

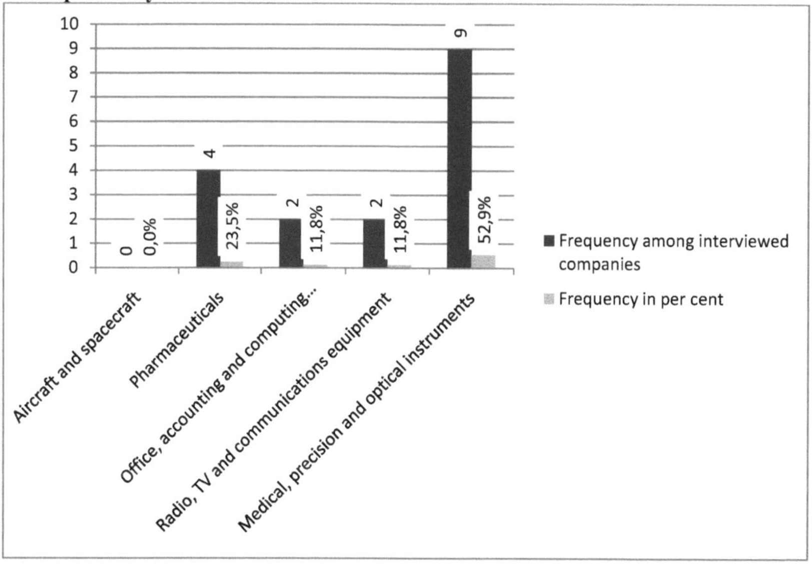

Note: one of the interviewed companies is active in two sectors: office, accounting and computing machinery and medical, precision and optical instruments. The number of companies is 16.
Source: own research.

Figure 12: Distribution of Croatian enterprises belonging to high-technology industries encompassed by interviews

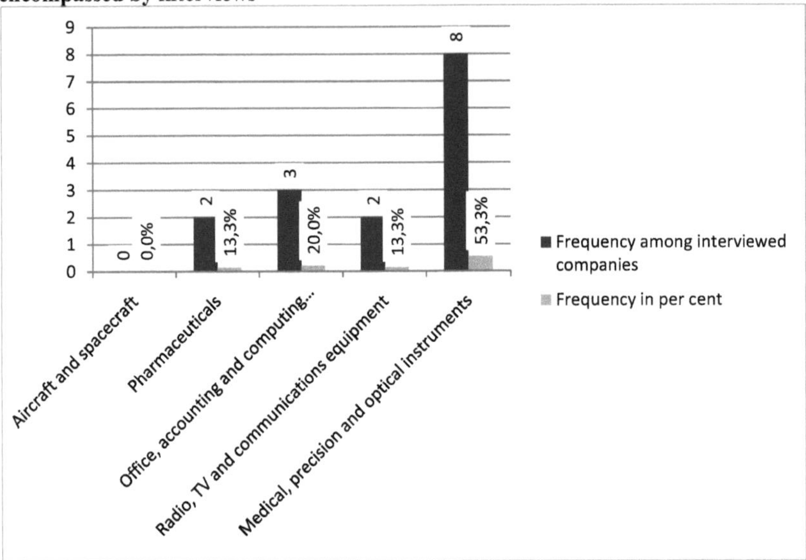

Note: the number of enterprises is 15.
Source: own research.

We analyse enterprises from all three countries in separate chapters and make observations based on mainly qualitative, but also quantitative[179] analysis in the light of the second hypothesis that we have advanced. We observe that the financing of innovations reflects the socio-economic characteristics of the enterprise itself and of its environment. We find support for the second hypothesis in cases from all three countries, while also noticing certain differences between them.

179 Enterprises will be compared to EU27 enterprises (data for 2005 for countries that today make up the 27 members of the EU). Micro EU27 enterprises had on average 1,151 million EUR turnover and 4 employees; Small EU27 enterprises made 3,657 million EUR of turnover and had 20 employees. Medium enterprises had a turnover of 16,847 million EUR and employed 100 persons (The Gallup Organization, 2007: 10). "Turnover" in the quoted source means annual sales. Below we compare these figures with figures on total revenue (and/or net sales where specified). Although total revenues should be higher than the value of turnover, this difference is usually negligible, and we shall disregard it here. For reasons of secrecy and/or discretion, exact data on individual enterprises from our sample cannot be presented, i.e. data on total revenues and number of employees.

5.2. Financing Innovations of SMEs in Hungary

Our field research covered 13 high-tech firms, 84,6% of them medium-sized, 15,4% small, and none micro (see Table 10). The small enterprises from the sample are, on average, bigger (indicated by total revenues) and more productive (indicated by the ratio of total revenues to employees) than an average small EU27 firm. But the medium-sized firms are smaller and less productive than the EU27 average. Seven enterprises in the sample are in mixed ownership, including at least 5 that have more than 50% foreign ownership (i.e. they are subsidiaries of foreign firms)[180].

Table 10: Distribution of Hungarian enterprises from the sample according to their size

Size of the enterprises	Number of enterprises (total is 13)	Share in total number of enterprises in %	Average number of employees	Average total revenues in million Euro
Micro enterprises	0	0%	-	-
Small enterprises	2	15,4%	24	5,1
Medium enterprises	11	84,6%	127	9,7

Note: For one company, we estimated all variables from the table on the basis of this company's website, as well as other internet sources. For another company, we only estimated the number of employees at the time of the survey. Otherwise, all data[181] are as at the end of 2008. Amounts originally in Forint were converted to Euro according to the exchange rate of the Hungarian Central Bank at the end of that year.
Source: HMJ 2009, Europages (2009), companies' websites, and an interview.

Out of the 13 high-tech firms covered in the interviews, only 3 (23,08%) did not have innovations in the past 3-5 years, while 10 (76,93%) had either incremental or radical innovations – as can be seen in Figure 13. Interestingly, three firms that reported no innovations were all subsidiaries of foreign (multinational) companies. Two of them produce no innovations because only their parent companies innovate. In one of them, financing is controlled by the parent company. They plan to introduce innovation soon. Other subsidiaries all had innovations.

180 For one firm, we only know that it is in mixed ownership. For data on ownership, see the source below Table 10.

181 I would like to thank Katalin Kasos-Seres for help with getting some data.

Figure 13: Kinds of innovation among the firms in Hungary

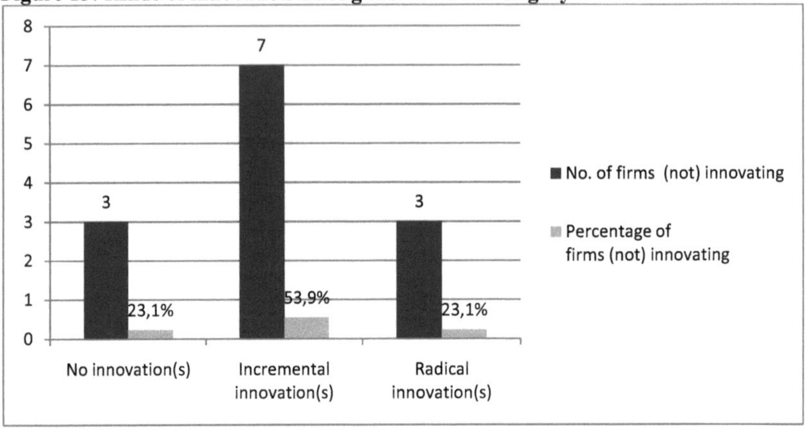

Source: own research.

The external sources for financing innovations that the firms innovating in the past 3-5 years would have used for financing current innovations are various, and we shall mention them in the order of diminishing frequency (multiple answers from each firm possible). Government funds obviously have an important role, since they would have been used by 6 (out of 10) firms. Banking credits were mentioned by 5 firms, and EU programmes by 3. Two firms mentioned they would have used financing sources from their partners. One firm mentioned grants, although it is unclear by whom. But it cannot go unnoticed that 8 (out of 10 firms) would have also used "own sources" or "retained earnings". Among those, there are two firms that would have used only own sources.

Evidence from some firms innovating in the past 3-5 years regarding their problems with financiers showed that they found banking credits expensive, and that it was hard to obtain large credits from a Hungarian bank. In addition, some firms complained about the state and EU programmes being low in value, overly bureaucratic, and entailing a time lag in getting the necessary funds. One firm found it hard to specify its activities for the next 2-3 years, for applications with the Hungarian government or EU programmes. As that company is very active with radical innovations, we presume this does not indicate a deficient strategy, but is no more than a general remark. Representatives of two companies remarked on VC and VC/PE, respectively. While the first claimed the VC's profit rate could endanger the company, the second felt the firm was pressed by demands of short-term profitability. However, both these firms are medium-sized, and their total revenue and productivity are close to the average presented in Table 10, i.e., much lower than in the EU27. Thus, their "projects" are probably not attractive to VC firms.

In order to test firms' openness and willingness to achieve growth, we asked each firm whether they had applied for funding by VC/PE firms. In spite of having a relatively developed VC market, only 2 out of 10 firms that provided an answer had applied for such funding (see Figure 14). When asked about the reasons they had not applied for funding by VC/PE firms, those firms that had innovated in the previous 3-5 years put forward several arguments. While three companies cited a number of "objective" reasons (either the fact that bank credit was cheaper than VC or limitations imposed by the parent company), four companies expressed a negative stance towards VC/PE. They either believe that investment by a VC/PE firm diverts the firm from achieving its goals, or that it is dangerous to strive for a high profit rate. Some were simply sceptical about investment by third parties, which could be indicative of a lack of social capital.

Figure 14: Number and percentage of Hungarian firms that applied for funding by VC/PE firms

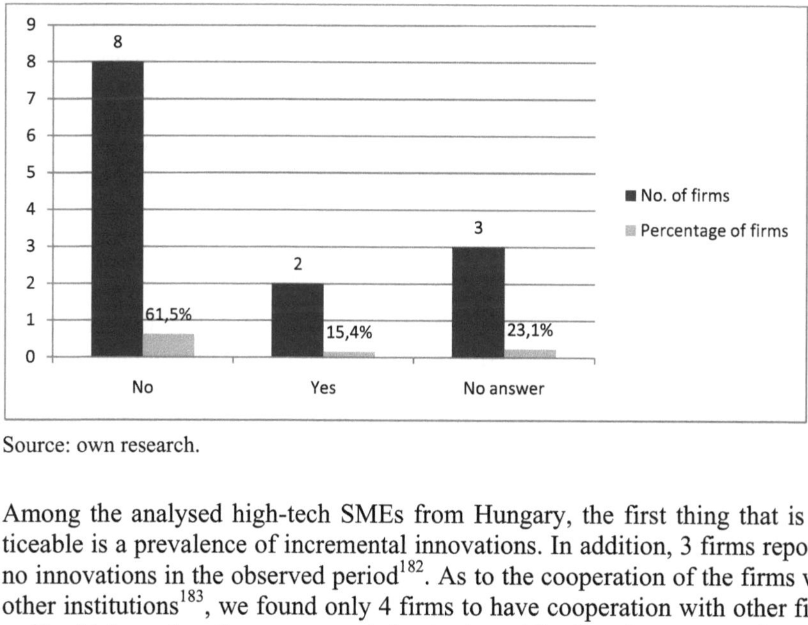

Source: own research.

Among the analysed high-tech SMEs from Hungary, the first thing that is noticeable is a prevalence of incremental innovations. In addition, 3 firms reported no innovations in the observed period[182]. As to the cooperation of the firms with other institutions[183], we found only 4 firms to have cooperation with other firms and/or higher education or research institutions. The fact that cooperation with

182 All three were subsidiaries of multinational firms.

183 The websites of those companies were searched for relevant information on possible cooperation.

other institutions and/or firms is rather infrequent is evident from the data in the table in Appendix III, as Hungary's innovative SMEs cooperate less than the EU average, while Slovenia's innovative SMEs outperform the EU average[184]. This points to the rather limited role of cooperation as a way to enhance innovation activities by way of communicating with potential knowledge sources. It further underlines the lower quality of innovation projects (i.e. innovations) among the analysed firms. Moreover, the vast majority of the firms in the sample is medium-sized, and making less total revenues than EU27 firms. In our view, the sources of innovation financing mentioned only reflect the nature of the innovation activities. Namely, government sources, EU programmes, and banking credits play a dominant role in this regard, while the role of VC/PE could be considered as no more than marginal. Because of the low asset specificity of the firms' projects, debt is among the most frequent sources. Furthermore, firms may prefer government sources and EU programmes not only because they supplement the supply from the financial sector, but also because they prefer financing options with fewer governance capabilities than VC/PE – in keeping with the pecking order theory. None of the firms would consider the latter source in case they needed funding for their innovations. Besides, a defensive stance towards VC/PE is noticeable with some companies. This, we believe, has to do with the underdeveloped equity culture in Hungary, in addition to the aforementioned considerations of transaction cost economics and the pecking order theory. Although the VC/PE option is not acceptable to all firms, it could be argued that, by refusing it, firms forego a chance for extra growth. This can be used as an argument supporting the assumption that not all financial sources are being tapped[185].

5.3. Financing Innovations of SMEs in Slovenia

The interviews covered 16 Slovenian companies. Their distribution according to size can be seen in Table 11. SMEs of all size classes are represented. Most of the enterprises from the sample are in domestic ownership[186]. Medium-sized firms have higher total revenue[187] than an average medium-sized EU27 enterprise. However, they are less productive.

---

184 The data are for 2004.

185 In addition, interviewees from 2 firms stated that their firms had no problems with financing (innovations).

186 There are no ownership data for one company. Out of the remaining 15, there are 11 domestically owned enterprises, and 4 with mixed (domestic and foreign) ownership (Pirs, 2009). However, in the quoted source there is no explanation as to the exact percentage of foreign ownership.

187 Or net sales, see table 11.

Table 11: Distribution of Slovenian enterprises from the sample according to their size

Size of the enterprises	Number of enterprises (total is 16)	Share in total number of enterprises in %	Average number of employees	Average net sales or total revenues in million Euro
Micro enterprises	8	50%	3,8	0,31
Small enterprises	5	31,3%	23,4	1,76
Medium enterprises	3	18,8%	121,8	20,01

Note: The data on total revenues or net sales refer to 2008 (13 enterprises), 2007 (2 enterprises) or to the April 2007-March 2008 period (one company). The number of employees for some enterprises was calculated as average, because original data were represented as size class (e.g. 5-9 employees). One of the companies merged with another in December 2008.
Source: Europages (2009), Pirs (2009) and Ajpes (2009).

Out of 16 enterprises, 4 (25%) had not had any innovations in the 3-5 years prior to the interview. The vast majority of other companies (11 or 68,75%) had had radical innovations, while only one company (6,25%) had had incremental innovations (see Figure 15). Out of the 4 firms that had had no innovations, we had 2 whose interviewees thought that there was no need for innovations, indicating a lack of intention to innovate and, consequently, a lack of innovation strategy[188]. Two other firms did not elaborate on their position.

188 Firm no. 3 trades in licensed products, which indicates the predominant influence of an external knowledge source.

Figure 15: Kinds of innovations among interviewed firms in Slovenia

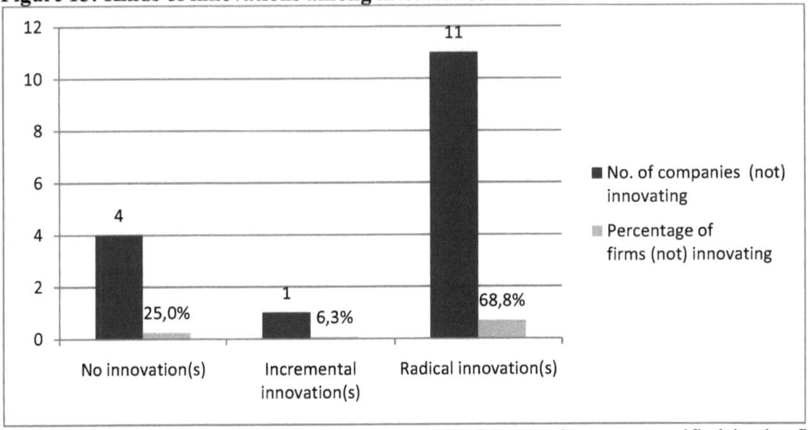

Note: Firms producing both incremental and radical innovations are specified in the figure only as producers of radical innovations.
Source: own research.

Firms that had innovated in the period of 3-5 years prior to the research had named various external financing sources that they would have used when wanting to finance innovations. Out of 12 innovating firms, 5 would have used some form of state assistance, while one additional firm would have applied for public tenders in order to get subsidies (which could also be government subsidies). The no. 2 spot on this list, at least according to the frequency with which some firms named a financing source, is occupied by different forms of EU financing sources. These were mentioned by 5 out of 12 companies. Only two firms would have asked their partners to either finance or cofinance innovation activities. Interestingly, banking credits were only explicitly mentioned by one firm. One firm that produced radical innovations would have considered only venture capital – both informal venture capital (business angels) and formal venture capital. Another firm is open to VC as well, among other sources. There is one firm that would have used own funds alone to finance innovations, and 5 firms that would have combined internal financing with other financing sources.

Firms that had innovated in the previous 3-5 years had several remarks on different financial sources. Some interviewees complained about bureaucratic procedures that required a lot of effort – such was the case with an interviewee complaining about European funds. Banking credits were criticized on the grounds that they were suited neither for financing innovations nor for SMEs. Regarding the latter, anecdotal evidence shows that the value of the required collateral needs to be quite high. Besides, the minimum size of the firm and business results have their role as well. Although 3 firms in our sample had indeed applied for funding by VC/PE firms (see Figure 16), the reluctance of the others (who answered the questions) was evident. This does seem to point to the fact that most of the entrepreneurs had not tried to attract all possible financing sources to finance their innovations, notwithstanding the opinion of an interviewee who claimed that investors in the European Union were much more conservative than investors in the USA[189]. Firms which did not consider the VC/PE option stated several reasons for refraining from it[190]. These reasons primarily reflected their mistrust towards VC/PE[191], but an interviewee was quite frank and said that their firm was not interested in having more innovations.

Figure 16: Number and percentage of Slovenian firms that applied for funding by VC/PE firms

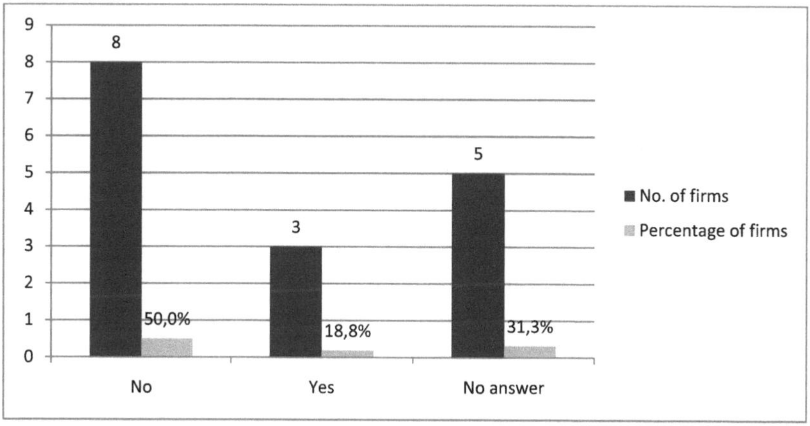

Source: own research.

Rather than financing problems, it is other problems that seem to be more acute to the innovating enterprises from our survey. Namely, interviewees from 7 innovating firms realised that financing innovations was not their worst problem, or that it was not a problem at all (see case study reports for details). There are other problems that were frequently mentioned. First of all, there is a problem with the lack of adequate or competent employees, as stated by interviewees from 4 firms. This is in line with the findings we reviewed for Slovenia in the

189 This argument did not serve as an "excuse", since this company's management indeed examined financing opportunities on both sides of the Atlantic. Besides, their product represents a radical innovation at the world level. For this quite revealing case, see enterprise no. 11 in the case study reports for Slovenia (Appendix II).

190 Only firms with innovations in the 3-5 years prior to this research were considered here.

191 Interviewees claimed VC/PE would want to buy their firm too cheaply, would like to force a high growth of the firm, would interfere with the firm's policy, etc.

chapter 4.5. An equally frequent problem concerns either too little innovation or the fact that innovations are not of sufficient quality – problems of this nature have been mentioned by 4 firms[192]. Second on the list of the most frequently mentioned problems concerns the products' market – it is either small in size and/or underdeveloped (3 firms).

With regard to the fact that firms from the sample mostly reported radical innovations, we would expect to see that this is positively connected with their cooperation and with openness to VC. Cooperation of innovative firms with other firms and/or research or higher education institutions seems to be frequent. Out of 12 innovators, we found that 10 cooperated with other organisations – 7 of which cooperated with both firms and research institutions or higher education institutions. That is not surprising, given the fact that 8 of the innovating companies are members of a technology park or a similar organisation in which such cooperation is encouraged. The data from the European Innovation Scoreboard (see table in Appendix III) confirm that frequent cooperation is no coincidence, as the percentage of Slovenia's innovative SMEs that cooperate with other institutions and/or firms lies above the EU average[193]. At first glance, the preferred financial sources do not differ much from those indicated by Hungarian companies, as state and EU sources dominate. As opposed to Hungarian enterprises, however, banking credit appears to be neglected in Slovenia as an innovation-financing source. And firms' acceptance of VC seems to be higher than in Hungarian firms, as two firms are willing to consider VC as a source for financing innovations, and 3 (out of the 11 that answered the specific question) had already applied for funding by VC/PE. Due to the similar entrepreneurial cultures in Slovenia and Hungary, this difference in financing sources can be attributed to the quality of innovation projects. A preponderant percentage (68,8%) of Slovenian firms from the field research had radical innovations and consequently a higher asset specificity. A ready-made solution for this is equity, as suggested by transaction costs economics. By the same token, debt (i.e., banking credit) is neglected.

5.4. Financing Innovations of SMEs in Croatia

In the case of Croatia, our interviews covered 14 companies and one trade, which adds up to a total of 15 case studies[194]. The distribution of enterprises according to their size can be seen in Table 12. Judging by the average total reve-

192 These problems concern stimulating innovativeness in an enterprise; the need to get good ideas, as well as to implement them; the need to have more innovations.

193 The data are for 2004.

194 For the sake of simplicity, we shall refer to them as enterprises or firms. Not least because of the fact that a trade can become a full-fledged company.

nues, the enterprises are quite small and definitely smaller than the EU27 average. Besides, there are no medium-sized enterprises. All enterprises from the sample are domestically owned[195].

Table 12: Distribution of Croatian enterprises from the sample according to their size

Size of the enterprises	Number of enterprises (total is 15)	Share in total number of enterprises in %	Average number of employees	Average total revenues in million Euro
Micro enterprises	5	33,3%	5,2	0,1
Small enterprises	10	66,7%	19,2	1,9
Medium enterprises	0	0	-	-

Note: Data refer to 2007. For one enterprise number of employees was from 2003. All enterprises are domestic.
Source: Poslovna Croatia (2009) and an interview; author's calculations.

All enterprises had innovations – most of the enterprises (13 or 86,67%) had incremental innovations, while 2 (13,33%) had radical innovations (see Figure 17).

Figure 17: Kinds of innovations among interviewed firms in Croatia

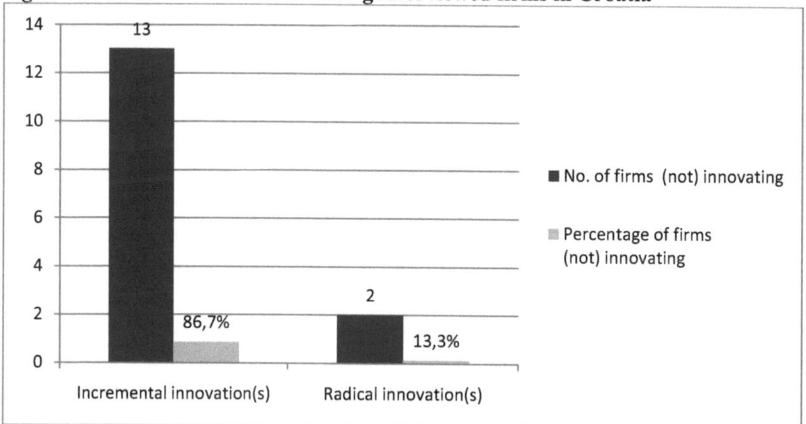

Note: A firm producing both incremental and radical innovations is specified in the figure only as producer of radical innovations.
Source: own research.

195 See source for this data under the table 12.

The enterprises from the sample would have normally used financing sources from various government authorities in order to finance innovations (e.g. Ministry of Economy, Labour and Entrepreneurship, county authorities, or local authorities, as well as the Croatian Bank for Reconstruction and Development) – 6 out of 15 enterprises. Nevertheless, 2 firms complain that these funds cover only a minor part of their expenses. Banking credit – quite expectedly – is the second most frequently reported financing source that the innovative enterprises found desirable (5 out of 15 enterprises). This confirms the earlier conclusions about Croatian enterprises' dependence on banks (see ch. 4.3.). Four of the 5 firms that would have liked to use banking credit would have used it as an exclusive option. Financing with the help of partners would have been an option for 2 enterprises. Two interviewees mentioned that their firms would have used solely internal sources to finance innovations. For one firm, VC would have been an exclusive option for financing innovations. As opposed to the Slovenian and Hungarian enterprises, only one Croatian enterprise would have tried EU funds as a financing source. Since Croatia is not in the EU, it cannot take part in all EU programmes, as opposed to the EU countries Slovenia and Hungary. The enterprises surveyed complained most of all about banking credits and venture capital funds. As for banking credits, some firms said they were expensive and administratively demanding, or even saw them as endangering their firm's independence. VC/PE was criticised by some interviewees on the grounds that it could interfere with their firms' business policies, endanger their independence, and that the investment process could become complicated and lengthy.

Only 3 out of the 15 surveyed enterprises (see Figure 18) had applied for VC/PE funding, which is similar to the firms surveyed in the other two countries. The reasons suggested by interviewees from the innovating firms that had not applied for VC/PE funds can be divided into four categories. Only one of the four can be said to comprise objective reasons for not wanting VC/PE. These enterprises reported a serious obstacle keeps them from expanding their business – they lack key employees. The other categories are as follows: 1) enterprises that did not want their independence endangered; 2) enterprises that have no ambition to expand their business; 3) enterprises that have not heard (a lot) about VC/PE. Hence neither in the case of Croatian SMEs can we confirm that they are open to all financing sources.

Figure 18: Number and percentage of Croatian firms that applied for funding by VC/PE firms

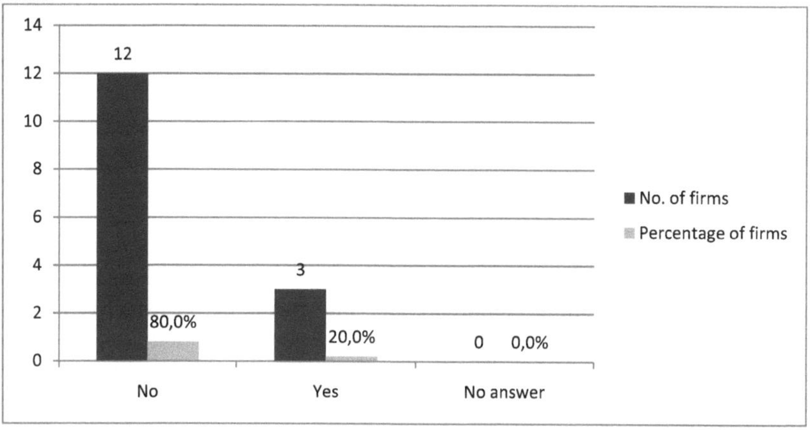

Source: own research.

For a total of 5 interviewees (firms), financing (innovations) was either explicitly not a problem (*per se*) or it was not the no. 1 problem at the level of their firm. There was only one firm whose representative indicated finance as the greatest problem their firm was faced with[196]. That qualification has to be seen in the light of other facts about this firm: it is understaffed, has no cooperation, and has only produced incremental innovations. The interviewee's arguments were that it was hard to get banking credit and VC. So the firm had not used all knowledge sources (since it had no cooperation) and was faced with a barrier in terms of being understaffed. Although it could be facing a problem with financing as well, its innovation projects are less interesting to VCs, since they are incremental (and not radical). Hence, the financing of innovations should be viewed in connection with these and other characteristics of the firm, as well as with some barriers to innovations that firms are faced with. Indeed, the interviewees mentioned various problems they were faced with in the innovation process. Namely, the absence of state support institutions was mentioned by 2 firms. A further 2 complained about lack of state support[197]. In view of the anecdotal evidence by two firms to which either a Ministry failed to make regular payments or a state owned enterprise refused to pay its own debt, the role of the state seems to be very problematic for the development of entrepreneurship in general, let alone entrepreneurs with innovations. Market-related barriers to innovation activities include the small size of the market (a remark made by 3

196 Firm no. 8.

197 Their remarks were perhaps more general: the state does not do enough to help, and the Republic of Croatia does not have an export strategy.

firms), and lack of qualified staff (mentioned by 2 interviewees). In addition to external problems, possibly influencing innovation activities are internal barriers to innovativeness: interviewees from 3 firms mentioned that they either had no ambition for development of their business or they did not want to incur more risk by developing their own product. All but 3 firms from the sample have some kind of cooperation with other partners: 4 with other firms, 2 with research institutions and/or higher education institutions, and 6 with both firms and scientific institutions.

Low asset specificity indicated by production of incremental innovations and small size of the enterprises can explain such willingness to use banking credit to finance innovations. These factors are also preventing enterprises from using a financial "instrument" with greater governance capabilities, that is, VC – even if they were willing to. The fact that finance is demand-constrained is indicated by the lack of motivation for development of business also noticed with some enterprises. However, the research has also identified other factors that represent barriers in the Croatian NSI. They prevent better innovative performance in general and have ramifications affecting financing innovations.

5.5. Summing Up

Based on the chapter on the NSIs in the studied countries, we know that the financing of innovations has been reported as the biggest problem in Hungary and in Croatia, whereas supply and expenses in connection with human resources figure as such in Slovenia (see above). By analysing the innovation financing of SMEs in the studied countries, we see that finance is complementary to innovations – not the primary or dominant factor that determines the outcome. In spite of relatively developed financial sectors, these countries have some firms that have voiced complaints about financing sources. But we saw that, when it comes to determining the quality of innovations produced, it is influenced by many factors both inside the firm and stemming from the national system of innovation. As regards the former, the most prominent are lack of cooperation and deficient motivation to further develop business. The latter is manifested in deficiencies of the national system of innovations, such as lack of support by state institutions and lack of adequate employees. Another conclusion from the empirical field research is that the quality of innovations (or "innovation projects") presupposes certain financing options. In the case of more radical innovations, as seen among Slovenian firms, acceptance of VC is higher. In the case of more incremental innovations, there is greater acceptance of banking credit, as seen in firms from Hungary and Croatia. We nevertheless saw that entrepreneurs in general may be reluctant about VC financing, which must be traced back to the bank-oriented culture and lack of social capital.

Our research supports the earlier findings for Slovenia, in that interviewees from the Slovenian SMEs identified the lack of adequate employees as their actual problem. In addition, we also identified problems related to their firms' innovativeness, as well as with small and/or underdeveloped market for their products. An evident problem for the Hungarian SMEs in our sample was their lack of cooperation. For the Croatian SMEs, barriers to innovative activities can be seen in the general lack of state support. Anecdotal evidence even suggests that there may be problems with regular payments by some state institutions. We did not find such evidence for the other two countries. Among other problems for the conduct of innovation activities in the surveyed SMEs, the small market size, lack of qualified staff, and lack of ambition stand out.

There are at least three things that we noticed in all three countries. First, the fact that state funds would have been the preferential funds for the financing of innovations in all three countries means that their NSIs are probably not as dependent on credit as one would expect from bank-based financial systems. This seems to be their advantage, since bank-based financial systems are more likely to support incremental innovations. Provided that government funds are abundant enough and that their conditionality affects firms' innovation activities positively (i.e., by increasing their innovative capabilities), that could be a way to change the NSIs. Second, financial instruments that are less interfering with corporate governance (e.g. subsidies and credits) are preferred over other instruments (e.g. VC/PE). That is a usual characteristic of bank-based financial systems. It is in accordance with both the pecking-order theory and the bank-oriented culture. Third, there is generally a reluctance to use VC/PE. This reluctance is most probably caused by the lack of equity culture and/or lack of social capital, as well as by other characteristics of bank-based financial systems – although there are some differences even in the acceptance of VC/PE in these countries.

6. Conclusion

Since innovations are crucial for economic growth, it is essential to understand how they are financed and how the financing of innovation activities can be improved. Our intention was to describe and explain the financing of innovation activities (in short: "innovation financing") in a systemic perspective, especially focusing on small and medium-sized enterprises in high-technology industries in such post-transition countries as Hungary, Slovenia and Croatia. We were motivated by the fact that there have only been very few studies of a conceptual/theoretical nature, or, for that matter, empirical studies in this field of economics. We, therefore, wanted to propose a novel approach to the financing of innovations. As far as the theoretical foundations of this enterprise are concerned, the task has proved demanding, not least because we do not yet have a theory of innovation financing. Our approach has been essentially institutional, in that we recognised that the financing of innovations is embedded in socio-economic systems, and that the latter largely influence the former.

In order to study innovation financing, in Chapter 2 we have integrated various concepts that were usually observed separately in the literature. The overarching concept is that of the national system of innovation, which we have used as a "common denominator" in the introduction of all of the other concepts. This is an important aspect of our approach, since elements of the national *systems* of innovation can only be changed in compliance with the other elements of this *system*. We see that the national system of innovation is itself ingrained in culture and social capital. Corporate governance (and its relations with the financial system) can be used to tackle the linkages between the financial and real sectors, since the financing and monitoring roles of the former have informational and strategic consequences for the latter – which again has implication for the financing of innovations. Furthermore, innovation activities are viewed in conjunction with entrepreneurial efforts which contribute to the dynamics of an economy – especially by founding new firms and influencing their growth paths by active management and risk-taking. However, since knowledge is dispersed, and innovation activities are time-consuming, knowledge-demanding and financially challenging innovative capabilities are not built solely inside the firm. Market-based and non-market-based relationships with other firms, customers, suppliers, research institutions and higher education institutions, government agencies, and others contribute to their sound foundations. Finally, notwithstanding the fact that competitiveness can be based on factors not related to innovativeness (like geographical location), producing innovations is the only way to build a competitive position in the market in the long run.

In order to derive principles of innovation financing, we have combined knowledge from the economics of innovation, theory of the firm, as well as financial theory, which can be seen in Chapter 3. Although mainstream economics theories do not consider the complex nature of innovation financing, we have nevertheless found the economics of information literature to be very helpful in terms of integration with our approach. Heterodox economics theories have provided us with "tools" to tackle the interrelatedness of the real and financial sectors. The institutional and post-Keynesian economics approaches led us to primarily concentrate on innovation (production) activities, and to observe their relations to financing. Hence, we have managed to prove the first hypothesis in light of the fact that the current approaches are either unsatisfactory or have insufficient explanatory power to explain innovation financing. In the concept of innovation financing that we advance, an essential role is given to the innovation capabilities of firms, which, in turn, influence innovation activities, the knowledge sources used, and the final output produced (incremental or radical innovations). While interactions with the financial sector are multiple, but particularly important before innovation activities have started, in the early stages of a firm or even before a firm is founded, it is the firm that eventually produces innovative product, not the financial institution[198]. The financial sector can be a facilitating and/or constraining influence on the real sector, but it does not have a primary role in the innovation activities of firms from the real sector. In our approach, the primary role is in the hands of the producer.

We believe this study makes a significant contribution to the literature on the economics of innovation by suggesting an approach to how innovation financing could be analysed. Implications for future theoretical work essentially concern the generally neglected link between the financial sector and the real sector. We need theories of the firm that focus more on the financial factors, and financial theories that focus more on the factors affecting production. There is a need for competing concepts of innovation financing, which can hopefully lead us towards a theory of innovation financing. To devise a theory of innovation financing is not an easy task and – as long as researchers believe that their theories should be as close to reality as possible – it is hard to believe that it could ever be simple and concise. The process of innovation financing is anything but simple – it involves interaction of the firm with financial institutions and/or private sources of finance (like family, relatives and friends) at different points in time, and it is systematically influenced by culture and social capital, as well as other socio-economic conditions.

198 This is not to say that financial institutions cannot be innovative – on the contrary, innovations in financial services and financial instruments can play a very important role for the real sector (as has already been suggested in this work).

The empirical part of this work comprises both the results of desk research and those of field research. It encompasses Hungary, Slovenia and Croatia, all post-transition countries of East (or South-East) Europe. As we elaborate in Chapter 4 on the national systems of innovation, these countries do not belong to the group of the largest innovators in Europe, because their NSIs have not yet been developed. Their social capital and entrepreneurial culture appear similar. However, their innovation performance differs – Slovenia is ahead of the other two countries, Hungary is in the middle, with Croatia lagging behind. In the fifth part on the innovation financing of SMEs in Hungary, Slovenia, and Croatia, we analysed the results of our field research, which corroborate our second hypothesis. Although financial factors often constrain firms in their innovation activities, we have found no reason, on the level of SMEs from each country, to believe that they are primary or decisive. Rather, insufficient innovative performance of firms, coupled with unwillingness to consider financing options with greater governance capabilities such as VC, and restraining non-financial conditions in the NSI are crucial. Besides, the quality of innovative output is largely determinant of financing options, in that incremental innovations are associated with banking credit, and radical innovations with venture capital. Needless to say, not all firms experience financial problems, and the innovative performance of firms, as well as the NSIs themselves, differ. We arrived at this result by pursuing largely qualitative research practices. Our analysis was based on the case studies described in the case study database (Appendix I) and case study reports (Appendix II). Therefore, it is not our intention to claim that these results are the absolute truth. Nonetheless, we believe that the findings are fairly robust and consistent with the theory and concepts employed.

Since the development of innovations in sectors of high technology is crucial for economic growth (in the countries that we have studied), public policies should properly address this issue. This requires formulation of a concept and strategy of socio-economic development at the state level, in order to make a substantial impact on the NSI. High-tech sectors are normally R&D intensive, which is not always the case in Hungary, Slovenia, and Croatia. Although R&D is just one of the innovation activities, it strongly influences the creation of new knowledge. Hence, enterprises conducting intramural R&D and those cooperating on R&D with other firms or research institutions should be of interest to policy makers. But raising the overall level of R&D expenditures is not enough, if innovative performance at country level is to be improved. We know that models suggesting a linearity in innovation development, such as the science-push model, the demand-pull model, or the money-first approach, can be rejected as being one-sided. Raising innovative performance involves increasing firms' innovation competencies and making structural changes in the NSI. Government can, and should, influence the overall structure of the NSI, in that it should set up efficient institutions (laws, by-laws, and other regulations regarding innovation activities) and organisations (such as technology transfer offices, research institutes, agencies for promoting competitiveness, etc.). In addition, linkages between state institutions, business, and research and higher education institutions should be enhanced. The role of the state should be particularly pronounced in building the innovation capabilities of firms, e.g., by promoting and supporting entrepreneurial education, conferences, seminars, and similar events that facilitate knowledge building and knowledge sharing. We have seen from this and other studies that entrepreneurs are often constrained by lack of information and knowledge. These, however, are the *sine qua non* of entrepreneurial activities. Building innovation capabilities (capacities) in SMEs means contributing to their competitiveness and innovativeness in the long run. This is indispensable for improving an NSI. These policy measures should be supplemented by attracting high-tech, R&D-performing FDI, and stimulating formal and informal knowledge transfer between firms in foreign ownership, and those in domestic ownership. There is another set of policies that ought to address innovation financing, and it should be pursued in parallel, not separately or exclusively. A national system of innovation is significantly determined by the characteristics of its financial system, that is, whether it is bank-based or capital-market-oriented. Since a bank-based system is more supportive of incremental innovations (and *vice-versa*), and capital-market-based system of radical innovations (and *vice-versa*), it is important to think about introducing a change in such a structure. All the more so because a bank-based system has its cognitive reflection in an entrepreneurial culture which is credit-oriented, and a capital-market-based system finds its cognitive counterpart in the equity culture. Hence, the government should focus on the particular way it influences innovation financing. This is especially important, since innovative high-tech SMEs from the empirical research prefer state-supplied financial sources for innovation financing. Such measures should be used to bridge the equity gap, especially for high-tech SMEs, which are particularly vulnerable in this regard. Kaufmann and Kokalj (1996: 59) essentially claim that programmes that just enable enterprises to take more credit, or obtain it on better terms, have positive feedback on both entrepreneurs' credit-based mentality, on their getting used to credit, as well as on the advantageous position of credit in the financial system. We may say that this could be dangerous for an NSI because it fixes its structure. Government programmes may concern other financial sources, not just credit ones, but, for example, also venture capital. By facilitating and supplementing the role of the financial sector in innovation financing, policy makers have a good chance to make it a coherent system with mutually supporting real and financial sectors. This can only be done if there are constant evaluations of measures undertaken, and if they are complementary to other elements of the NSI.

References

Acs, Z., C. O'Gorman, L. Szerb, and S. Terjesen (2006), "Could the Irish Miracle Be Repeated in Hungary?", Max-Planck-Gesellschaft Discussion Papers on Entrepreneurship, Growth and Public Policy, No. 3305.

Adam, F., M. Makarovič, B. Rončević, and M. Tomšič (2005), *The Challenges of Sustained Development. The Role of Socio-Cultural Factors in East-Central Europe*, Budapest: CEU Press.

Afuah, A. (2003), *Innovation Management. Strategies, Implementation, and Profits*, New York: Oxford University Press.

Allen, F. and D. Gale (2000), *Comparing Financial Systems,* Cambridge, MA, London, England: The MIT press.

Allen, F., L. Bartiloro, and O. Kowalewski (2007), "The Financial System of the EU-25", in Liebscher, K., J. Christl, P. Mooslechner, and D. Ritzberger-Grünwald (eds) (2007), *Financial Development, Integration and Stability. Evidence from Central, Eastern and South-Eastern Europe*, Cheltenham: Edward Elgar, pp. 80-104.

Amable, B. (2000), "Institutional Complementarity and Diversity of Social Systems of Innovation and Production", *Review of International Political Economy*, 7(4): 645–687.

Andritzky, J. R. (2007), "Capital Market Development in a Small Country: The Case of Slovenia", IMF Working Paper, WP/07/229.

Antoniou, A., Y. Guney, and K. Paudyal (2008), "The Determinants of Capital Structure: Capital Market-Oriented Versus Bank-Oriented Institutions", *Journal of Financial and Quantitative Analysis*, 43(1): 59-92.

Aralica, Z. (2007), "Utjecaj tehnologije na izvoznu konkurentnost Republike Hrvatske", Doktorska disertacija, Ekonomski fakultet Zagreb, Zagreb.

Ardalan, K. (2003), "Money and Academic Finance: The Role of Paradigms", *International Journal of Social Economics*, 30(5-6): 720-740.

Ardalan, K. (2004), "On the Theory and Practice of Finance", *International Journal of Social Economics*, 31(7-8): 684-705.

Arestis, P. (1992), *The Post-Keynesian Approach to Economics. An Alternative Analysis of Economic Theory and Policy*, Aldershot: Edward Elgar.

Arnsperger, C. and Y. Varoufakis (2006), "What is Neoclassical Economics?", *Post-Autistic Economics Review*, 38. Available at: http://www.paecon.net/PAEReview/issue38/ArnspergerVaroufakis38.htm , accessed 8 August 2009.

Audretsch, D. B. and E. E. Lehmann (2004), "Financing High-Tech Growth: The Role of Banks and Venture Capitalists", *Schmalenbach Business Review*, 56(4): 340-357.

Avnimelech, G. and M. Teubal (2004), "Targeting Venture Capital: Lessons from Israel's Yozma Program", in Bartzokas, A. and S. Mani (eds)

(2004), *Financial Systems, Corporate Investment in Innovation, and Venture Capital*, Cheltenham: Edward Elgar, pp. 85-116.
Bajuk, J., V. Cvijanović, T. Dadić, D. Račić and M. Veršić Marušić (2005), *Korporacijsko upravljanje u javnim dioničkim društvima u Republici Hrvatskoj*, Zagreb: Odvjetnički ured Veršić Marušić, Ekonomski institut, Zagreb, Socius (Working Materials).
Bakker, M.-R. and A. Gross (2004), "Development of Non-Bank Financial Institutions and Capital Markets in European Union Accession Countries", World Bank Working Paper, No. 28.
Bank of Slovenia (2008), *Financial Stability Review*, May. Available at: http://www.bsi.si/library/includes/datoteka.asp?DatotekaId=3223, accessed 1 July 2009.
Bank of Slovenia (2009), "Monthly Bulletin", *Monthly Bulletin of Bank of Slovenia*, 18(5). Available at: http://www.bsi.si/library/includes/datoteka.asp?DatotekaId=3490, accessed 24 June 2009.
Bank of Slovenia (2009a), *Poročilo o finančni stabilnosti*, May. Available at: http://www.bsi.si/library/includes/datoteka.asp?DatotekaId=3482, accessed 1 July 2009.
Bank of Slovenia (2009b), *Price Stability Report*, March. Available at: http://www.bsi.si/library/includes/datoteka.asp?DatotekaId=3470, accessed 1 July 2009.
Bartlett, W. (2002), "Prepreke razvoju malih i srednjih poduzeća u Bosni i Hercegovini, Makedoniji i Sloveniji: komparativna analiza", in Čengić, D. and M. Vehovec (eds) (2002), *Poduzetništvo, institucije i sociokulturni capital*, Zagreb: Institut društvenih znanosti Ivo Pilar, pp. 95-110.
Berger, A. N. and G. F. Udell (1998), "The Economics of Small Business Finance: The Roles of Private Equity and Debt Markets in the Financial Growth Cycle", *Journal of Banking and Finance*, 22: 613-673.
Berglöf, E. (1997), "Reforming Corporate Governance: Redirecting the European Agenda", *Economic Policy*, 12(24): 91-123.
Berglöf, E. and P. Bolton (2002), "The Great Divide and Beyond: Financial Architecture in Transition", *Journal of Economic Perspectives*, 16(1): 77-100.
Berthélemy, J.-C. and A. Varoudakis (1996), *Financial Development Policy and Growth*, Paris: Development Centre of the Organisation for Economic Co-Operation and Development.
Bigus, J. (2003), *Zur Theorie der Wagnisfinanzierung. Informations- und Anreizprobleme vor dem Börsengang*, Wiesbaden: Deutscher Universitäts-Verlag.
Bojnec, Š. (1999), "Privatisation, Restructuring and Management in Slovene Enterprises", *Comparative Economic Studies*, 41(4): 71-102.

Bojnec, Š. (2001), "Business and Managerial Start-Ups, R&D, and Product Innovation in Slovenia", *Eastern European Economics*, 39(4): 53-89.
Borrás, S. (2004), "System of Innovation Theory and the European Union", *Science and Public Policy*, 31(6): 425-433.
Bouchoux, D. E. (2001), *Protecting Your Company's Intellectual Property. A Practical Guide to Trademarks, Copyrights, Patents & Trade Secrets*, New York: AMACOM.
Boyd, R. and P. J. Richerson (2005), *The Origin and Evolution of Cultures*, Oxford: Oxford University Press.
Boynton, R. S. (2004), "The Tyranny of Copyright?", *The New York Times*, 25 January. Available at http://www.nytimes.com/2004/01/25/magazine/25COPYRIGHT.html?pagewanted=1&th, accessed 25 July 2009.
Bradley, M., G. A. Jarrell, and E. H. Kim (1984), "On the Existence of an Optimal Capital Structure: Theory and Evidence", *The Journal of Finance*, 39(3): 857-878.
Brealey, R. A. and S. C. Myers (2003), *Principles of Corporate Finance*, Boston: McGraw-Hill.
Brennan, M. J. (1995), "Corporate Finance Over the Past 25 Years", *Financial Management*, 24(2): 9-22.
Brouwer, E. and A. Kleinknecht (1997), "KEYNES-PLUS? Effective Demand and Changes in Firm-Level R&D. An Empirical Note", Research Memorandum 1997-38.
Brown, T. E. and J. Ulijn (eds) (2004), *Innovation, Entrepreneurship and Culture. The Interaction Between Technology, Progress and Economic Growth*, Cheltenham: Edward Elgar.
Bučar, M. and M. Stare (2002), "Slovenian Innovation Policy: Underexploited Potential for Growth", *Journal of International Relations and Development*, 5(4): 427-448.
Bygrave, W. D. and J. A. Timmons (1992), *Venture Capital at the Crossroads*, Boston, MA: Harvard Business School Press.
Callahan, J. and S. Muegge (2003), "Venture Capital's Role in Innovation: Issues, Research and Stakeholder Interests", in Shavinina, L. V. (ed) (2003), *The International Handbook on Innovation*, Oxford: Elsevier Science, pp. 641-663.
Caprio, Jr., G. and A. Demirgüç-Kunt (1997), "The Role of Long Term Finance: Theory and Evidence", Policy Research Department, The World Bank, February.
Carayannis, E. G. and E. Gonzales (2003), "Creativity and Innovation = Competitiveness? When, How and Why", in Shavinina, L. V. (ed) (2003), *The International Handbook on Innovation*, Oxford: Elsevier Science, pp. 587-606.

Casper, S. and F. van Waarden (2005), "Introduction: Scanning Literature on Institutions, Organizations and Innovation", in Casper, S. and F. van Waarden (eds) (2005), *Innovation and Institutions. A Multidisciplinary Review of the Study of Innovation Systems*, Cheltenham: Edward Elgar, pp. 3-18.

Čengić, D. (2002), "Sociološki aspekti razvoja malih i srednjih poduzeća u Međimurskoj županiji", in Čengić, D. and M. Vehovec (eds) (2002), *Poduzetništvo, institucije i sociokulturni capital*, Zagreb: Institut društvenih znanosti Ivo Pilar, pp. 175-212.

Chang, H.-J. (2002), "Kicking Away the Ladder", *Post-Autistic Economics Review*, 15. Available at: http://www.btinternet.com/~pae_news/review/issue15.htm, accessed 15 August 2007.

Christensen, J. L. (1992), *The Role of Finance in Industrial Innovation – Aspects of Financial Systems*, Aalborg: Institute of Production, Aalborg University Press.

Christensen, J. L. (1997), "Financing Innovation: ISE Sub-Project 3.2.3", TSER Project: Innovation Systems and Europe – An Integration (ISE), Aalborg University.

CNB (Croatian National Bank) (2007), *Makrobonitetna analiza*, No. 6. Available at: http://www.hnb.hr/publikac/makrobonitetna-analiza/h-mba-06.pdf?tsfsg=d1d3a7aa79a60d07ae7eac3a5fd6b4e1, accessed 28 December 2008.

CNB (Croatian National Bank) (2009a), *Bilten 148*, Vol. XIV, May. Available at: http://www.hnb.hr/publikac/bilten/arhiv/bilten-148/hbilt148.pdf?tsfsg=1b2c58eac0d71488647bfa523b81adb8, accessed 27 June 2009.

CNB (Croatian National Bank) (2009b), *Godišnje izvješće 2008*. Available at: http://www.hnb.hr/publikac/godisnje/2008/h-god-2008.pdf?tsfsg=509adea54b5ae609f0e0fd85957cce63, accessed 27 June 2009.

Cohen, W. M. and D. A. Levinthal (1989), "Innovation and Learning: The Two Faces of R&D", *The Economic Journal*, 99(397): 569-596.

Cohen, W. M. and D. A. Levinthal (1990), "Absorptive Capacity: A New Perspective on Learning and Innovation", *Administrative Science Quarterly*, 35(1): 128-152.

Cohen, W. M. and R. C. Levin (1989), "Empirical Studies of Innovation and Market Structure", in Schmalensee, R. and Willig, R. D. (eds) (1989), *Handbook of Industrial Organization*, Vol. II, Amsterdam: North Holland, pp. 1059-1107.

Cornell, B. and A. C. Shapiro (1993), "Financing Corporate Growth", in Chew, D. H., Jr. (ed) (1993), *The New Corporate Finance. Where Theory Meets Practice*, New York: McGraw-Hill, pp. 195-211.
Crespi, F. (2004), "Notes on the Determinants of Innovation: A Multi-Perspective Analysis", Working Paper 42.2004, The Fondazione Eni Enrico Mattei Note di Lavoro Series Index.
Cressy, R. and C. Olofsson (1997), "European SME Financing: An Overview", *Small Business Economics*, 9(2): 87-96.
Cvijanović, V. (1999), "Ekonomske paradigme", diplomski rad, Ekonomski fakultet Zagreb, Zagreb.
Cvijanović, V. (2007), "Croatia", in Bölling, A. K. (ed) (2007), *"Mittelstand" and Trade in South Eastern Europe. Social Market Economy as a Compass of Development*, Sankt Augustin/Berlin: Konrad Adenauer Stiftung e.V., pp. 63-73.
Cvijanović, V. and S. Kušić (2002), "Izravna strana ulaganja kao izvori financiranja investicija: usporedna analiza tranzicijskih ekonomija s primjerom Republike Hrvatske", *Financijska teorija i praksa*, 26(4): 879-893.
Dalić, M. (2002), "Usporedna analiza hrvatskog financijskog sustava i financijskih sustava naprednih tranzicijskih zemalja", *Privredna kretanja i ekonomska politika*, 12(92): 27-52.
Danis, W. M. and A. V. Shipilov (2002), "A Comparison of Entrepreneurship Development in Two Post-Communist Countries: The Cases of Hungary and Ukraine", *Journal of Developmental Entrepreneurship*, 7(1): 67-94.
Demirguc-Kunt, A. and R. Levine (eds) (2001), *Financial Structure and Economic Growth: A Cross Country Comparison of Banks, Markets and Development*, Cambridge, MA: The MIT Press.
Demsetz, H. (1997), "The Firm in Economic Theory: A Quiet Revolution", *The American Economic Review*, 87(2): 426-429.
Dequech, D. (2000), "Fundamental Uncertainty and Ambiguity", *Eastern Economic Journal*, 26(1): 41-60.
Dequech, D. (2003), "Cognitive and Cultural Embeddedness: Combining Institutional Economics and Economic Sociology", *Journal of Economic Issues*, 37(2): 461-470.
Djankov, S. and P. Murrell (2002), "Enterprise Restructuring in Transition: A Quantitative Survey", *Journal of Economic Literature*, 40(3): 739-792.
Dosi, G. (1988), "The Nature of the Innovative Process", in Dosi, G., C. Freeman, R. Nelson, G. Silverberg, and L. Soete (eds) (1988), *Technical Change and Economic Theory*, London and New York: Pinter Publishers, pp. 221-238.
Dosi, G. and L. Orsenigo (1988), "Coordination and Transformation: an Overview of Structures, Behaviours and Change in Evolutionary Environments", in Dosi, G., C. Freeman, R. Nelson, G. Silverberg, and L. Soete

(eds) (1988), *Technical Change and Economic Theory*, London and New York: Pinter Publishers, pp. 13-37.
Dow, S. C. (1991), "The Post-Keynesian School", in Mair, D. and A. G. Miller (eds) (1991), *A Modern Guide o Economic Thought. An Introduction to Comparative Schools of Thought in Economics*, Aldershot: Edward Elgar, pp. 176-206.
Dow, S. C. (2005), *Ekonomska metodologija. Istraživanje*, Zagreb: Politička kultura.
Drahos, P. (2005), "Intellectual Property Rights in the Knowledge Economy", in Rooney, D., G. Hearn, and A. Ninan (eds) (2005), *Handbook on the Knowledge Economy*, Cheltenham: Edward Elgar, pp. 139-151.
Drucker, P. (1983), "Modern Prophets: Schumpeter and Keynes". Available at: www.peterdrucker.at/en/texts/p_drucker_proph_en.pdf, accessed 2 December 2008.
Dunn, S. P. (2000), "Wither Post Keynesianism", *Journal of Post Keynesian Economics*, 22(3): 343-364.
Dunn, S. P. (2002), "A Post Keynesian Approach to the Theory of the Firm", in Dow, S. C. and J. Hillard (eds) (2002), *Post Keynesian Econometrics, Microeconomics and the Theory of the Firm. Beyond Keynes, Volume One*, Cheltenham: Edward Elgar, pp. 60-80.
Eagleton, T. (2000), *The Idea of Culture*, Malden: Blackwell Publishing.
Edquist, C. (2004), "Reflections On The Systems Of Innovation Approach", *Science and Public Policy*, 31(6): 485-489.
Edquist, C. and L. Hommen (1999), "Systems of Innovation: Theory and Policy For the Demand Side", *Technology in Society*, 21: 63-79.
Egert, B., P. Backe, and T. Zumer (2007), "Private-Sector Credit in Central and Eastern Europe: New (Over)shooting Stars?", *Comparative Economic Studies*, 49(2): 201-231.
Eisenberg, M. A. (1999), "The Conception That the Corporation is a Nexus of Contracts, and the Dual Nature of the Firm", *Journal of Corporation Law*, 24(4): 819-836.
ERAWATCH (2009a), *ERAWATCH Research Inventory Report: Croatia*, Bruxelles: European Communities. Available at: http://cordis.europa.eu/erawatch/index.cfm?fuseaction=ri.downloadCountryReport&filename=country_profile_HR.pdf, accessed 21 June 2009.
ERAWATCH (2009b), *ERAWATCH Research Inventory Report: Hungary*, Bruxelles: European Communities. Available at: http://cordis.europa.eu/erawatch/index.cfm?fuseaction=ri.downloadCountryReport&filename=country_profile_HU.pdf, accessed 21 June 2009.
ERAWATCH (2009c), *ERAWATCH Research Inventory Report: Slovenia*, Bruxelles: European Communities. Available at:

http://cordis.europa.eu/erawatch/index.cfm?fuseaction=ri.downloadCountryReport&filename=country_profile_SI.pdf, accessed 21 June 2009.
Estrin, S. (2002), "Competition and Corporate Governance in Transition", *The Journal of Economic Perspectives*, 16(1): 101-124.
European Commission (2007a), *INNO-Policy TrendChart – Policy Trends and Appraisal Report. Croatia 2007*, Bruxelles: European Commission. Available at: http://www.proinno-europe.eu/index.cfm?fuseaction=country.showCountry&topicID=108&parentID=52&ID=46, accessed 7 August 2009.
European Commission (2007b), *INNO-Policy TrendChart – Policy Trends and Appraisal Report. Hungary 2007*, Bruxelles: European Commission. Available at: http://www.proinno-europe.eu/index.cfm?fuseaction=country.showCountry&topicID=108&parentID=52&ID=20, accessed 7 August 2009.
European Commission (2007c), *INNO-Policy TrendChart – Policy Trends and Appraisal Report. Slovenia 2007*, Bruxelles: European Commission. Available at: http://www.proinno-europe.eu/index.cfm?fuseaction=country.showCountry&topicID=108&parentID=52&ID=19, accessed 7 August 2009.
European Commission (2008), "European Innovation Scoreboard 2007. Comparative Analysis of Innovation Performance", PRO INNO Europe paper, No. 6.
Eurostat (2007), *European Business. Facts and Figures*, Luxembourg: Office for Official Publications of the European Communities.
Eurostat (2008), *Key Figures on Europe. 2009 Edition*, Luxembourg: Office for Official Publications of the European Communities.
Fama, E. F. and K. R. French (1998), "Taxes, Financing Decisions, and Firm Value", *The Journal of Finance*, 53(3): 819-843.
FINA (2008), *Analiza financijskih rezultata poduzetnika Republike Hrvatske u 2007. godini. Prema obrađenim podacima godišnjih financijskih izvještaja zaprimljenih u Registar godišnjih financijskih izvještaja*, Zagreb: Financijska agencija.
Fischer, M. M. (2001), "Innovation, Knowledge Creation and Systems of Innovation", *The Annals of Regional Science*, 35: 199-216.
Fogel, G. and A. Zapalska (2001), "A Comparison of Small and Medium-Size Enterprise Development in Central and Eastern Europe", *Comparative Economic Studies*, 43(3): 35-68
Foray, D. (2004), *The Economics of Knowledge*, Cambridge, MA: The MIT Press.
Foster, J. (1991), "The Institutionalist (Evolutionary) School", in Mair, D. and A. G. Miller (eds) (1991), *A Modern Guide to Economic Thought. An In-*

troduction to Comparative Schools of Thought in Economics, Aldershot: Edward Elgar, pp. 207-232.
Fox, G. (1997), *Reason and Reality in the Methodologies of Economics. An Introduction*, Cheltenham: Edward Elgar.
Francis, J. C. (1991), *Investments. Analysis and Management*, New York: McGraw-Hill.
Franičević, V. (2003), "Douglass C. North – Bilješke uz intelektualnu biografiju", Mimeo. Available at: http://web.efzg.hr/dok/ETE/FRANICEVIC/NORTH_Pogovor_2003.pdf, accessed 18 May 2009.
Frankfurter, G. M. and E. G. McGoun (1999), "Ideology and the Theory of Financial Economics", *Journal of Economic Behavior & Organization*, 39: 159-177.
Freel, M. S. (1999), "The Financing of Small Firm Product Innovation within the UK", *Technovation*, 19: 707-719.
Freeman, C. (1982), *The Economics of Industrial Innovation*, London: Frances Pinter (Publishers).
Freeman, C. (1988), "Introduction", in Dosi, G., C. Freeman, R. Nelson, G. Silverberg, and L. Soete (eds) (1988), *Technical Change and Economic Theory*, London and New York: Pinter Publishers, pp. 1-8.
Freeman, C. (2006), "'Catching Up' and Innovation Systems: Implications for Eastern Europe", in Piech, K. and S. Radosevic (eds) (2006), *The Knowledge-based Economy in Central and Eastern Europe. Countries and Industries in a Process of Change*, Houndmills, Basingstoke: Palgrave Macmillan, pp. 13-30.
Frielinghaus, A., B. Mostert, and C. Firer (2005), "Capital Structure and the Firm's Life Stage", *South African Journal of Business Management*, 36(4): 9-18.
Futo, P., P. Hoggett, and L. Kallay (1997), "Small Firms and Economic Transformation in Hungary", *The European Journal of Social Sciences*, 10(2): 171-184.
Gate 2 Growth (2002), *A Guide to Financing Innovation*, Bruxelles: European Communities.
Geels, F. W. (2005), *Technological Transitions and System Innovations. A Co-Evolutionary and Socio-Technical Analysis*, Cheltenham: Edward Elgar.
Gelenčer, G. (2008), "Kreditne garancije: male banke naizloženije rizicima", *Lider*, 2 May: 8-9.
Gnos, C. (2006), "French Circuit Theory", in Arestis, P. and M. Sawyer (eds) (2006), *A Handbook of Alternative Monetary Economics*, Cheltenham: Edward Elgar, pp. 87-104.

Goldsmith, R. E. and G. R. Foxall (2003), "The Measurement of Innovativeness", in Shavinina, L. V. (ed) (2003), *The International Handbook on Innovation*, Oxford: Elsevier Science, pp. 321-330.
Goldsmith, R. W. (1969), *Financial Structure and Development*, New Haven, CT: Yale University Press.
Grant, S., S. King, and B. Polak (1996), "Information Externalities, Share-Price Based Incentives and Managerial Behaviour", *Journal of Economic Surveys*, 10(1): 1-21.
Greenwald, B., J. E. Stiglitz, and A. Weiss (1984), "Informational Imperfections in the Capital Market and Macroeconomic Fluctuations", *The American Economic Review*, 74(2): 194-199.
Grootaert, C., D. Narayan, V. Nyhan Jones, and M. Woolcock (2004), "Measuring Social Capital. An Integrated Questionnaire", World Bank Working Paper, No. 18.
Grupp, H. (1998), *Foundations of the Economics of Innovation. Theory, Measurement and Practice (New Horizons in the Economics of Innovation)*, Cheltenham: Edward Elgar.
Guinet, J. (1995), "Financing Innovation", *OECD Observer*, 194: 10-16.
Gullander, S. and G. Napier (2003), *Handbook in Business Angel Networks – The Nordic Case*, Stockholm: Stockholm School of Entrepreneurship.
Günter, L. and H. Frommann (1998), *Eigenkapital für den Mittelstand: Venture Capital im In- und Ausland*, München: Beck.
Hackethal, A. and R. H. Schmidt (2000), "Finanzsystem und Komplementarität", in Francke, H.-H., E. Ketzel, and H.-H. Kotz (eds) (2000), *Finanzmärkte im Umbruch (Beihefte zu Kredit und Kapital, Heft 15)*, Berlin: Duncker & Humblot, pp. 53-102.
Hall, P. A. and D. Soskice (2001), "An Introduction to Varieties of Capitalism", in Hall, P. A. and D. Soskice (eds) (2001), *Varieties of Capitalism. The Institutional Foundations of Comparative Advantage*, Oxford: Oxford University Press, pp. 1-68.
Hampden-Turner, C. and F. Trompenaars (1995), *The Seven Cultures of Capitalism. Value Systems for Creating Wealth in the United States, Britain, Japan, Germany, France, Sweden, and the Netherlands*, London: Piatkus.
Hare, P., T. Révész, P. Aven, G. Oblath, and H.-W. Sinn (1992), "Hungary's Transition to the Market: The Case against a 'Big-Bang'", *Economic Policy*, 7(14): 228-264.
Harris, M. and A. Raviv (1991), "The Theory of Capital Structure", *The Journal of Finance*, 46(1): 297-355.
Havas, A. and L. Nyiri (eds) (2007), *National System of Innovation in Hungary. Background Report for the OECD Country Review 2007/2008*, Budapest: National Office for Research and Technology.

HCSO (Hungarian Central Statistical Office) (2009), *Economy and Society, January – March 2009*. Available at: http://portal.ksh.hu/pls/ksh/docs/eng/xftp/gyor/jel/ejel20903.pdf, accessed 27 June 2009.
Hellmann, T. and J. Stiglitz (2000), "Credit and Equity Rationing in Markets With Adverse Selection", *European Economic Review*, 44: 281-304.
Hemmings, P. (2005), "Hungarian Innovation Policy: What's the Best Way Forward?", OECD Economics Department Working Papers, No. 445.
Hielscher, U., G. Dorn, and G. Lampe (1982), *Innovationsfinanzierung mittelständischer Unternehmungen*, Stuttgart: Poeschel Verlag.
Hirschey, M. (2003), "Organization Structure and Corporate Governance: A Survey", in Hirschey, M., K. John, and A. K. Makhija (eds) (2003), *Corporate Governance and Finance*, Kidlington, Oxford: Elsevier Science Ltd., pp. 65-112.
Hisrich, R. D. and G. Fulop (1995), "Hungarian Entrepreneurs and Their Enterprises", *Journal of Small Business Management*, July: 88-94.
Hodgson, G. M. (1998), "Evolutionary and Competence-Based Theories of The Firm", *Journal of Economic Studies*, 25(1): 25-56.
Hodgson, G. M. (2001), *Evolution and Institutions. On Evolutionary Economics and the Evolution of Economics*, Cheltenham: Edward Elgar.
Hodgson, G. M. (2006), "What are Institutions?", *Journal of Economic Issues*, 40(1): 1-25.
Hollingsworth, J. R. (2000), "Doing Institutional Analysis: Implications for the Study of Innovations", *Review of International Political Economy*, 7(4): 595-644.
Hutchinson, R. W. (1995), "The Capital Structure and Investment Decisions of the Small Owner-Managed Firm: Some Exploratory Issues", *Small Business Economics*, 7: 231-239.
Iliev, I. P. (2000), "Financing Innovation: Focus on Central and East Europe", Thesis for the degree of master of commerce in economics, University of the Witwatersrand, Johannesburg.
Iliev, I. P. (2001), "Changing Boundaries: The Importance of Including Measures of Innovation Finance in Innovation Surveys", a submission to the joint DACST/OECD seminar on innovation measurement, 28-29 March.
Iliev, I. P. (2006), "Barriers to Venture Capital Investment in Innovative Small and Medium Enterprises in Central and Eastern Europe: Causes and Policy Implications", in Piech, K. and S. Radosevic (eds) (2006), *The Knowledge-based Economy in Central and Eastern Europe. Countries and Industries in a Process of Change*, Houndmills, Basingstoke: Palgrave Macmillan, pp. 127-144.
IMF (International Monetary Fund) (2009), *Republic of Croatia: 2009 Article IV Consultation—Staff Report; Public Information Notice on the Executive*

Board Discussion; And Statement by the Executive Director for the Republic of Croatia. Available at: http://imf.org/external/pubs/ft/scr/2009/cr09185.pdf, accessed 27 June 2009.
Inzelt, A. (2003), "Innovation Process in Hungary", in Shavinina, L. V. (ed) (2003), *The International Handbook on Innovation*, Oxford: Elsevier Science, pp. 859-872.
Inzelt, A. (2004), "The Evolution of University-Industry-Government Relationships During Transition", *Research Policy*, 33: 975-995.
Inzelt, A. (2008), "Private Sector Involvement in Science and Innovation Policy-Making in Hungary", *Science and Public Policy*, 35(2): 81-94.
Ivičić, L., M. Dumičić, A. Burić, and I. Huljak (2008), "Rezultati petoga HNB-ova anketiranja banaka", pregledi, P-24.
Jaklič, A., J. P. Damijan, and M. Rojec (2008), "Innovation Cooperation and Innovation Activity of Slovenian Enterprises", Paper presented at the conference Innovation, Competitiveness and Growth, Zagreb, 27 November 2008.
Jambrović, M. (2007), "Malo i srednje poduzetništvo. Pokretači razvoja", *Eukonomist* (posebni prilog: Banke i mali te srednji poduzetnici), 10: 32-38.
Janeway, W. (2007), "Foreword", in Lamoreaux, N. R. and K. L. Sokoloff (eds) (2007), *Financing Innovation in the United States, 1870 to the Present*, Cambridge, MA: The MIT Press, pp. ix-xii.
Jensen, M. C. (1986), "Agency Costs of Free Cash Flow, Corporate Finance, and Takeovers", *The American Economic Review*, 76(2): 323-329.
Jensen, M. C. and W. H. Meckling (1976), "Theory of the Firm: Managerial Behavior, Agency Costs and Ownership Structure", *Journal of Financial Economics*, 3(4): 305-360.
Karajić, N. (1998), "Javnost u hrvatskoj tranziciji – između države i tržišta?", in Meštrović, M. and A. Štulhofer (eds) (1998), *Sociokulturni kapital i tranzicija u Hrvatskoj*, Zagreb: Hrvatsko sociološko društvo, pp. 173-204.
Karsai, J., M. Wright, and I. Filatotchev (1997), "Venture Capital in Transition Economies: The Case of Hungary", *Entrepreneurship: Theory and Practice*, 21(4): 93-110.
Kaufmann, F. and Lj. Kokalj (1996), *Risikokapitalmärkte für mittelständische Unternehmen*, Stuttgart: Schäffer-Poeschel.
Keasey, K., S. Thompson, and M. Wright (1997), "Introduction: The Corporate Governance Problem - Competing Diagnoses and Solutions", in Keasey, K, S. Thompson, and M. Wright (eds) (1997), *Corporate Governance: Economic and Financial Issues*, Oxford: Oxford University Press, pp. 1-17.
Keller, W. (2004), "International Technology Diffusion", *Journal of Economic Literature*, 42: 752-782.

Khan, A. (2000), "The Finance and Growth Nexus", *Business Review*, January/February: 3-14.
King, R. G. and R. Levine (1993), "Finance and Growth: Schumpeter Might be Right", *The Quarterly Journal of Economics*, 108(3): 717-737.
Király, J., M. Nagy, and V. E. Szabó (2008), "Contagion and the Beginning of the Crisis – Pre-Lehman Period", Magyar Nemzeti Bank Occasional Papers, No. 76.
Kiseljak, L. (2007), "Čišćenje burze: ostaje samo 15 tvrtki iz JDD kotacije", *Lider*, 23 November: 6-8.
Klapper, L. F., V. Sarria-Allende, and V. Sulla (2002), "Small- and Medium-Size Enterprise Financing in Eastern Europe", World Bank Policy Research Working Paper, No. 2933.
Knell, M. and M. Rojec (2007), "The Economics of Knowledge and Knowledge Accumulation: A Literature Survey", Draft of 20 February 2007, Mimeo, U-KNOW Project. Understanding the Relationship between Knowledge and Competitiveness in the Enlarging EU.
Knight, F. H. (1921), *Risk, Uncertainty, and Profit*, Boston, MA: Hart, Schaffner & Marx, Houghton Mifflin Company. Available at: http://www.econlib.org/library/Knight/knRUP6.html, accessed 29 August 2006.
Kochhar, R. (1996), "Explaining Firm Capital Structure: The Role of Agency Theory vs. Transaction Cost Economics", *Strategic Management Journal*, 17: 713-728.
Kornai, J., E. Maskin, and G. Roland (2003), "Understanding the Soft Budget Constraint", *Journal of Economic Literature*, 41(4): 1095-1136.
Krahnen, J. P. (1998), "Where Do We Stand in the Theory of Finance? A Selective Overview With Reference To Erich Gutenberg", CFS Working Paper, No. 98/09.
Kulicke, M. (1997), "Die Finanzierung technologieorientierter Unternehmensgründungen", in Koschatzky, K. (ed) (1997), *Technologieunternehmen im Innovationsprozeß. Management, Finanzierung und regionale Netzwerke*, Heidelberg: Physica-Verlag, pp. 127-152.
Kurik, S., R. Lumiste, E. Terk, and A. Heinlo (2002), *Innovation in Estonian Enterprises 1998-2000*, Tallinn: Estonian Institute for Futures Studies.
Kušić, S. (2001), *Privatisierung im Transformationsprozess. Das Beispiel der Republik Kroatien*, Wiesbaden: Deutscher Universitäts-Verlag GmbH.
Kušić, S. (ed) (2005), *Path-Dependent Development in the Western Balkans. The Impact of Privatization*, Frankfurt am Main: Peter Lang.
Kušić, S. and M. König (1999), "Zwei ungleiche Brüder auf dem Weg nach Europa. Die Entwicklung Sloweniens und Kroatiens seit der staatlichen Unabhängigkeit und die Annäherung an die Europäische Union", *Südosteuropa Mitteilungen*, 39(3): 230-246.

Kutlača, Đ., D. Ristić, M. Andevski, D. Semenčenko, O. Delić, M. Vukadinović, T. Buta, V. Holodkov, V. Radun, S. Arsenijević, and I. Šćepanović (2006), *Inovaciona kultura u Srbiji – mit ili realnost*, Novi Sad: Fakultet za menadžment.

La Porta, R., F. Lopez-de-Silanes, A. Shleifer, and R. W. Vishny (1997), "Legal Determinants of External Finance". *The Journal of Finance*, 52(3): 1131-1150.

La Porta, R., F. Lopez-de-Silanes, A. Shleifer, and R. W. Vishny (1998), "Law and Finance", *Journal of Political Economy*, 106(6): 1113-1155.

Laffont, J.-J. and D. Martimort (2002), *The Theory of Incentives – The Principal-Agent Model*, Princeton and Oxford: Princeton University Press.

Lamoreaux, N. R. and K. L. Sokoloff (2007), "Introduction: The Organization and Finance of Innovation in American History", in Lamoreaux, N. R. and K. L. Sokoloff (eds) (2007), *Financing Innovation in the United States, 1870 to the Present*, Cambridge, MA: The MIT Press, pp. 1-37.

Landry, R., N. Amara, and M. Lamari (2000), "Does Social Capital Determine Innovation? To What Extent?", Paper presented at the 4th International Conference on Technology, Policy and Innovation, Curitiba, Brazil, 28-31 August 2000.

Landry, R., N. Amara, and M. Lamari (2001), "Social Capital, Innovation and Public Policy", *Printemps*, Spring: 73-79.

Lawson, T. (2001), "Back to Reality", *Post-Autistic Economics Newsletter*, 6. Available at: http://www.paecon.net/PAEReview/wholeissues/issue6.htm, accessed 21 June 2007.

Lawson, T. (2002), "Social Explanation and Popper", Lecture given at the Popper Anniversary Conference, University of Galway, September 2002. Available at http://www.econ.cam.ac.uk/faculty/lawson/PDFS/Popper.pdf, accessed 24 June 2007.

Lawson, T. (2003), *Reorienting Economics*, London and New York: Routledge.

Lee, F. S. (2002), "Theory Creation and the Methodological Foundation of Post Keynesian Economics", *Cambridge Journal of Economics*, 26: 789-804.

Lerner, J. (2000), *Venture Capital and Private Equity. A Casebook*, New York: John Wiley & Sons, Inc.

Lerner, J. (2007), "The Governance of New Firms: A Functional Perspective", in Lamoreaux, N. R. and K. L. Sokoloff (eds) (2007), *Financing Innovation in the United States, 1870 to the Present*, Cambridge, MA: The MIT Press, pp. 405-432.

Levine, R. and S. Zervos (1998), "Stock Markets, Banks and Economic Growth", *The American Economic Review*, 88(3): 537-558.

Lopez-Gracia, J. and C. Aybar-Arias (2000), "An Empirical Approach to the Financial Behaviour of Small and Medium Sized Companies", *Small Business Economics*, 14(1): 55-63.
Lorentzen, A. (1999), "Industrial Development, Technology Change, and Regional Disparity in Hungary", *European Planning Studies*, 7(4): 463-482.
Lundvall, B.-Å. (2007), "National Innovation Systems – Analytical Concept and Development Tool", *Industry and Innovation*, 14(1): 95-119.
Lundvall, B.-Å., B. Johnson, E. S. Andersen, and B. Dalum (2002), "National Systems of Production, Innovation and Competence Building", *Research Policy*, 31: 213-231.
Major, E. and M. Cordey-Hayes (2003), "Encouraging Innovation in Small Firms Through Externally Generated Knowledge", in Shavinina, L. V. (ed) (2003), *The International Handbook on Innovation*, Oxford: Elsevier Science, pp. 667-679.
Mantzavinos, C. (2001), *Individuals, Institutions and Markets*, Cambridge: Cambridge University Press.
Mateljić, H. (2007), "Poslujete li s najbrže rastućim tvrtkama?", *Lider*, 29 June: 24-29.
Medić, Đ. Š. and B. Markov (2003), "Institutions and Evolutionary Theory of the Firm", The Fifth International Conference Enterprise in Transition (Extended Abstract; Full text on CD ROM), Faculty of Economics Split, Split-Tučepi, 22-24 May 2003: 16-20.
Medić, Đ. Š. and V. Cvijanović (2001), "Post-Keynesian Analysis of Financial Markets: Basic Characteristics and Possibilities for Use in Croatia", in Kandžija, V., L. D. Bernard, E. Claessens, O. Gabrovec-Mei, J. Kröger and A. Kumar (eds) (2001), *Economic System of European Union and Accession of the Republic of Croatia* (Third International Conference), Lovran: University of Rijeka, Faculty of Economics Rijeka, Universitá degli studi di Trieste, Facolta di Economia, Universiteit Antwerpen, Europacentrum Jean Monnet, pp. 251- 264.
Meeus, M. and L. Oerlemans (2005), "National Innovation Systems", in Casper, S. and F. van Waarden (eds) (2005), *Innovation and Institutions. A Multidisciplinary Review of the Study of Innovation Systems*, Cheltenham: Edward Elgar, pp. 51-67.
Melody, W. A. (1987), "Information: An Emerging Dimension of Institutional Analysis", *Journal of Economic Issues*, 21(3): 1313-1339.
Mencinger, J. (2002), "The Benefits of Ignoring IMF", *Ekonomski pregled*, 53(3-4): 391-403.
Mencinger, J. (2004), "Transition to a National and a Market Economy: A Gradualist Approach", in Mrak, M., M. Rojec, and C. Silva-Jáuregui (eds) (2004), *Slovenia. From Yugoslavia to the European Union*, Washington, D.C.: The World Bank, pp. 67-82.

Ministry of Foreign Affairs and European Integration (2009), "Pregled stanja procesa pregovora". Available at: http://www.eu-pregovori.hr/Download/2008/12/30/Tabela_pregled_stanja_pregovora_23_12_08-hrv.pdf, accessed 10 May 2009.
Minniti, M. and L. Polutnik (1999), "Financial Development and Small Firms Financing in Slovenia", *Comparative Economic Studies*, 41(2/3): 111-133.
MNB (Magyar Nemzeti Bank) (2008a), "Senior Loan Officer Survey on Bank Lending Practices. Summary of the Aggregate Results of the Survey on 2007 H2". Available at: http://english.mnb.hu/Resource.aspx?ResourceID=mnbfile&resourcename=slo_2007_ii_en_20080313_final, accessed 26 June 2009.
MNB (Magyar Nemzeti Bank) (2008b), "Senior Loan Officer Survey on Bank Lending Practices. Summary of the Aggregate Results of the Survey on 2008 H1". Available at: http://english.mnb.hu/Resource.aspx?ResourceID=mnbfile&resourcename=slo_2008_i_en_20080910_final, accessed 26 June 2009.
MNB (Magyar Nemzeti Bank) (2009a), "Senior Loan Officer Survey on Bank Lending Practices. Summary of the Aggregate Results of the Survey on 2008 H2". Available at: http://english.mnb.hu/Resource.aspx?ResourceID=mnbfile&resourcename=slo_2008_ii_en_final_20090316, accessed 26 June 2009.
MNB (Magyar Nemzeti Bank) (2009b), "Senior Loan Officer Survey on Bank Lending Practices. Summary of the Aggregate Results of the Survey on 2009 Q1". Available at: http://english.mnb.hu/Resource.aspx?ResourceID=mnbfile&resourcename=slo_2009_i_en_final_20090516, accessed 22 June 2009.
Modigliani, F. and M. H. Miller (1963), "Corporate Income Taxes and the Cost of Capital: A Correction", *The American Economic Review*, 53(3): 433-443.
Myers, S. (1993), "The Search for Optimal Capital Structure", in Chew, D. H., Jr. (ed) (1993), *The New Corporate Finance. Where Theory Meets Practice*, New York: McGraw-Hill, pp. 142-150.
Myers, S. C. (1984), "The Capital Structure Puzzle", *The Journal of Finance*, 39(3): 575-592.
Nelson, R. (1988), "Institutions Supporting Technical Change in the United States", in Dosi, G., C. Freeman, R. Nelson, G. Silverberg, and L. Soete (eds) (1988), *Technical Change and Economic Theory*, London and New York: Pinter Publishers, pp. 312-329.
Nelson, R. R. (1994), "Firm, Theory of the (II)", in Hodgson, G. M., W. J. Samuels, and M. R. Tool (eds) (1994), *The Elgar Companion to Institu-*

tional and Evolutionary Economics (A-K), Aldershot: Edward Elgar, pp. 241-246.
Nelson, R. R. (1997), "Why Do Firms Differ, and How Does it Matter?", in Foss, N. J. (ed) (1997), *Resources, Firms and Strategies. A Reader in the Resource-Based Perspective*, Oxford: Oxford University Press, pp. 257-267.
Nelson, R. R. (ed) (1993), *National Innovation Systems. A Comparative Analysis*, New York: Oxford University Press.
Nelson, R. R. and S. G. Winter (1977), "In Search of Useful Theory of Innovation", in Martin, B. R. and P. Nightingale (eds) (2000), *The Political Economy of Science, Technology and Innovation*, Cheltenham: Edward Elgar, pp. 547-586 [Reprint from Research Policy Vol. 6, No. 1, 1977: 36-76].
Niederöcker, B. (2002), *Finanzierungsalternativen in kleinen und mittleren Unternehmen. Eine neo-institutionalistische Analyse unter besonderer Berücksichtigung der Innovationsfinanzierung*, Wiesbaden: Deutscher Universitäts-Verlag.
Nielsen, K. (2000), "Social Capital and Systemic Competitiveness", Research Papers Network Institutional Theory, Research Paper, No. 2/2000.
North, D. C. (1990), *Institutions, Institutional Change and Economic Performance*, Cambridge: Cambridge University Press.
Noteboom, B. (2004), *Inter-firm Collaboration, Learning and Networks. An Integrated Approach*, New York: Routledge.
Nowak, P. and J. Murrow (2006), "Letting the Investment Funds Flow: Issues and Challenges in Slovenia: A Transitioning Economy", *Investment Management and Financial Innovations*, 3(2): 9-17.
O'Sullivan, M. (2000), "The Innovative Enterprise and Corporate Governance", *Cambridge Journal of Economics*, 24: 393-416.
OECD (1995), *National Systems for Financing Innovation*, OECD.
OECD (1997), *The Measurement of Scientific and Technological Activities: Proposed Guidelines for Collecting and Interpreting Technological Innovation Data (Oslo Manual)*, Paris: OECD.
OECD (2004), *OECD Principles of Corporate Governance*, Paris: OECD.
OECD (2005), *OECD Science, Technology and Industry Scoreboard 2005*, Paris: OECD.
OECD (2009a), "OECD Economic Outlook. Hungary". Available at: http://www.oecd.org/dataoecd/6/17/20211580.pdf, accessed 24 June 2009.
OECD (2009b), "OECD Economic Outlook. Slovenia". Available at: http://www.oecd.org/dataoecd/39/11/41713038.pdf, accessed 24 June 2009.

Paldam, M. and G. T. Svendsen (2000), "Missing Social Capital and the Transition in Eastern Europe", University of Aarhus, Aarhus School of Business, Department of Economics Working Papers, No. 00-5.
Palley, T. I. (1996), *Post Keynesian Economics. Debt, Distribution and the Macro Economy*, London: MacMillan.
Perez, C. (2002), *Technological Revolutions and Financial Capital. The Dynamics of Bubbles and Golden Ages*, Cheltenham: Edward Elgar.
Peteraf, M. A. (1997), "The Cornerstones of Competitive Advantage: A Resource-Based View", in Foss, N. J. (ed) (1997), *Resources, Firms and Strategies. A Reader in the Resource-Based Perspective*, Oxford: Oxford University Press, pp. 187-203.
Pistor, K., M. Raiser, and S. Gelfer (2000), "Law and Finance in Transition Economies", European Bank for Reconstruction and Development Working Paper, No. 48.
Pleschak, F. (1997), "Technologie- und Gründerzentren als Instrument der regionalen Wirtschaftsförderung", in Koschatzky, K. (ed) (1997), *Technologieunternehmen im Innovationsprozeß. Management, Finanzierung und regionale Netzwerke*, Heidelberg: Physica-Verlag, pp. 229-246.
Poljanec-Borić, S. (2002), "Poduzetnički vrijednosni sklop i problem konkurentskoga razvoja Međimurske županije", in Čengić, D. and M. Vehovec (eds) (2002), *Poduzetništvo, institucije i sociokulturni capital*, Zagreb: Institut društvenih znanosti Ivo Pilar, pp. 213-240.
Porter, M. E. (1990), "The Competitive Advantage of Nations", *Harvard Business Review*, March-April: 73-93.
Porter, M. E. (1996), "What Is Strategy", *Harvard Business Review*, November-December: 59-79.
Powell, W. W. (1990), "Neither Market Nor Hierarchy: Network Forms of Organization", *Research in Organizational Behavior*, 12: 295-336, reset [Reprinted in Grabher, G. and W. W. Powell (eds) (2004), *Networks Volume I*, Cheltenham: Edward Elgar, pp. 69-103].
Prahalad, C. K. and G. Hamel (1997), "The Core Competence of the Corporation", in Foss, N. J. (ed) (1997), *Resources, Firms and Strategies. A Reader in the Resource-Based Perspective*, Oxford: Oxford University Press, pp. 235-256.
Proleksis enciklopedija (various editors and authors) (2009), entry "teorija". Available at: http://enciklopedija.carnet.hr, accessed 8 August 2009.
Quéré, M. (2004), "National Systems of Innovation and National Systems of Corporate Governance: A Missing Link?", *Economics of Innovation and New Technology*, 13(1): 77-90.
Račić, D. and N. Podrug (2004), "Corporate Governance as a Learning Process: Organisational and Societal Levels", Mimeo, Paper prepared for the

European Academy of Management Meeting, St. Andrews, 5-8 May 2004.
Račić, D. and V. Cvijanović (2005), "Privatization, Institution Building and Market Development: The Case of Croatia", in Kušić, S. (ed) (2005), *Path-Dependent Development in the Western Balkans. The Impact of Privatization*, Frankfurt am Main: Peter Lang, pp. 43-77.
Račić, D., V. Cvijanović, and Z. Aralica (2008), "The Effects of the Corporate Governance System on Innovation Activities in Croatia", *Sociological Review*, 39(1-2): 101-114.
Račić, D., Z. Aralica, and T. Dadić (2005), "The Determinants of SME Growth in Croatia", in Raffa, M. and L. Iandoli (eds) (2005), *Entrepreneurship Competitiveness and Local Development*, Napoli: Edizioni Scientifiche Italiane, pp. 1-20 (CD ROM).
Račić, D., Z. Aralica, K. Bačić, Lj. Božić, G. Buturac, V. Cvijanović, M. Čenić, T. Dadić, I. Đokić, Z. Marić, N. Podrug, E. Rajh, I. Rašić Bakarić, D. Redžepagić, and S. Slijepčević (2004), *Inovacije u hrvatskim poduzećima, 2001.-2003.* (Working Material – Project Study), Zagreb: The Institute of Economics.
Radosevic, S. (2006), "The Knowledge-based Economy in Central and Eastern Europe: An Overview of Key Issues", in Piech, K. and S. Radosevic (eds) (2006), *The Knowledge-based Economy in Central and Eastern Europe. Countries and Industries in a Process of Change*, Houndmills, Basingstoke: Palgrave Macmillan, pp. 31-53.
Radosevic, S. and D. E. Yoruk (2004), "Industrial Networks in Central and Eastern Europe at the Firm Level: Summary and Overview of Ten Case Studies", in McGowan, F., S. Radosevic, and N. von Tunzelmann (eds) (2004), *The Emerging Industrial Structure of the Wider Europe*, London and New York: Routledge, pp. 57-76.
Ragin, C. C. (2004), "Comparative Methodology – The Logic of Case Oriented Research" (Background for the course taught at Oslo Summer School in Comparative Social Science Studies 2004). Available at: http://www.sv.uio.no/oss/Museum/2004/Ragin.html, accessed 19 June 2007.
Raiser, M. (1999), "Trust in Transition", European Bank for Reconstruction and Development Working Paper, No. 39.
Raiser, M., A. Rousso, and F. Steves (2003), "Trust in Transition: Cross-Country and Firm Evidence", European Bank for Reconstruction and Development Working Paper, No. 82.
Ratz, K. (1988), *Innovationsfinanzierung für kleine und mittlere Unternehmen*, Wien: Signum Verlag.

Realfonzo, R. (2006), "The Italian Circuitist Approach", in Arestis, P. and M. Sawyer (eds) (2006), *A Handbook of Alternative Monetary Economics*, Cheltenham: Edward Elgar, pp. 105-120.
Ritzer, G. (ed) (2005), *Encyclopedia of Social Theory*, Volume I, Thousand Oaks: Sage Publications.
Rojec, M. and A. Jaklič (2004), "Integration of Slovenia into EU and Global Industrial Networks", in McGowan, F., S. Radosevic, and N. von Tunzelmann (eds) (2004), *The Emerging Industrial Structure of the Wider Europe*, London and New York: Routledge, pp. 158-174.
Roland, G. (2004), "Understanding Institutional Change: Fast-Moving and Slow-Moving Institutions". Available at: http://www.econ.berkeley.edu/~groland/pubs/gr3.pdf, accessed 8 August 2009.
Rončević, B. (2002), "Path from the (Semi)Periphery to the Core: On the Role of Socio-Cultural Factors", *IES Proceedings 1.1*, November: 1-26.
Rosenberg, N. (1992), "Economic Experiments", *Industrial and Corporate Change*, 1(1) [Reprinted in Dosi, G., D. J. Teece, and J. Chytry (eds) (2005), *Understanding Industrial and Corporate Change*, Oxford: Oxford University Press, pp. 287-308.]
Rus, A. (2002), "Društveni kapital i razvoj malih i srednjih poduzeća u jugoistočnoj Europi", in Čengić, D. and M. Vehovec (eds) (2002), *Poduzetništvo, institucije i sociokulturni capital*, Zagreb: Institut društvenih znanosti Ivo Pilar, pp. 37-72.
Sabadinowitsch, T. (2000), *Risikokapital für mittelständische Wachstumsunternehmen*, Diskussionsbeiträge zur Bankbetriebslehre, Bd. 19, Frankfurt: Bankakad.-Verl.
Sabatini, F. (2006), "The Empirics of Social Capital and Economic Development: A Critical Perspective", Fondazione Eni Enrico Mattei Working Paper, 15.2006.
Sala-I-Martin, X., J. Blanke, M. Drzeniek Hanouz, T. Geiger, I. Mia, and F. Paua (2008), "The Global Competitiveness Index: Prioritizing the Economic Policy Agenda", in World Economic Forum (2008), *The Global Competitiveness Report 2008-2009*, Geneva: World Economic Forum, pp. 3-41.
Samuelson, P. A. and W. D. Nordhaus (2001), *Economics*, Irwin: McGraw-Hill.
Santarelli, E. (1995), *Finance and Technological Change. Theory and Evidence*, Houndmills: MacMillan Press Ltd.
Scherer, F. M. (1999), *New Perspectives on Economic Growth and Technological Innovation*, Washington: Brookings Institution.
Schertler, A. and M. Stolpe (2000), "Venture Mania in Europe: Its Causes and Consequences", Kiel Discussion Papers, No. 358.
Scott, W. R. (1995), *Institutions and Organizations*, London: Sage.

Searle, J. R. (1996), *The Construction of Social Reality*, London: Penguin Books.
Searle, J. R. (2005), "What is an Institution?", *Journal of Institutional Economics*, 1(1): 1-22.
Setterfield, M. (1993), "A Model of Institutional Hysteresis", *Journal of Economic Issues*, 27(3): 755-774.
Sik, E. (1994), "From the Multicoloured to the Black-and-White Economy: The Hungarian Second Economy and the Transformation", *Journal of Urban and Regional Research*, 18: 46-70.
Singer, S. (2007), *What Makes Croatia an Entrepreneurial Country? Results of GEM Croatia Research 2006*, Zagreb: CEPOR.
Singer, S. and B. Lauc (2004), "Razvoj sektora malih i srednjih poduzeća – Hrvatska", Support to Promotion of Reciprocal Understanding of Relations and Dialogue Between the European Union and the Western Balkans, Specific Grant Agreement RELEX 1-2 190202 reg 4-14.
Singer, S., N. Šarlija, S. Pfeifer, Đ. Borozan, and S. Oberman Peterka (2006), *What Makes Croatia an Entrepreneurial Country? Results of GEM Croatia Research 2002 – 2005*, Zagreb: CEPOR.
Singer, S., S. Pfeifer, Đ. Borozan, N. Šarlija, and S. Oberman (2003), *What Makes Croatia a(n) (Non) Entrepreneurial Country? GEM 2002 Results for Croatia*, Zagreb: CEPOR.
Souitaris, V. (2003), "Determinants of Technological Innovation: Current Research Trends and Future Prospects", in Shavinina, L. V. (ed) (2003), *The International Handbook on Innovation*, Oxford: Elsevier Science, pp. 513-528.
Srholec, M. (2006), "Global Production Systems and Technological Catching Up: Thinking Twice About High-tech Industries in Emerging Countries", in Piech, K. and S. Radosevic (eds) (2006), *The Knowledge-based Economy in Central and Eastern Europe. Countries and Industries in a Process of Change*, Houndmills, Basingstoke: Palgrave Macmillan, pp. 57-78.
Stanford Encyclopedia of Philosophy (2003), entry "Philosophy of Economics". Available at: http://plato.stanford.edu/entries/economics/#4.3, accessed 24 June 2007.
Stiglitz, J. and A. Weiss (1981), "Credit Rationing in Markets with Imperfect Information", *The American Economic Review*, 71(3): 393-410.
Stiglitz, J. E. (2000), "The Contributions of the Economics of Information to Twentieth Century Economics", *Quarterly Journal of Economics*, 115(4): 1441-1478.
Stojić, H. and M. Barać (2008), "Trendovi u Hrvatskoj i u regiji", in Cvijanović, V., M. Marović, and B. Sruk (eds) (2008), *Financiranje malih i srednjih poduzeća*, Zagreb: HVCA and Binoza press, pp. 174-185.

Stratmann, A. W. (1998), *Die Finanzierung von Innovationen. Eine theoretische und empirische Untersuchung unter besonderer Berücksichtigung konjunktureller Einflüsse*, Materialien zur Wissenschaftsstatistik, Heft 9, Februar 1998, Essen: Wissenschaftsstatistik GmbH im Stifterverband für die Deutsche Wissenschaft.

Štulhofer, A. (1998), "Što kultura ima s tim? Sociokulturni kapital, civilnost i hrvatsko gospodarstvo", in Meštrović, M. and A. Štulhofer (eds) (1998), *Sociokulturni kapital i tranzicija u Hrvatskoj*, Zagreb: Hrvatsko sociološko društvo, pp. 161-172.

Stulz, R. M. (2000), "Financial Structure, Corporate Finance and Economic Growth", *International Review of Finance*, 1(1): 11-38.

Sundbo, J. (1998), *The Theory of Innovation. Entrepreneurs, Technology and Strategy*, Cheltenham: Edward Elgar.

Švarc, J. (2009), *Hrvatska u društvu znanja. Prijepori i perspektive inovacijske politike*, Zagreb: Školska knjiga.

Svendsen, G. T. (2005), "Social Capital in Western and Eastern Europe", in Höhmann, H.-H. and F. Welter (eds) (2005), *Trust and Entrepreneurship. A West-East Perspective*, Cheltenham: Edward Elgar.

Szalavetz, A. (2007), "Knowledge-Based Entrepreneurship in Hungary", *Romanian Economic and Business Review*, 2(4): 64-74.

Szelenyi, I. and B. Szelenyi (1994), "Why Socialism Failed: Toward a Theory of System Breakdown – Causes of Disintegration of East European State Socialism", *Theory and Society*, 23: 211-231.

Szerb, L., G. Rappai, Z. Makra, and S. Terjesen (2007), "Informal Investment in Transition Economies: Individual Characteristics and Clusters", *Small Business Economics*, 28: 257-271.

Teece, D. J. (2000), *Managing Intellectual Capital*, Oxford: Oxford University Press.

Temple, P. (1999), *Private Equity. Examining the New Conglomerates of European Business*, Chichester: John Wiley & Sons.

The Gallup Organization (2007), "Observatory of European SMEs. Analytical Report", Flash EB Series #196.

Thomas, A. S. and S. L. Mueller (2000), "A Case for Comparative Entrepreneurship: Assessing the Relevance of Culture", *Journal of International Business Studies*, 31(2): 287-301.

Tidd, J., J. Bessant, and K. Pavitt (2001), *Managing Innovation. Integrating Technological, Market and organizational Change*, Chichester: John Wiley & Sons, Ltd.

Torrance, T. S. (1991), "The Philosophy and Methodology of Economics", in Mair, D. and A. G. Miller (eds) (1991), *A Modern Guide to Economic Thought. An Introduction to Comparative Schools of Thought in Economics*, Aldershot: Edward Elgar, pp. 21-39.

Tremblay, D.-G. (2004), "Innovation, Strategic Management and Economics: How Does Economic Theory Explain Innovation Within Enterprises?", Research Note, No. 2004-04.
Tschach, I. E. (2003), "The Theoretical Derivation of Credit Market Segmentation as the Result of a Free Market Process", J.W. Goethe-University Frankfurt am Main Working Paper Series Finance and Accounting, No. 103.
Ulijn, J. and T. E. Brown (2004), "Innovation, Entrepreneurship and Culture, a Matter of Interaction Between Technology, Progress and Economic Growth? An Introduction", in Brown, T. E. and J. Ulijn (eds) (2004), *Innovation, Entrepreneurship and Culture. The Interaction Between Technology, Progress and Economic Growth*, Cheltenham: Edward Elgar, pp. 1-38.
UNDP (United Nations Development Programme) (2002), *Human Development Report 2002. Deepening Democracy in a Fragmented World*, New York: Oxford University Press.
Vadnjal, J., M. Letonja, and J. Berginc (2007), "An Equity Gap: Does it Really Exist in Slovenia", Seventh International Conference on "Enterprise in Transition" (Conference Proceedings), Faculty of Economics Split, Bol – Split, 24-26 May 2007: 1-12.
Van Auken, H. E. (2001), "Financing Small Technology-Based Companies: The Relationship Between Familiarity with Capital and Ability to Price and Negotiate Investment", *Journal of Small Business Management*, 39(3): 240-258.
Van Ees, H. and H. Garretsen (1993), "Financial Markets and the Complementarity of Asymmetric Information and Fundamental Uncertainty", *Journal of Post Keynesian Economics*, 16(1): 37-48.
van Waarden, F. (2005), "A Prototypical Institution: Law, Regulation and Innovation", in Casper, S. and F. van Waarden (eds) (2005), *Innovation and Institutions. A Multidisciplinary Review of the Study of Innovation Systems*, Cheltenham: Edward Elgar, pp. 229-262.
Vercellone, C. (2007), "Uvod", in Vercellone, C. (ed) (2007), *Kognitivni kapitalizam. Znanje i financije u postfordističkom razdoblju*, Zagreb: Politička kultura, pp. 9-12.
Vincensini, C. (2001), "Is Path Dependence a Useful Concept to Analyse the Evolution of Ownership Structures in Central Europe? A Theoretical and Empirical Discussion", Paper prepared for the conference "Institutions in Transition", organised by IMAD, Otocec, Slovenia, 13-14 July 2001.
Vincentz, V. (2002), "Entwicklung und Tendenzen der Finanzsysteme in Osteuropa", Osteuropa-Institut München Working Papers, No. 237.

Vitols, S. (2004), "Changes in Germany's Bank-Based Financial System: A Varieties of Capitalism Perspective", Wissenschaftszentrum Berlin Discussion Paper, SP II 2004 – 03.
von Tunzelmann, N. (2004), "Network Alignment in the Catching-Up Economies of Europe", in McGowan, F., S. Radosevic, and N. von Tunzelmann (eds) (2004), *The Emerging Industrial Structure of the Wider Europe*, London and New York: Routledge, pp. 23-37.
Wachtel, P. (2001), "Growth and Finance: What Do We Know and How Do We Know it?", *International Finance*, 4(3): 335-362.
Walker, W. (1993), "National Innovation Systems: Britain", in Nelson, R. R. (ed) (1993), *National Innovation Systems. A Comparative Analysis*, New York: Oxford University Press, pp. 158-191.
Watson, R. and N. Wilson (2002), "Small and Medium Size Enterprise Financing: A Note on Some of the Empirical Implications of a Pecking Order", *Journal of Business Finance & Accounting*, 29(3 & 4): 557-578.
WB and IFC (World Bank and the International Finance Corporation) (2007), *Doing Business 2008*, Washington: The International Bank for Reconstruction and Development and the World Bank.
Webster's (1996), *Webster's Encyclopedic Unabridged Dictionary of the English Language*, New York/Avenel: Gramercy Books.
Wernerfelt, B. (1997), "A Resource-Based View of the Firm", in Foss, N. J. (ed) (1997), *Resources, Firms and Strategies. A Reader in the Resource-Based Perspective*, Oxford: Oxford University Press, pp. 117-130.
Westlund, H. (2006), *Social Capital in the Knowledge Economy. Theory and Empirics*, Berlin: Springer.
Westlund, H. and E. Nilsson (2003), "Measuring Enterprises' Investments in Social Capital – A Pilot Study", Paper prepared for presentation at the 43rd Congress of the European Regional Science Association, Jyväskylä, Finland, 27-30 August 2003.
Whitley, R. (2000), *Divergent Capitalisms. The Social Structuring and Change in Business Systems*, New York: Oxford University Press.
Williamson, O. E. (1988), "Corporate Finance and Corporate Governance", *The Journal of Finance*, 43(3): 567-591.
Wilson, P. and S. Bates (2003), *The Essential Guide to Managing Small Business Growth*, Chichester: John Wiley & Sons, Ltd.
Wolfson, M. H. (1996), "A Post Keynesian Theory of Credit Rationing", *Journal of Post Keynesian Economics* 18(3): 443-470.
World Economic Forum (2006), *The Lisbon Review 2006. Measuring Europe's Progress in Reform*, Geneva: World Economic Forum.
World Economic Forum (2008), *The Global Competitiveness Report 2008-2009*, Geneva: World Economic Forum.

Wray, L. R. (2006), "Banking, Finance, and Money: A Socioeconomics Approach", The Levy Economics Institute of Bard College Working Paper, No. 459.

Young, A. E. and V. Cvijanović (2006), "The Market for Venture Capital in Croatia", in W. Besler (ed) (2006), *Börsen, Banken und Kapitalmärkte. Festschrift für Hartmut Schmidt zum 65. Geburtstag*, Berlin: Duncker & Humblot, pp. 571-582.

Zagoršek, H. and M. Štembergar (2007), "Culture and its Influence on Business Performance", in Prašnikar, J. and A. Cirman (eds) (2007), *New Emerging Economies and Their Culture*, New York: Nova Science Publishers, Inc., pp. 57-71.

Ženko, Z., M. Mulej, and J. Marn (2004), "Innovation Before Entry into the EU: The Case of Slovenia", *Post-Communist Economies*, 16(2): 169-189.

Žugaj, M. (1989), *Osnove znanstvenog i stručnog rada*, Samobor: "Zagreb".

Other sources

Ajpes (Agency of the Republic of Slovenia for Public Legal Records and Related Services) (2009), Various Data. Available at: www.ajpes.si.

BSE (Budapest Stock Exchange) (2009), Various Data. Available at: www.bse.hu.

Business Croatia (Database "Business Croatia", Version 2007) (2007), Various Data.

CROSTAT (Republic of Croatia, Central Bureau of Statistics) (2009), *Statistical Information 2009*, Zagreb: Republic of Croatia, Central Bureau of Statistics.

EBRD (European Bank for Reconstruction and Development) (2008), Structural Change Indicators. Available at: http://www.ebrd.com/country/sector/econo/stats/sci.xls.

EBRD (European Bank for Reconstruction and Development) (2009), Structural Change Indicators. Available at: http://www.ebrd.com/country/sector/econo/stats/sci.xls.

Enterprise Surveys (2008), Various Data. Available at: http://www.enterprisesurveys.org/ExploreTopics/?topicid=7.

Europages (The European B2B Search Engine) (2009), Various Data. Available at: www.europages.co.uk.

Eurostat (2009), Concepts and Definitions. Eurostat's Concepts and Definitions Database (CODED). Available at: http://ec.europa.eu/eurostat/ramon/nomenclatures/index.cfm?TargetUrl=LST_NOM_DTL_GLOSSARY&StrNom=CODED2&StrLanguageCode=EN&CFID=17373998&CFTOKEN=7f8983fca99c7c31-5FE4DC61-FBD0-AECF-34BA12AFF850ED06&jsessionid=f900e7f319604319c374.

Eurostat Database (2009), Eurostat Database on Monetary and other Financial Statistics. Available at: http://epp.eurostat.ec.europa.eu/portal/page/portal/monetary_financial_statistics/data/database.

Eurostat Database (2009a), Eurostat Database on Short-Term Business Statistics. Available at: http://epp.eurostat.ec.europa.eu/portal/page/portal/short_term_business_statistics/data/database.

Eurostat Database (2009b), Eurostat Database on Science, Technology and Innovation. Available at: http://epp.eurostat.ec.europa.eu/portal/page/portal/science_technology_innovation/data/database.

Exporters Database (Database "Središnja baza hrvatskih izvoznika", BETA version) (2006), Various Data.

HMJ (Hungarian Ministry of Justice and Law Enforcement, Database of the Ministry of Justice and Law Enforcement) (2009), Various Data. Available at: http://www.e-beszamolo.irm.hu/default.aspx.
JSI (Jožef Stefan Institute) (2009), Information on Jožef Stefan Institute. Available at: http://www.ijs.si/ijsw/JSI.
LJSE (Ljubljana Stock Exchange) (2009), Various Data. Available at: www.ljse.si.
Pirs (Business Directory of Republic of Slovenia) (2009), Various Data. Available at: www.pirs.si.
Poslovna Croatia (Database "Poslovna Croatia", Web Version 2009) (2009), Various Data. Available at: www.poslovnahrvatska.hr.
Rudjer Innovations (Website of Rudjer Innovations) (2008), Various Information. Available at: http://www.r-i.hr/en/index.php.
SORS (Statistical Office of the Republic of Slovenia) (2009), Various Data. Available at: http://www.stat.si/eng/indikatorji.asp?ID=12.
Technology Network ICT (2006), Various Information on Slovenian ICT Network. Available at: http://ict-slovenia.net/index.php?lang=en.
Technology Network PCT (2006), Various Information on Slovenian Network on Process Control Technology. Available at: http://www.tvp.si/index.php?id=81.
United Nations Statistics Division (2009), Various Data for Hungary, Slovenia and Croatia from National Accounts Main Aggregates Database. Available at: http://unstats.un.org/unsd/snaama/SelectionCountry.asp.
USPTO (U.S. Patent and Trademark Office) (2008), "Patenting By Geographic Region (State and Country), Breakout By Organization. Count of 1969 - 2007 Utility Patent Grants, By Calendar Year of Grant. Croatia". Available at: http://www.uspto.gov/go/taf/asgstca/hrx_ror.htm.
WFE (World Federation of Exchanges) (2009), Various Data on Newly Listed and Delisted Companies. Available at: http://www.world-exchanges.org/statistics/annual.
World Values Survey (2009), Various Data. Available at: www.worldvaluessurvey.org.
ZSE (Zagreb Stock Exchange) (2009), Pregled trgovine u 2008. godini. Available at: http://www.zse.hr/UserDocsImages/reports/ZSE-2008.pdf.
ZSE (Zagreb Stock Exchange) (2009a), Various Data. Available at: www.zse.hr.

APPENDIX I: Case Study Database

HUNGARY

Table 13: Case study database of high-tech SMEs in Hungary

Entrepreneurs No. (*below*) / Characteristics (*right*)	Has your company produced innovations in the past 3-5 years? If not, why?	What kind of innovations do you normally come up with – incremental or radical?	If you have an idea about an innovation, but no money to finance it, who do you turn to for funding? Why?	What are the greatest problems of financiers with whom you apply for funding?	Have you applied for funding by VC/PE firms? Why (not)?
1[199]	Yes.	Incremental.	Own sources and to a lesser extent (10%-20%) Economic Competitiveness Operative Programme of the Hungarian Government (GVOP).	N/a	No. The firm does not want financing from "anyone from outside", no matter whether it is a VC/PE firm or any other entity. The reason stated is that they can solve financing by themselves, since the amount needed to be invested is not so huge.

199 Pharmaceutical company.

2[200]	Yes.	Radical.	The firm would have mostly (over 50%) used state funds, mainly available through Economic Competitiveness Operative Programme of the Hungarian Government (GVOP) as well as through National Radio and Television Commission (ORTT). The rest of the funds they would have liked to get through banking credit and retained earnings.	N/a	No. They do not want participation in equity from third parties. They prefer "own financing", since they believe "other shareholders" could divert them from attaining long-term development goals.

200 Active in medical, precision and optical instruments.

3[201]	Yes.	Incremental.	The firm would have mostly used banking credits (up to 70% of needed money). However, they would have liked to use 30% of own money (retained earnings). They would have applied for financing by the various programmes offered by the EU.	Venture capital firms set too high a profit rate that needs to be accomplished - which could be "dangerous for the company".	No. Venture capital participation could be "dangerous for the company" because the profit rate it wants to achieve is often too high.
4[202]	Yes.	Incremental.	The firm would have used various sources, like government funds, refund of innovation tax[203] as well as own funds. Banking credit is an option as well.	There are no problems with financiers.	Yes. A part of the company had been bought by a governmental venture capital fund, only later to be sold back to the original owners.

201 Active in medical, precision and optical instruments.
202 Medical, precision and optical instruments.
203 It is stipulated by the law that each company in Hungary needs to pay a tax of 1,5%. Eventually it gets refunded if the company invests into R&D.

5[204]	No.	-	The firm has to turn to the parent company. This is because they are not allowed by the parent company to apply for any other resources, not even the ones from the EU.	-	No. See the answer to the third question.
6[205]	No, because they are just assembling parts. Besides, they do not have an R&D department. Only the parent company introduces innovations.	-	-	-	-

204 Pharmaceutical industry.
205 Medical, precision and optical instruments.

7[206]	Yes.	Incremental.	They mostly finance their innovations from own funds. Too high costs of other funding sources, or too much administration for some EU or governmental funds prevent them from considering other sources.	As for the EU or governmental sources – the amount they could get is pretty low and it could cost them a lot of time to administer. Besides, aforementioned sources can provide funding only two years after they really needed finance. Banking credits cost too much.	No. The reason stated is that they have enough money at disposal and that they are a stock company whose ownership structure is well defined (there are only Hungarian shareholders and they do not want any third party to have a share).

206 Pharmaceutical industry.

8[207]	Yes.	Incremental.	Sources of external finance most extensively used are banking credits. This is logical since they have developed good connections with banks. Other sources of funding which they consider are EU and governmental funds.	N/a	No. It probably would not be tolerated by the parent company (from Italy).
9[208]	Yes.	Incremental.	Mostly own sources of financing.	N/a	N/a

207 Pharmaceutical company.
208 Pharmaceutical company.

10[209]	Yes.	Incremental.	In most cases they use own sources of financing (up to 1/3) and funds from the foreign strategic partner (up to 2/3 of the total sum). However, recently they also got money from their applications to Economic Competitiveness Operative Programme of the Hungarian Government (GVOP).	High administration costs (GVOP), delay in transferring GVOP funds. As for VC/PE firms, their main problem was that they wanted to push for profitability in much shorter period of time.	Yes. They have tried to apply but the main problem was that VC/PE firms expected and pushed for shorter return period.
11[210]	No. They only produce products developed in their mother company in Asia.	-	-	-	-

209 Pharmaceuticals as well as medical, precision and optical instruments.

210 Pharmaceutical industry.

12[211]	Yes.	Radical.	They are mostly using retained earnings. However, they also apply for funds elsewhere: Economic Competitiveness Operative Programme of the Hungarian Government (GVOP), Ányos Jedlik Programme[212] or FP7.	As regards applications for various projects from the Hungarian government or from the EU, they find it hard to specify exactly what they are going to do in the next 2-3 years.	No.

211 Pharmaceutical industry.

212 The National Office for Research and Technology is in charge of this programme. It provides a total of almost 30 million EUR for successful projects.

13[213]	Yes.	Radical.	They (would have) mostly use(d) bank loans and different grants. There is a solid share of contractual work as well. They have been considering venture capital as a financing instrument, but have never really applied for equity investment.	The interviewee explicitly said financing did not cause any limitations to his firm. However, a problem worth mentioning was difficulty in obtaining large loans with Hungarian banks.	No. The reason for such decision lies in the competitive advantage of a bank loan, which is cheaper than venture capital.

Source: author.

SLOVENIA

Table 14: Case study database of high-tech SMEs in Slovenia

Entrepreneurs No. (*below*) / Characteristics (*right*)	Has your company produced innovations in the past 3-5 years? If not, why?	What kind of innovations do you normally come up with – incremental or radical?	If you have an idea about an innovation, but no money to finance it, who do you turn to for funding? Why?	What are the greatest problems of financiers with whom you apply for funding?	Have you applied for funding by VC/PE firms? Why (not)?

213 Pharmaceutical industry.

1[214]	Yes.	Radical.	The firm would have used different sources. From internal financing (10% of total revenues is invested into research and development) to different funding sources of the state, European funds (EUREKA) as well as other subsidies.	Bureaucracy and too much paper work (unrealistic demands from financiers like projections about use value of innovations). In addition, funds offered are too little.	No and they would not do that "because the ownership structure is clear".
2[215]	No. There is no need for innovation, because the size of the enterprise is too small and there is not enough finance.	-	N/a	N/a	N/a
3[216]	No. There is no need for innovation because they only implement products of another firm.	-	N/a	N/a	N/a

214 Sector of medical, precision and optical instruments.
215 Sector of radio, TV and communications equipment.

4[217]	Yes.	Incremental.	To all kinds of possible financiers - partners, state and EU funds. They would have also used internal funds.	They are connected with size of demanded collateral and minimum size of the company.	Yes. The firm has competed for funds of a domestically-owned venture capital firm, though without success.
5[218]	Yes.	Radical.	They (would have) only use(d) own funds.	N/a	No.
6[219]	Yes.	Radical.	Only banking credits.	Regarding banking credits – problems are connected with high value of collateral and sound business results that would be desirable.	No. Because they are not interested in producing more innovations, but in getting the current ones onto market.
7[220]	Yes.	Radical.	The only option is applying to tenders for various subsidies.	The greatest problem consists of finding the appropriate financiers as well as to fit into the framework they impose.	No.
8[221]	No.	-	N/a	N/a	N/a

216 Sector of office, accounting and computing machinery.
217 Sector of office, accounting and computing machinery, as well as medical, precision and optical instruments.
218 Pharmaceutical company.
219 Sector of medical, precision and optical instruments.
220 The firm is active in the sector of medical, precision and optical instruments.
221 Sector of medical, precision and optical instruments.

9[222]	No.	-	N/a	N/a	N/a
10[223]	Yes.	Radical.	EU-funding through projects, and state institution for roads.	N/a	N/a
11[224]	Yes.	Radical.	Informal venture capital (business angels), formal venture capital. In principle, there are no other options because informal and formal VC are the only sources of funds that are adequate and sufficient.	The greatest problem of financiers in European Union is their conservatism, regardless of the type of financial institution. Opposite can be said of the financial institutions in the USA, which are much more direct and ready to invest even in young and small firms, but with a good idea.	Yes. VC is the optimal source for funding innovations of this firm.

222 Pharmaceutical company.
223 Sector of radio, TV, and communications equipment.
224 Sector of medical, precision and optical instruments.

12[225]	Yes.	Radical.	Self-financing (more than 50%), EU projects (1/3 of total financing); Various subsidies. Financing from these sources offers the most elegant way to stay independent and develop your own ideas.	Certain problems regarding banking credits have been due to the implementation of the Basel II framework from January 1, 2008: It is not as easy as it used to be to get a credit from a bank. Also, financiers like VC or PE firms do not necessarily care about development of the business, but rather how to make money. Financiers' procedures are often very formal, whereby bureaucrats with insufficient knowledge about the final product make decisions. In addition, due to relatively small value of their assets they cannot demand lot of money from anyone.	No. The interviewee is against selling his firm "below reasonable price", which VC firms would want.

225 Pharmaceutical company.

13[226]	Yes.	Radical.	The firm is open to all kinds of financing sources, because this activity takes a lot of financing to be developed.	No problems.	Yes. VC is a suitable financing vehicle since a lot of money can be invested in a firm this way. Besides, the firm was founded 1997 with a VC firm's share.
14[227]	Yes.	Radical.	The firm considers only co-financing by other firms (business partners).	No problems (with current financing sources).	No. The enterprise cannot grow as fast as a VC firm would like to have it. Hence it is not sensible to apply for VC (yet).

226 Pharmaceutical company.
227 Medical, precision and optical instruments industry.

15[228]	Yes.	Incremental and radical.	They normally finance their innovation processes from internal sources, but also use European (structural) funds.	There are several problems in connection with the European funds they compete for, and they are connected with complicated administrative procedures that require a lot of energy. Banks do not have loans for innovations, so they use banking credits only for the working capital.	No. They have applied neither by VC firms nor by business angels. The reason is that both the former and the latter have different goals than their firm. Their only goal is to "sell a firm".
16[229]	Yes.	Incremental and radical.	The company normally receives money from state programmes. In addition, they finance innovation processes from internal sources.	The company started its operations with small bank loan. However, banks have not supported their growth. Only now they would like to give them loans (when they have money). VC might negatively influence company's policy.	No. Being financed at this stage of company's development by VC "does not make sense". VC firm might have strong influence on company's policy.

Source: author.

228 Medical, precision and optical instruments.
229 Medical, precision and optical instruments.

CROATIA

Table 15: Case study database of high-tech SMEs in Croatia

Entrepreneurs No. (*below*) / Characteristics (*right*)	Has your company produced innovations in the past 3-5 years? If not, why?	What kind of innovations do you normally come up with – incremental or radical?	If you have an idea about an innovation, but no money to finance it, who do you turn to for funding? Why?	What are the greatest problems of financiers with whom you apply for funding?	Have you applied for funding by VC/PE firms? Why (not)?
1[230]	Yes.	Radical.	They do not turn to anybody, because nobody really offers financing of (radical) innovations, or at least of a quality that they produce. However, in the past years they have utilised funds that both the city and the state offer.	They do not finance innovations (not enough finance and no continuous financial support); VC/PE funds impose unacceptable conditions which can ruin their independence and provide a serious obstacle to the business policy they have conducted.	Yes. The firm has had several contacts with VC/PE firms (through initiative of the latter), but none was successful, since VC/PE firms impose unrealistic criteria that can endanger this firm's business policy.

230 The firm produces medical, precision and optical instruments for a specific industry.

2[231]	Yes.	Incremental.	They (would have) finance(d) the innovation activities by themselves or through partners. All other financing possibilities are not acceptable for their firm, since they do not have their own product, but only produce products jointly with other enterprises.	They would be best off with a credit for working capital. However, banking credits are expensive and take a lot of paper work they do not have enough time to do. They are below the radar of VC/PE firms.	No. The firm has had contacts with VC/PE firms (through initiative of the latter), but the firm does not seem to be attractive to them because of its relatively small value. On the firm's managers'/owners side, they do not want their entrepreneurial freedom to be endangered.
3[232]	Yes.	Incremental and radical.	In order to sustain their independence, they have not been open to neither banks nor venture capital firms. Hence they prefer financing innovation activities by themselves and with help of their partners.	Banking credits and venture capital could endanger firm's independence.	No, because a venture capital firm might derogate their independence.

231 It is a producer of office, accounting and computing machinery in a cluster of domestic and foreign firms and research institutions.

232 Active in office, accounting and computing machinery.

4[233]	Yes.	Incremental.	Banks, Croatian Bank for Reconstruction and Development, and own financing sources[234]. Namely these are the most appropriate financing sources for their production and innovation activities – Croatian Bank for Reconstruction and Development offers some programmes for financing innovations.	External financing is generally very expensive. Venture capital firms can put their independence at risk.	No, because they believe they might lose grip of the firm.

233 The firm produces pharmaceutical products.

234 In the "language of entrepreneurs" this might stand for a variety of financing sources, from retained earnings to entrepreneur's credit cards and sources from family, friends.

5[235]	Yes.	Incremental.	Mostly their own financing sources, although they also turn to the Ministry of Economy, Labour and Entrepreneurship (which covers a minor part of costs of their innovation activities).	Banking credits are not suitable for them (because of the irregular receipt of payments to accounts receivable that is with serious violations of the established timeframe).	No and they do not have any interest in contacting them because of the possibility of losing independence.
6[236]	Yes.	Mostly incremental.	Venture capital firm.	N/a	Yes, they are currently (at the time of the interview – mid December 2007) competing for funds from a VC firm.
7[237]	Yes.	Incremental.	Banking credits.	Banking credits are too expensive.	No, there is no such ambition. This may possibly endanger their independence.

235 The firm produces pharmaceutical products and medical, precision and optical instruments.
236 Producer of office, accounting and computing machinery.
237 Active in radio, TV and communications equipment industry.

8[238]	Yes.	Incremental.	Banking credits.	Banking credits are too expensive and procedures to obtain them are too lengthy. Venture capital funds are also too complicated and lengthy to acquire.	Yes, but have not been successful.
9[239]	Yes.	Incremental.	Internal (own) sources – as they have no ambition nor need for larger financing sources.	N/a	No – as they have no ambition nor need for larger financing sources.
10[240]	Yes.	Incremental.	Banking credits.	N/a	No – because there is no ambition for extraordinary expansion (as there are not enough employees for production development).
11[241]	Yes.	Incremental.	Internal (own) sources – as there are no business plans to further expand their innovation activities.	N/a	No – as there are no business plans to further expand their innovation activities.

238 Active in medical, precision and optical instruments.
239 Active in medical, precision and optical instruments.
240 Active in medical, precision and optical instruments.

12[242]	Yes.	Incremental.	Banking credit.	N/a	No – because they lack competent staff for further development of their business.
13[243]	Yes.	Incremental.	They would have not only used internal sources, but would have also applied by some regional and state institutions for subsidies and credit. They are about to try to apply for some EU-funds.	They believe bureaucrats do not understand innovations they often have to judge.	No. The interviewee does not know about availability of venture capital in Croatia. Hence, they have not considered these questions so far. There has been a try from a large Croatian company to buy their company, however they have offered too little money for that transaction.
14[244]	Yes.	Incremental.	Either own sources or funds from the Ministry of Economy, Labour and Entrepreneurship or from the county.	Generally, funds are too little and quite fragmented – that makes it hard to invest into the innovation process.	No, not yet. However, they would be willing to.

241 Active in radio, TV and communications equipment.
242 Producer of medical, precision and optical instruments.
243 Producer of medical, precision and optical instruments.
244 Producer of medical, precision and optical instruments. This is not a firm, but a trade.

15[245]	Yes.	Incremental.	They usually turn to the Ministry of Economy, Labour and Entrepreneurship and to the city of Rijeka. Since they have a problem with small market size their revenues are not certain and therefore cannot use other financing instruments. Besides, they have not been acquainted with functioning of VC firms.	Financing is not a problem but rather product placement.	No. They have not heard a lot about it and are not sure as to how VC firms function.

Source: author.

245 Medical, precision and optical instruments.

APPENDIX II: Case Study Reports

HUNGARY

Firm 1

This is a firm from the pharmaceutical industry, founded in 1992. However, the company has a history of over 100 years, since its first legal predecessor was established in 1905. Nowadays, the company has a strong in-house R&D department. Their work focuses largely on developing and producing basic and more complex substances for pharmaceutical products. It has been exporting to both the USA and Western Europe.

The innovations have been mainly incremental. Innovation activities for the development of simpler products cost around 4.000 EUR, which can rise up to 12.000 EUR for more complex products. In order to finance innovations, they prefer own funds. However, they have also got some money from the Economic Competitiveness Operative Programme of the Hungarian Government (GVOP). The budget received from the latter amounts to up to 20% of the cost of their innovation activities. Because of the relatively small need for investment, they believe they can finance their innovation activities without financial sources from third parties.

Firm 2

This company operates in the industry of medical, precision and optical instruments. It has produced radical innovations. Due to high equipment costs, their innovation processes cost up to 3 million Euro.

They have diversified sources of funds to finance their investment projects. They mainly include funds from certain state agencies – the Economic Competitiveness Operative Programme of the Hungarian Government (GVOP) and the National Radio and Television Commission (ORTT). External finance also comes from bank loans and, occasionally, subscription fees. An important source of funds also comes from their retained earnings.

The already fixed ownership structure and their hitherto successful strategy make the decision makers in the company reluctant to accept equity investment from third parties. They believe that would seriously negatively affect the level of motivation of today's owners for continuous development of the company.

Firm 3

This firm is active in the industry of medical, precision and optical instruments. It was founded in 1992, and remained in Hungarian ownership. Although it does not cooperate with research institutions, it does have a cooperation agreement with the University of Győr. In spite of this agreement, the company has not found convenient people to cooperate with on the operational level. It is a member of the Pannon Mechatronics Cluster[246].

The interviewee stated they did not base their firm on research. However, developing cooperation with universities is a definite aim. So far, they have not had any success. Furthermore, they are in the process of searching for quality employees and partners. The firm's CEO is quite satisfied with the financial situation within the firm. They have had excellent cooperation with a Hungarian bank for the past 14 years. Bank personnel even visit the firm in order to better assist its development. If they could not get EU funds, they would have taken banking credit to cover 70% of innovation activity costs, while the remaining 30% would have been financed from retained earnings. They currently take part in an EU project which has brought them around 100.000 EUR (30% of the project's total costs).

However, some financial problems are evident at the moment, as the firm would like to invest in robots, and only has half the money needed for that investment. On the other hand, they have not contacted any VC firms, due to what they perceive to be extremely high profit expectations. Such a situation would be particularly dangerous for the firm.

Firm 4

This company's history started in 1964 in the town of Zalaegerszeg, which is still home to the firm's headquarters. After the old state-owned company was privatised, several companies were founded, along with this one. The company belongs to the industry of medical, precision and optical instruments. The owners are 100% Hungarian. The company employs around 200 people and has a turnover of slightly less than 20 million EUR, of which 98% comes from exports to Western European countries. This firm is a member of the Pannon Mechatronics Cluster.

The company produces customised products as specified by the(ir) contract(s). Its innovations have been only incremental. It was not until a year ago (2007)

246 This is a regional cluster which covers companies that are mostly from Zala and Vas counties.

that they started to focus their intentions on innovations with greater intensity. The interviewee stressed that they had no problems with financing. The company can have a bank loan whenever it is needed. Moreover, the relationship with the bank is deemed to be excellent. These financing sources have been combined with own funds as well as with innovation tax refund. Two years ago, they built a new production plant. In order to finance it, they had applied for funds from the government, although without success.

The company even had an experience with equity investment, since a part of the company had been bought by a development agency, only to later be sold back to the original owners. They no longer want to consider such investments, since their liquidity is good.

The only problem mentioned by the interviewee was the small size of the Hungarian market.

Firm 5

This firm is a subsidiary of a large multinational pharmaceutical company. It has not had any innovations in the last 3-5 years, although they do plan to introduce an innovation to the market soon. The financial side of their business is strongly controlled by the parent company. They are not allowed to look for financial sources themselves, not even for EU funds.

Firm 6

This is a company from the medical, precision and optical instruments industry. It is a subsidiary of a multinational company from Germany. It has not had any innovations, because they have just been doing the assembly. Besides, they do not have an R&D department. The interviewee believes they "do not have the right to innovations".

Firm 7

This middle-sized company belongs to the pharmaceutical industry. It has been quite successful, producing products for (Western) European markets, their turnover being several million Euros. The innovations it has produced have been incremental.

The innovation process cost around 40.000 EUR. Internal funds have been the primary source of their financing. Other financing instruments are either coupled with excessive costs, or are insufficient for their needs. That is even the case with EU and governmental sources that may grant and lend them money up to two years after they really need it. The interest rate for a bank loan is too high. The company's management does not consider private equity and/or venture capital, since they have enough money. Besides, equity investments are not wanted because their ownership structure has already been defined (the owners are all Hungarian). However, when they plan to buy equipment that is particularly expensive, they do consider EU or governmental grants. In that case, the grants have to be substantial.

Firm 8

This is a pharmaceutical company that was founded in 1993. They have done only incremental innovations in the past 3-5 years. They believe that the development of innovations is only dependent on the market demand. The interviewee emphasised that there had not been any recent demands for serious changes to their products.

Their primary financial source has been bank credit. The company has developed good connections with banks. Other sources considered have been governmental and EU funds. Any involvement of venture capital would probably not be allowed by the parent company from Italy. Hence, they have not taken it into consideration.

Firm 9

This is a pharmaceutical company established in 1993. It has been producing incremental innovations, for which it mostly uses own sources of financing.

Firm 10

This is a company active both in the pharmaceutical industry and in medical, precision and optical instruments. The modern company was established in 1993, although its predecessor dates from 1959. Nowadays, the company exports to numerous countries around the world. It cooperates with research institutions.

Innovation activities cost around 200.000 EUR per year, with incremental innovations as output. Thanks to a foreign strategic partner, their share of financing is only 1/3 of the total investment, while the partner participates with 2/3 of the investment amount. Equally important is the foreign partner's marketing capacity, which helps the company place their products on new markets. They have recently had an experience with applications to the Economic Competitiveness

Operative Programme of the Hungarian Government (GVOP). A significant share of the money won from that application actually came from EU funds. In connection with this application, they experienced substantial administration costs and delays in financing. It was only after this transaction that they realised they could have utilised other financing instruments that would have been cheaper, such as leasing.

The company has also had initial negotiations with VC (or PE) firms, although without any results. The biggest problem standing in the way of these transactions was that VC firms demanded return on investment over a period that was much shorter than usual for this industry. The firm also faces problems with regard to market research. Because they export around the world, it is hard to estimate any future demand for their products. This, of course, is essential when planning for investments and their financing sources.

Firm 11

This is a company from the pharmaceutical industry which does not produce innovations. It is their parent company from Asia that is in charge of innovating.

Firm 12

This is a pharmaceutical company founded in 2002 in Budapest. It has been very successful, producing and exporting to over 20 countries on 4 continents. It produces radical innovations, which is a result of intensive in-house R&D. Every year, they place 5-10 new products on the market. Innovation activities per product cost from 20.000 EUR to approximately 80.000 EUR per year. Innovation processes total several hundred thousand Euro per year.

They finance innovations mostly from retained earnings, but also apply to some other programmes: the Economic Competitiveness Operative Programme of the Hungarian Government (GVOP), the Ányos Jedlik Programme, the FP7. The problem that arises in this context is the inability to correctly specify (predict) the situation 2-3 years ahead.

The interviewee mentioned that they did not use venture capital (private equity), without providing any explanation as to why this was the case. When considering financing a future innovation, they usually turn to partner companies, which are also their customers. This is especially advantageous because of the trust that exists between them. The benefit for the partner company providing finance is that it gets the final products cheaper than it normally would. The advantage for the firm in focus is that it can freely sell its products to other companies as well.

Firm 13

This is another company from the pharmaceutical industry. It was founded in Pécs, and later opened a subsidiary in the USA. This Hungarian company is a member of the MEDIPOLIS Knowledge Centre. The firm cooperates with other firms and with at least one research institution (University of Pécs) – as a member of MEDIPOLIS.

Its products are distributed by one of the largest companies in the field. Its innovations can be considered radical, as they have no competition in this market niche at the world level. In order to finance their innovation activities, they combine two primary financing instruments: bank loans and grants. They also produce products on a contractual basis. The interviewee was convinced that financing was not the limiting factor for this type of business activity in general, or, indeed, for their firm in particular.

The only problem related to financing is that Hungarian banks are reluctant to approve big loans. They have been considering venture capital, but have never competed for those funds. The reason is that they find bank loans cheaper and safer than venture capital.

SLOVENIA

Firm 1

The firm is active in the sector of medical, precision and optical instruments. It exports most of its products all around the world and has representative offices on three continents. The firm is a member of a high-technology product manufacturers' cluster (VT-Cluster) which currently consists of 10 firms and 5 support institutions. They have been producing radical innovations. The innovation process costs approximately several million Euro.

They are acquainted with different sources of financing and use some of them. Internal financing accounts for approximately 10% of investments, while the rest is covered through different subsidies available in Slovenia, as well as through some European funds (EUREKA). However, they have not shown any interest in venture capital.

They believe that unnecessary bureaucracy and too much paperwork are the most significant factors contributing to financing problems. Besides, financiers often have unrealistic demands, such as those related to projections about the use value of innovations. In addition, the funds offered are not sufficient.

The only barrier to the innovative process is stimulating innovativeness inside the firm.

Firm 2

This firm is active in the sector of radio, TV, and communications equipment. It is one of 13 firms in the eAliansa Cluster of Information Technologies. This cluster also includes 6 support institutions, one of which is a development agency (however, there are no scientific institutions involved). The cluster was formed only in 2003, following an initiative of the Ministry of Economic Affairs. Since their current business is quite successful, they feel that there is no need for innovations. Other factors contributing to such decision are the small size of the firm and insufficient finance.

Firm 3

Sector of office, accounting, and computing machinery. A member of the same cluster as enterprise No. 2. It has a history of 30 years (from 1978) – it used to exist as a part of a larger enterprise in the former Yugoslavia. It started as a private enterprise in 1989, as soon as this became possible under the federal law.

While they used to be a highly innovative firm in the late 1970s and throughout the 1980s, nowadays they feel that it is not necessary to innovate, as they mostly only implement licensed products.

Firm 4

The firm was founded in 1992, but has only been operational from the beginning of 1993 – in the field of office, accounting and computing machinery and medical, precision and optical instruments. The firm cooperates with other companies.

The firm has produced incremental innovations (in the past 3-5 years). An average innovation development process costs around 200.000 EUR, and the final product around 20.000 EUR. They have already applied for financing with a domestic venture capital firm. Unfortunately, their application was refused, since the VC firm did not recognise the value of their product. They would have intended to seek funding for innovations from a number of external financing sources – state and EU funds, their partner companies.

The biggest problem they see with the financiers is that they demand too high a value in terms of collateral. In addition, the small size of their firm has been a disadvantage. As regards other problems in developing innovations, the most acute one is the lack of adequate employees in Slovenia.

Firm 5

This is a pharmaceutical company. The interviewee claims that they have produced radical innovations in the past 3-5 years. The firm has cooperated with firms and research institutions from both Slovenia and abroad. In terms of financing sources, they (would have) only consider(ed) internal finance, and would not have let a VC firm invest into their equity.

Firm 6

This firm is from the sector of medical, precision and optical instruments. It has cooperation with domestic and foreign companies. It also cooperates with numerous research institutions both in Slovenia and abroad.

In the past 3-5 years, the firm has produced radical innovations. While the products cost from 1.500 EUR to 2.500 EUR, the innovation process costs from 20.000 EUR to 40.000 EUR. In terms of financing sources, they have only considered banking credits, and would never let a VC/PE firm invest into their equity. Obtaining banking credits is burdened with some problems, as they need to have a sound balance sheet, as well as high-quality collateral. Financing innovations is not the biggest problem the firm faces. The interviewee believes that producing innovations is not the primary task of their enterprise, but rather to place already existing products on the market. The biggest problems they encounter in the innovation process are the lack of competent employees and the limited time frame in product development.

Firm 7

This firm was founded in 1989 and has produced high-quality products since. They have also had successful cooperation with the University of Ljubljana. Their innovations have been radical and have reached both western and eastern markets. The innovation process costs from 20.000 EUR to 25.000 EUR.

In order to finance their innovation processes, they apply to public tenders. They have not contacted VC/PE firms. The problem they usually experience with financiers is how to fit their innovations into financiers' requirements. There are no special problems regarding innovations, other than how to get good ideas.

Firm 8

The firm is active in the sector of medical, precision and optical instruments. It has not had any innovations in the past 3-5 years.

Firm 9

This is another pharmaceutical company. However, the interviewee claims that it does not produce any innovations.

Firm 10

The company is active in the field of radio, TV, and communications equipment. It has a history of just over a year. It is a spin-off company founded by researchers from the University of Ljubljana, another Slovenian Institute, and an Institute from Austria. However, because of its huge potential and a lucrative FP7 project they got[247], the firm was admitted as a member of the Technology Park Ljubljana. It collaborates with a number of domestic and foreign institutions and firms.

This is a case of a radical innovation – it is radical not only in Slovenia, but also in the whole of Europe. Because this firm has only recently begun its activities, and also because of the FP7 project it got, the only other source of financing is from the public tenders of a Slovenian state institution.

Firm 11

This is another young company, founded only in 2005. Early in 2006, they joined the Styrian Technology Park. It produces medical, precision and optical instruments. Its radical innovations have been recognised at exhibitions at home and abroad. These innovations can be considered highly innovative even in a global context, since they also have a product that only three other products in the world can compete with. None of those three products are in Europe, but in the USA and Australia.

As the interviewee explained, the company represents a textbook example of a "garage company", since the founders started with no more than bare enthusiasm. Their first business plan was written in a "naive" way, as they thought the investors would readily give them money. They believe investors in the European Union to generally be rather conservative, as opposed to those in the USA.

---

247 The project is worth around 4,5 million EUR.

Even European VC firms, which should cope with extreme risk almost by definition, do not behave that way. During their fundraising trip to the USA, they learned that financial institutions over there were much more willing to invest something in their ideas. However, because Slovenia is across the Atlantic Ocean from them (which increases uncertainty and risk), these investors offered the company's founders to start a company in the USA. The founders refused to do that, since they did not want to move that far away from home. Hence, they finally decided to stay in their homeland – Slovenia. They soon discovered a business angel who invested around 30.000 EUR into their firm, despite the fact that they were only students at the time. That contact led to another, so they managed to raise some finance with their family and friends as well. This year (2008) they are focused on expanding their market share. They plan to invest around 2 million EUR, and are looking for a VC firm willing to finance them.

As for problems limiting them in their intentions to expand their business, they mention the small size of their market.

Firm 12

This is a relatively young company, founded only in 2006. It belongs to the pharmaceutical industry, especially biotechnology. It is a member of the Business Incubator Network Pomurje in Murska Sobota. It cooperates with other enterprises and the University of Ljubljana.

This firm's work is organised on a project basis, which means they work on various projects that last for 1-3 years. The costs for innovation activities amount to 100.000 EUR per year. They have produced radical innovations and have financed them mainly from own funds (more than 50%), EU projects (1/3 of the investment amount), as well as through subsidies. The interviewee explicitly stated that financing was not the primary problem his firm faced. Instead, it is essential to have a novel idea and to know what to do with it.

Problems relating to the financiers' side include more difficult access to banking credit (because of Basel II, which is applied in Slovenia since January 1, 2008), and the exceedingly formal procedures that SMEs are required to follow. Bureaucrats cannot know for sure which firms are good and which are not. The small size of the firm limits the amount of credit and/or equity it can receive, the interviewee believes. Moreover, this firm is not open to receiving VC/PE because its owner fears that such transactions could result in a draining of the firm's resources. In essence, his explanation comes down to him not being willing to work for someone else in return for what he perceives as much lower pay or profit.

Problems above and beyond those connected with financing are evident in the lack of fundamental research on the level of the country.

Firm 13

This firm has a history of 11 years, having been founded in 1997. Since the beginnings of its activities, they have had a clear focus on the pharmaceutical industry and biotechnology. Their intentions have been reflected in a strong inhouse R&D department. The company is a member of the Technology Park Ljubljana and strongly cooperates with other firms and research institutions.

The company stems from an R&D project that was developed at a Slovenian University. It was founded with a VC firm as a shareholder. After the opening of their subsidiary in Austria in 2001 (in order to be present on the EU market), the firm went through a rough period, which nearly ended in a bankruptcy. The firm was then bought by a Slovenian holding company, which has remained their parent company until today. Although the holding's focus is not on the pharmaceutical industry, the firm is free to make its business decisions and to look for alternative financing sources.

They have recently focused strongly on the development of the market and expansion of their market share. This means they have to look for markets abroad. They would especially like to expand to the market of South-Eastern Europe. Parallelly with that, they want to attract VC financing. They are open to all financing options because they need to invest a lot of money to develop these specific business and innovation activities.

There are three main problems faced by this firm (as stated by the interviewee): 1) because of the specific nature of their products, it is hard to export their products to countries far from Slovenia; 2) medical practice is quite rigid and does not accept new ideas very easily; 3) the market for their products will take another 5 years to develop.

Firm 14

This is an enterprise from the sector of medical, precision and optical instruments. Judging solely on the basis of the number of full time staff (around 35), it can be considered a small enterprise. However, its full time employees work closely together with more than 80 students, as well as with several research institutions and other firms. They have been producing products that have been exported to developed countries in Europe and the world (USA, Japan, Australia). The firm is a member of the Technology Park Ljubljana.

The history of the firm dates back only to 2001, although they made an informal start in 1996, working on projects of the Jožef Stefan Institute. Although not a spin-off in the narrow sense of the word, core ideas have indeed been created while working on projects at or of the Jožef Stefan Institute.

As regards innovation financing, the only sources considered are their partner companies and research institutions. Moreover, they often work on a contractual basis. This firm does not experience any problems with financing. The interviewee recognised that finance was not primary. The interviewee is fully acquainted with venture capital and even acts as a consultant for a Slovenian business angel network. However, he has not considered venture capital as a source of finance for his enterprise as yet. The reason he stated was that he could not work any harder than now. However, this is not because of quality and/or quantity of finance or their ideas, but because of their two main problems, which are explained in detail below.

This company's problems are related to finding quality employees and to the size and level of development of the market. Even when they find employees with adequate technical capabilities, it remains uncertain whether they could fit into the corporate culture. The latter was the case recently, when they employed a few new people who never managed to fit into the company's culture. Since they have been working with a pool of students, the latter are potential recruits as well. The other problem mentioned was related to the size of the market for the high-tech products they produce. This issue is twofold: first of all, the markets for their products are scattered around the world, and are not big (developed); secondly, their national market is rather small and gravitates towards the markets of South-Eastern Europe, which are still underdeveloped.

Firm 15

This company is the first spin-off company of the Jožef Stefan Institute, which founded it in February 1987. It has operated in its current legal form since 1991, when the company was privatised. However, the seeds of the future company were sown in the late 1970s and were rooted in projects of the Jožef Stefan Institute. Its focus is on certain branches of the industry of medical, precision and optical instruments. It cooperates with many different firms at home and abroad, as well as with research institutions in Slovenia. It is present in South-Eastern Europe through its subsidiaries. The firm is a member of the Technology Park Ljubljana.

The company produces incremental and radical innovations. Radical innovations are aimed at specific market niches. For the purpose of constant innovation, the enterprise has an annual budget of around 250.000 EUR to 400.000 EUR. The budget mainly comes from internal sources. In addition, they have recently also had a chance to apply for European (structural) funds.

This company does not consider any other sources of innovation financing. Bank loans are suitable for financing working capital, but not for financing innovations, as banks do not offer such programmes. They believe that neither business angels nor VC firms are desirable funding sources, as their only aim is to "sell a firm". However, there are problems with European funds caused by huge administration requirements (which leads to further costs).

As for the other problems the company faces, they result from insufficient innovation produced.

Firm 16

This is another company formed by former workers of the Jožef Stefan Institute. A few of them established this company in 1997 and have since maintained a close relationship with the Institute. Besides, the company has cooperation with a research institute from the USA. Almost 100% of their revenue comes from exports to various countries of the world, mostly highly developed ones. The company operates in the sector of medical, precision and optical instruments within the Technology Park Ljubljana.

The company has produced both incremental and radical innovations. Its products are based on in-house R&D, for which 15% of the total revenues are reserved. Innovation processes are primarily financed through state programmes, but also through the company's internal funds.

The problems mentioned in relation to financing are connected with bank loans, as well as with VC. The interviewee claims that banks give you money when you already have it. Although they started their business with a bank loan, it was not banks that have financed their expansion. Now that they have a sound financial situation, banks would like to give them loans. As for VC, it might negatively influence the company's policy, and that is why they do not want to have VC firms in their ownership structure.

However, financing is not the primary problem the firm faces. The biggest problems concern the very making of a product (innovation). For a good product, an employee needs to have 10 years experience. Furthermore, it takes a long time from product development to placement on the market. The difficulty lies in selling the product, that is, finding distribution channels.

CROATIA

Firm 1

This firm has produced more than 600 different products (hardware and software for a specific industry) since it was founded in 1994. Each of them are authentic and innovative. Since this company's only advantages are knowledge and know-how, it is logical that it has relied on producing innovations to be competitive. Because of the small size of the Croatian market, it has sold its products worldwide and has spread to 4 continents. Many of the aforementioned innovations are radical, which made it possible for the company to win international tenders, beating some large, established enterprises in the process. Much of all this has been possible because they had organised a cluster of almost 40 co-operators. The firm also has intensive cooperation with research institutions. The firm operates in the premises of the the Technology Park Zagreb.

In order to finance innovations, they mostly rely on their internal funds. Namely, years of experience in this sector have taught them that nobody on the market offers the kind of innovation financing that is suited to them. However, they do apply for certain governmental funds, as well as for some offered by the city in which they are headquartered. However, these funds cover only a small percentage of their total costs (e.g., they recently got roughly 7.000 EUR from the city, and approximately the same amount from the Ministry of Economy, Labour and Entrepreneurship). Comparing these grants to the price of their recent product (approximately 20.000 EUR), the costs covered in this way are only minor. On top of that, such grants can only be obtained sporadically (and on competitive terms), which makes them quantitatively inadequate ("symbolic", as the interviewee stated). Surprisingly, the company was turned down by the biggest governmental innovation-funding agency BICRO, although the company came to compete with a product that represented the very best solution currently available on the market. As for the VC/PE firms that could provide large enough funds for their innovations, these firms tend to interfere with business policy in a negative way, as they expect to make too high a profit in a short-to-medium term, disregarding the long-term potential of the company they buy (a share in). Besides the financing aspect as such, another large problem this company has faced was getting paid for the products and solutions delivered to a state owned enterprise (SOE), as this SOE refused to pay its debt.

The company's chief manager (our interviewee) believes that financing *per se* is not a problem, but rather the way the whole innovation system operates. Firstly, the state lacks a general orientation towards financing products of a specific innovativeness level. Instead of concentrating on many different products with low value added, the state should concentrate its funds on a few products distinguished by great innovativeness, which could have a huge exporting potential. Secondly, the state has totally disregarded some other solutions that a successful innovation system should include, like certifying institutions, lawyers and consultants/specialists for international tenders, etc. In the words of our interviewee, that would provide bigger support than just giving away funds, as has been the case so far. Thirdly, there is a shortage of qualified engineers of the kind the company would like to recruit.

Firm 2

This firm has had an impressive history of producing high quality high-tech products since it was founded in 1993. They constantly produce innovations in the high-tech field, which encompass both own hardware and software solutions. The price of their final product varies, depending on the type and quality of the product. The bulk of their recent revenues have come from hardware solutions, which they develop in close cooperation with their domestic and foreign partners (firms and research institutions). Some 70%-80% of their revenues have been obtained abroad.

Innovations are financed with their own funds/retained earnings or through their project partners. Their strategic orientation has mostly been on projects which are done in cooperation (consortium) with other, usually foreign, partners (that is, hardware products). Hence, they do not have "their own" hardware products, but these products belong to the foreign company that finances them. As for the development of hardware, they have adapted perfectly to this way of doing business. The other part of their business activities is concentrated on software development (their own software) or software customisation (software licensed by another firm). The latter part of their activities is much less demanding in terms of financial and human resources invested in it, as well as in terms of the risk incurred.

Other answers on innovation financing logically follow from the structure and strategic positioning of this company. Namely, they are not interested in VC/PE firms, since they could endanger the owners' influence on the company. Although they claim that their firm is not interesting to venture capitalists, their intention to maintain the status quo and not to venture into unknown territory seems obvious. Thus, they emphasise that they do not want to incur more risk by developing their own (hardware) product. Within the confines that they have adapted to, they would prefer banking credit for working capital on favourable terms (but banking credit is currently too expensive).

Furthermore, they point out some problems with financiers – banking credits are too expensive and demand too much paperwork. Their bank (one of the largest in Croatia) has been rather passive and has not recognised their need for liquidity. Aside from VC/PE firms, which they do not perceive as advantageous for their current strategy and structure, clear financial support from the state is missing.

Although the firm has never had serious financial problems – either financial obstacles or insufficient financial support –, it is not only financing that worries the owners/managers interviewed. The interviewees complain in no uncertain terms about the lack (or passivity) of governmental support institutions. For a small firm of their size and structure (they mostly employ engineers), it has been extremely hard to cope with international regulations, information, and advice when entering foreign markets. In this sense, they feel that the economic diplomacy of the Republic of Croatia is not doing its job. In spite of the small firm size, their managers and employees have to deal with technical things regarding their core job, as well as with legal regulations and some marketing functions, such as promotion. Outsourcing the latter business function would require too much money. Besides, intellectual property rights protection has been too expensive. For some of those obstacles to innovation, they blame the small size of the Croatian market, which is, therefore, not suitable for the economies of scale. For the aforementioned reasons, the owners/managers do not see any clear advantages of their doing business in the Republic of Croatia.

Firm 3

Since it was founded in 1994, the firm has produced chips (processors) and similar electrical components. The market value of their products is around 10.000 EUR. They have cooperation with other firms, which also includes partial financing. They have no cooperation with research institutions, and are not willing to resort to a banking credit or to venture capital funds. They are convinced that credits and/or venture capital could jeopardize their independence.

Firm 4

This firm has been a producer and developer of pharmaceutical products since 1991. Besides medicaments, it has also produced food supplements and cosmetical products, products for human care, as well as insecticides, repellents, and other. It is an environmentally and socially conscious and responsible firm. Hence, its products are as natural as they can be – there are around 500 of them, with an average price of 4 EUR. They have cooperation with research institutions in Croatia, as well as abroad, and are committed to constantly gaining new knowledge.

This firm prefers financing through banking credits, state funds, and own sources. Although they are not willing to open up for talks with a venture capital firm, they feel there are no cheap and adequate financing sources.

Firm 5

This is a producer of medical, precision and optical instruments, as well as pharmaceutical products, that has had incremental innovations. The prices of their products go up to 40 EUR. They do not have cooperation with other firms and research institutions. At the time, they are satisfied with their own research and development capacities.

As regards innovation financing, they have managed with their internal (own) sources, although, in 2007, they received a minor amount of support from the Ministry of Economy, Labour and Entrepreneurship amounting to 2.700 EUR. They would have been even more successful, had they had their accounts receivable paid on time. Although this is a common problem for SMEs in Croatia, their principal customer is the Ministry of Health and Social Security of the Republic of Croatia[248], which very much aggravates the problem and makes it fundamental. The average time it takes to receive payment is 3-5 months. The aforementioned problem is reflected in their being unable to take banking credit (since they simply do not know when they will be able to pay it back). Despite this problem, they are not interested in having any venture capital firm invest into their equity, since they might lose their independence.

Firm 6

This firm is an exceptionally successful one in terms of speed of development and in terms of being export-oriented. It was founded in 1995 and has produced products belonging to the category of office, accounting and computing machinery, which it has exported to the Western European, US, Japanese, Thai, and other markets. The firm has offices in Germany and Japan. Exports have accounted for almost 100% of their total revenues. Their solutions cost between 10.000 EUR and 50.000 EUR. They have cooperation both with other firms and a domestic university.

248 Actually, the Ministry of Health and Social Security is only indirectly the principal customer (or debtor) of this firm, since they work with pharmacies and health institutions, such as hospitals. However, these institutions directly depend on the Ministry, since they are part of a state-operated social security scheme.

In order to finance their innovation activities and development of their company, they are in the midst of a screening process by a venture capital firm.

Firm 7

They were founded in 1992 and have managed to penetrate the Russian and Montenegrin market – not only with their products (from the radio, TV, and communications equipment industry), but also with branches. They also export to markets in other countries of the former Yugoslavia. They have cooperated with other firms, although with a very limited number. Their products range from 670 EUR to 13.500 EUR.

They believe that credit is the only option for their innovation activities. Although they use banking credit, they are very cautious in that respect because of interest rates being too high. They reject the possibility of partnership with a venture capital firm because any loss of independence is deemed too high a price.

Firm 8

This firm is the leading manufacturer of its kind in Croatia (medical, precision and optical instruments). They currently only export to Bosnia and Herzegovina. However, Western Europe was an important market for them a few years ago. Innovations have been their constant output, and are priced from 30 EUR to 600 EUR. There has been no cooperation on innovation activities with other firms or with research institutions, as they believe that they are professional enough to develop their products alone.

They are convinced that their greatest problem is financing. Namely, it is very hard to get a credit from domestic banks – because they are foreign-owned and, as a result, have no interest in financing domestic entrepreneurs (SMEs). Bureaucratic procedure for obtaining a credit is another factor (financially) limiting them. In addition, venture capital funds are also hard to obtain (lengthy and complicated), and they have given up contacting them. This is in part due to the lack of staff capable of preparing the firm for a venture capital investment.

Firm 9

They were founded in 1990 and have had success with export activities to some European countries and the USA. Exports amount to 35% - 50% of their total revenues. They have developed their own range of products and have taken care that the products are periodically innovated. The final products' prices vary from

20 EUR to 20.000 EUR. They were particularly keen to highlight their own R&D and production – having enough workers for the products they make. There is only one firm they have cooperation with, because it has the necessary equipment for testing their products.

In accordance with the sound business foundations that they have built over the years (payment of accounts receivable has also been on time) and their current ambition, they have no need for credit or venture capital, nor is it their intention to look into either. Moreover, they have sufficient internal (own) funds.

Firm 10

This is a producer of medical, precision and optical instruments founded in 1993. It exports almost 100% of its products to Western European and US markets. They cooperate with 2 research institutions in their home town.

In order to finance their innovations, they would have gone to a bank and apply for a credit (which they currently use). However, they explicitly state they have no ambition or sufficient human resources for further development. This also explains their unwillingness to use venture capital funds.

Firm 11

This firm was founded in 1994 and has been producing radio, TV, and communications equipment. They do not export their products. Cooperation with other firms and scientific institutions is sporadic. The value of their products could be from 0,5 EUR to 340 EUR.

During the interview, they pointed out that they currently had no ambitions of developing their business. In addition, they are satisfied with incremental innovations, for which they claim they have enough resources – both human and financial. In the absence of expansion plans, they are reluctant to use either banking credit and/or venture capital.

Firm 12

This is a producer of medical, precision, and optical instruments for a particular industry. They cooperate with other firms (however, not with research institutions).

They are willing to finance further innovations with banking credit (and are, as a matter of fact, currently using one), but have not been seriously considering venture capital as a financial source. Their most important problem is the lack of competent staff for further development.

Firm 13

This firm was founded in 1990 in Zagreb and has been active in the sector of medical, precision and optical instruments ever since. The founders used to work for the research institute of one of the largest and most successful companies in its sector in the former Yugoslavia (Končar). They are members of various organisations and cooperate with numerous research institutions in Croatia. Furthermore, this firm's managers (owners) organised the first high-tech cluster in Croatia, which comprises firms from the Technology Park Zagreb (which is where their headquarters is) and others.

Although they have exported to Greece, Slovenia, and Macedonia, their market is currently rather small, so they cannot expand and develop their production. The small market size and their difficulties in expanding to other markets make up the largest problems this firm faces. The innovation process costs around 100.000 EUR, but they only manage to get around 13.000 EUR through various subventions by the City of Zagreb and the state. They plan to apply for credit with the Business Innovation Centre of Croatia, and have considered applying for some EU funds as well. The largest amount for investment comes from within the firm, as 40%-60% of their revenues are invested into development.

As regards equity investment, they were offered some money from a Croatian state-owned firm. However, since they thought that the amount was too low, they refused the offer. Investment by VC or PE firms has not been considered so far because the interviewee was not acquainted with the functioning of such investors.

The interviewee perceives the lack of export strategy of the Republic of Croatia as the No. 1 problem for the development of innovations in the country. The only mentioned problem with financiers was that the bureaucrats did not understand innovations and could therefore not competently decide which innovation to finance.

Firm 14

This is a one-man-trade from the sector of medical, precision and optical instruments. Its headquarters are located in the Technology-Innovation Centre of Rijeka. In principle, they produce only after orders are taken from their buyers.

The interviewee (the owner) cooperates with other institutions, but mainly with the Croatian Institute "Ruđer Bošković".

The firm has mainly produced incremental innovations. An average innovation process costs a few thousand Euro. The interviewee normally finances his innovations directly from customers who order his products. He is also considering financing by the Ministry of Economy, Labour and Entrepreneurship and awards from the county. The main problem with financiers is that they are not effective, since too little money is offered. Although he knows what VC firms do and how they work, he has not so far heard about VC in Croatia. However, he believes the firm is too small for VC firms.

Besides problems with financiers, the interviewee thinks that the state does not help enough. Furthermore, he is the company's only employee and, as a consequence, manages all the necessary things by himself.

Firm 15

This is another one-man firm based in the Technology-Innovation Centre of Rijeka, active as a producer of medical, precision and optical instruments. The innovations that this firm has produced have mainly been incremental.

The interviewee, who is both owner and director, explicitly states that financing is not a problem, but product placement is. This is the largest problem this enterprise faces. As the interviewee explains, only when you can place your products (and therefore make revenue) can you repay your debts. Moreover, he can get financing from the City of Rijeka or from a Ministry. This enterprise does not have a stable market, nor can the market absorb a lot of its products. The innovation process costs around 20.000 EUR on average.

As for the VC/PE firms, the interviewee has barely heard about these organisations and does not know exactly how they function.

APPENDIX III: Innovation indicators and Selected Competitiveness Indicators

Table 16: Indicators from the European Innovation Scoreboard

			Hungary		Slovenia		Croatia		EU	
Indicator no.	Indicator name	Explanation of the indicator	Value	Year	Value	Year	Value	Year	Value	Year
1	S&E graduates	New science and engineering graduates per 1000 population aged 20-29	5,1	2005	9,8	2005	5,7	2005	12,9	2005
2	Population with tertiary education	Population with tertiary education per 100 population aged 25-64	17,7	2006	21,4	2006	16,2	2006	23	2006
3	Broadband penetration rate	Number of broadband lines per 100 population	7,5	2006	11,4	2006	—	—	14,8	2006
4	Participation in life-long learning	Participation in life-long learning per 100 population aged 25-64	3,8	2006	15	2006	2,1	2005	9,6	2006
5	Youth education attainment level	Percentage of population aged 20-24 having completed at least higher secondary education	82,9	2006	89,4	2006	93,8	2005	77,8	2006
6	Public R&D expenditures	Public R&D expenditures as percentage of GDP	0,5	2005	0,35	2005	0,7	2004	0,65	2005
7	Business R&D expenditures	Business R&D expenditures as percentage of GDP	0,41	2005	0,87	2005	0,51	2004	1,17	2005
8	Share of medium-high/high-tech R&D	Share of medium-high/high-tech R&D as percentage of manufacturing R&D expenditures	90,9	2002	89,3	2004	—	—	85,2	2004

9	Enterprises receiving public funding for innovation	Share of innovative enterprises (in total number of both innovating and non-innovating enterprises according to Community Innovation Survey) receiving public funding for innovation	5,7	2004	—	—	—	—	9	2004
10	SMEs innovating in-house	Percentage of SMEs innovating in-house (in total number of SMEs according to Community Innovation Survey)	9,3	2004	—	—	—	—	21,6	2004
11	Innovative SMEs co-operating with others	Innovative SMEs co-operating with others, as percentage of SMEs (in total number of SMEs according to Community Innovation Survey)	6,6	2004	10,5	2004	—	—	9,1	2004
12	Innovation expenditures	Innovation expenditures as percentage of turnover (in total turnover for all enterprises, according to Community Innovation Survey)	1,16	2004	—	—	—	—	2,15	2004
13	Early-stage venture capital	Early-stage venture capital (defined as private equity raised for investment in companies) as percentage of GDP	0,005	2006	—	—	—	—	0,053	2006
14	ICT expenditures	Total expenditures for information and communication technology (ICT) as percentage of GDP	8,1	2005	5,4	2005	—	—	6,4	2005

15	SMEs using organisational innovation	Percentage of SMEs using organisational innovation (in total number of SMEs according to Community Innovation Survey)	19,1	2004	—	—	—	—	34	2004
16	Employment in high-tech services	Number of employed persons in the high-tech services sectors (post and telecommunications (NACE64), information technology including software development (NACE72) and R&D services (NACE73)) as percentage of total workforce	3,37	2006	2,87	2006	2,18	2006	3,26	2006
17	Exports of high technology products	Exports of high technology products (according to SITC Rev. 3 codes), as share of total exports - measured in national currency and current prices	20,2	2006	4,5	2006	6,8	2006	16,7	2006
18	Sales of new-to-market products	Sales of new-to-market products as percentage of total turnover of all enterprises, according to Community Innovation Survey	4,2	2004	7,4	2004	—	—	7,3	2004
19	Sales of new-to-firm products	Sales of new-to-firm products as percentage of total turnover of all enterprises, according to Community Innovation Survey	2,5	2004	6,9	2004	—	—	6,2	2004

20	Employment in medium-high/high-tech manufacturing	Employment in medium-high/high-tech manufacturing as percentage of total workforce (which includes all manufacturing and service sectors)	8,41	2006	8,5	2006	4,87	2006	6,63	2006
21	EPO patents per million population	European Patent Office (EPO) patents per million population (by year of filing)	18,9	2003	50,4	2003	18,2	2003	128	2003
22	USPTO patents per million population	United States Patent and Trademark Office (USPTO) patents per million population (by year of grant)	3,5	2003	7	2003	3,1	2000	52,2	2003
23	Triad patents per million population	Triad patents per million population (a patent is a triad patent only in the case if it is filed with the EPO, the Japanese Patent Office (JPO) and is granted by the USPTO)	1,8	2005	2,7	2005	0,7	2000	20,8	2005
24	Community trademarks per million population	Community trademarks (they offer a uniform protection in all EU countries) per million population	20,5	2006	30,5	2006	1,6	2006	108,2	2006
25	Community industrial designs per million population	Community industrial designs per million population (they offer a uniform protection in all EU countries)	11,3	2006	51,5	2006	1,8	2006	109,4	2006

Note: Explanation of the values in the table (European Commission, 2008: 39): "For the EU the average value shown is that of the EU27, except, due to missing data for EU27 respectively EU25, EU25 for indicators" 3, 14, 22 and 23 "and EU15 for indicator" 13. "For the indicators based on CIS-4 data, EU averages are not available from Eurostat. The EU averages for indicators" 9, 10, 11, 12,

15, 18 and 19 "are weighted estimates based on CIS-4 country data available from Eurostat. The EU averages for these CIS indicators are thus not official Eurostat estimates".
Source: European Commission (2008: 39-50).

Figure 19: Selected competitiveness indicators for Hungary, Slovenia and Croatia

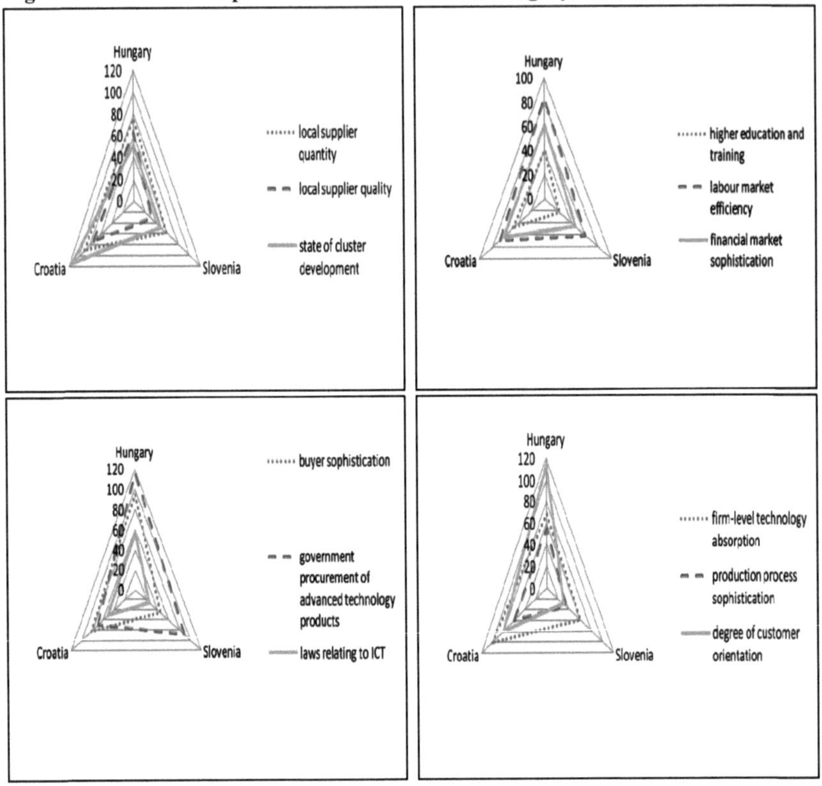

Note: the upper left graphic shows selected indicators for supporting industries; the upper right graphic shows some "efficiency enhancers" that can be used to represent a part of factor conditions; the lower left graphic depicts selected indicators for demand conditions (ICT is an abbreviation for information and communication technologies); and the lower right graphic depicts indicators described as "strategy and operational effectiveness" of enterprises – in the source listed below (World Economic Forum, 2008). In that report, 134 countries are given grades and, as a rule, the closer to the starting point in the graphic a country is ranked, the better it performs.
Source: World Economic Forum (2008).

Sozio-ökonomische Perspektiven in Südosteuropa
Socio-Economic Perspectives in South-Eastern Europe

Herausgegeben von Siniša Kušić, Slavko Kulić und Vladimir Cvijanović

www.peterlang.de